Studies in Computational Intelligence

Volume 988

Series Editor

Janusz Kacprzyk, Polish Academy of Sciences, Warsaw, Poland

The series "Studies in Computational Intelligence" (SCI) publishes new developments and advances in the various areas of computational intelligence—quickly and with a high quality. The intent is to cover the theory, applications, and design methods of computational intelligence, as embedded in the fields of engineering, computer science, physics and life sciences, as well as the methodologies behind them. The series contains monographs, lecture notes and edited volumes in computational intelligence spanning the areas of neural networks, connectionist systems, genetic algorithms, evolutionary computation, artificial intelligence, cellular automata, self-organizing systems, soft computing, fuzzy systems, and hybrid intelligent systems. Of particular value to both the contributors and the readership are the short publication timeframe and the world-wide distribution, which enable both wide and rapid dissemination of research output.

Indexed by SCOPUS, DBLP, WTI Frankfurt eG, zbMATH, SCImago.

All books published in the series are submitted for consideration in Web of Science.

More information about this series at http://www.springer.com/series/7092

S. Chakraverty
Editor

Soft Computing in Interdisciplinary Sciences

Springer

Editor
S. Chakraverty
Department of Mathematics
National Institute of Technology Rourkela
Rourkela, India

ISSN 1860-949X ISSN 1860-9503 (electronic)
Studies in Computational Intelligence
ISBN 978-981-16-4712-3 ISBN 978-981-16-4713-0 (eBook)
https://doi.org/10.1007/978-981-16-4713-0

This Springer imprint is published by the registered company Springer Nature Singapore Pte Ltd.
The registered company address is: 152 Beach Road, #21-01/04 Gateway East, Singapore 189721, Singapore

Preface

Soft computing is the recent development about the computing methods which include fuzzy set theory/logic, evolutionary computation (EC), probabilistic reasoning, artificial neural networks and other machine learning, etc. Soft Computing is the connection of computational and artificial intelligence with other science and engineering disciplines, which investigates governing models and analyzes related problems from different inter- and multidisciplinary areas. Traditional computing in general deals with approximate numerical methods whereas soft computing can handle imprecision, uncertainty, partial truth, and approximations.

Interdisciplinary sciences address different problems of science and engineering utilizing more than one or more areas. Correspondingly, increased dialog among these disciplines has resulted in this new book. The aim of this book has been to target the present and future requirements for the interaction between various science and engineering areas on the one hand and different branches of soft computing methods on the other.

In view of the above, this book includes a total of 12 chapters. Interval and fuzzy uncertainty-related issues are addressed in Chaps. 1–5. Machine intelligence problems are included in Chaps. 6–8. Chapters 9–11 include related problems of Internet of Things and causality. Lastly, hybrid evolutionary computing and dual parametric-based approach are targeted, respectively, in Chaps. 12 and 13.

As regards, Chap. 1 is contributed by A. M. C. H. Attanayake and S. S. N. Perera where interval regression has been used for predicting dengue outbreaks. This chapter discusses theories of interval regression procedures; Center method, Center and Range method, Constrained Center and Range method, interval regression based on interval least squares algorithm, and fuzzy regression techniques.

In general, the material and geometrical properties in the dynamic analysis of structural systems are assumed to be in crisp (or exact) form. However, due to several errors and insufficient or incomplete information of data, the uncertainties are assumed to be present in the material and geometrical properties. These uncertain material and geometrical properties may be modeled through convex normalized fuzzy sets. In this regard, a fuzzy-affine approach is developed by S. Rout and S.

Chakraverty in Chap. 2. This chapter deals with the dynamic analysis of various structural systems, viz., multi-degrees-of-freedom spring-mass system and multi-story frame structural system, etc. by adopting the proposed approach.

Developing a weather index is a quite sophisticated task due to the uncertainty of the factors associated with the phenomena. As such, Chap. 3 includes the fuzzy-based weather index problem by I. T. S. Piyatilake and S. S. N. Perera. The uncertainty is handled here by fuzzy theories to identify the weather-related risk in different regions.

Type-2 fuzzy linear eigenvalue problems are considered by D. Mohapatra and S. Chakraverty in Chap. 4 along with application in dynamical problems of structures. Triangular perfect quasi-type-2 fuzzy numbers are used and four parameters are utilized to solve the problems in parametric form.

In Chap. 5, fuzzy dynamical system in alcohol-related health risk behaviors and beliefs is addressed by Maranya M. Mayengo, Moatlhodi Kgosimore, and S. Chakraverty. A mathematical model has been developed for alcohol-related health risks incorporating fuzziness in uncertainties associated with individual risk behavior and induced death rate.

Artificial Neural Network (ANN) has been convincingly used in the past few years for finding the solution of differential equations (DEs). Accordingly in Chap. 6, Arup Kumar Sahoo and S. Chakraverty addressed an alternate learning method, viz., Curriculum Learning-based ANN model for solving differential equations. Different problems have been solved to illustrate the proposed training method, and analytical results have been compared with neural results.

Drowsiness is one of the main causes of decreasing strength and alertness, which may lead to an increase of accidents in personal or professional activities such as driving a vehicle, operating a crane, working with heavy machinery in large industries such as steel plants, mine blasts, and so on. In this regard, Venkata Phanikrishna B and Suchismita Chinara contributed a machine learning approach for Analysis of EEG signal for drowsiness detection in Chap. 7.

In Chap. 8, Deepti Moyi Sahoo and S. Chakraverty proposed an interval neural network-based strategy for the simultaneous identification of mass, stiffness, and damping of multi-storey shear buildings. The present model is validated by considering various example problems of different multi-storey shear structures.

Uddalak Chatterjee and Sangram Ray discussed the security issues on IoT communication and evolving solutions in Chap. 9. An extensive description of security threats and challenges across the different layers of the architecture of IoT systems is presented. The issues related to IoT cloud is also highlighted.

Causal analysis supports the study of causes and effect as they are observed, and their underlying relationship directing such trends is analyzed for predictive modeling. Accordingly, Pramod Kumar Parida gives an overview of causality and its applications in Chap. 10. The criteria for causal influence analysis in deep neural models are also discussed with examples.

In Chap. 11, Ganghishetti Pradeep, Vadlamani Ravi, and Gutha Jaya Krishna investigated an interesting problem of hybrid evolutionary computing-based association rule mining. This chapter proposes three novel association rule mining algorithms based on hybrid evolutionary techniques, which obviate the necessity of pre-specifying the minimum support and minimum confidence unlike Apriori and FP-Growth.

Finally, Chap. 12 written by Aanshi Rustagi, Annapurna Jonnalagadda, Aswani Kumar Cherukuri, and Amir Ahmad addresses the improvement of the accuracy of sarcasm detection and to better understand the context by proposing a model which integrates the ratings, reviews, and emojis. Detection of sarcasm is very crucial in today's world where social media become a major platform of expressing emotions. The performance shows that integrating more features enhances the accuracy by a considerable margin as compared to previously defined methodologies.

It is worth mentioning that the diversity of application problems via soft computing has made this book a very useful source for different subject areas. Graduate/postgraduate students, teachers, and researchers in colleges, universities, and industries in the fields of various engineering such as computer science/engineering, mechanical, civil, aerospace, electrical, and other sciences such as mathematics/applied mathematics, statistics, and physics will certainly be benefited. This book will be a handy and important asset to handle their problems. The multidisciplinary areas as said will surely be helpful to each and every institute/university/industry throughout the globe. The editor does appreciate the efforts of all the contributors for writing their important chapters and submitting those on time for the success of this book. Lastly, the Editor is very much thankful to the whole team of Springer for their help and support for publishing this challenging book on time.

Editor

Rourkela, India S. Chakraverty

Contents

Editor and Contributors

About the Editor

Prof. S. Chakraverty has 29 years of experience as a researcher and teacher. Presently, he is working in the Department of Mathematics (Applied Mathematics Group), National Institute of Technology Rourkela, Odisha, as a senior (Higher Administrative Grade) professor. Prior to this, he was with Central Building Research Institute (CSIR), Roorkee, India. After completing Graduation from St. Columba's College (Ranchi University), his career started from the University of Roorkee (now, Indian Institute of Technology Roorkee) and did M. Sc. (Mathematics) and M. Phil. (Computer Applications) from there securing the first positions in the university. Dr. Chakraverty received his Ph. D. from IIT Roorkee in 1992. Thereafter, he did his post-doctoral research at the Institute of Sound and Vibration Research (ISVR), University of Southampton, U.K., and at the Faculty of Engineering and Computer Science, Concordia University, Canada. He was also a visiting professor at Concordia and McGill universities, Canada, during 1997–1999 and a visiting professor of University of Johannesburg, South Africa, during 2011–2014. He has authored/co-authored/Edited 23 books, published 388 research papers (till date) in journals and conferences, two more books are in Press, and two books are ongoing. He is in the Editorial Boards of various International Journals, Book Series and Conferences. Professor Chakraverty is the chief editor of "International Journal of Fuzzy Computation and Modelling" (IJFCM), Inderscience Publisher, Switzerland, an associate editor of "Computational Methods in Structural Engineering, Frontiers in Built Environment" and happens to be the editorial board member of "Springer Nature Applied Sciences", "IGI Research Insights Books", "Springer Book Series of Modeling and Optimization in Science and Technologies", "Coupled Systems Mechanics (Techno Press)", "Curved and Layered Structures (De Gruyter)", "Journal of Composites Science (MDPI)", "Engineering Research Express (IOP)" and "Applications and Applied Mathematics: An International Journal". He is also the reviewer of around 50 national and international journals of repute, and he was the president of the Section of Mathematical Sciences (including Statistics) of "Indian Science Congress" (2015–2016) and was the vice president—"Orissa Mathematical Society" (2011–2013).

Professor Chakraverty is a recipient of prestigious awards, viz. Indian National Science Academy (INSA) nomination under International Collaboration/Bilateral Exchange Program (with the Czech Republic), Platinum Jubilee ISCA Lecture Award (2014), CSIR Young Scientist Award (1997), BOYSCAST Fellow. (DST), UCOST Young Scientist Award (2007, 2008), Golden Jubilee Director's (CBRI) Award (2001), INSA International Bilateral Exchange Award ([2010–2011 (selected but could not undertake), 2015 (selected)], Roorkee University Gold Medals (1987, 1988) for first positions in M. Sc. and M. Phil. (Comp. Appl.), etc. He has already guided nineteen (19) Ph. D. students, and twelve are ongoing. He is in the list of 2% world scientists recently (2020) in "Artificial Intelligence & Image Processing" category based on an independent study done by Stanford University scientists. His world rank is 1862 out of 215114 researchers throughout the globe.

Professor Chakraverty has undertaken around 17 research projects as a principle investigator funded by international and national agencies totaling about Rs.1.6 crores. He has hoisted around 8 international students with different international/national fellowships to work in his group as PDF, Ph.D., visiting researchers for different periods. A good number of international and national conferences, workshops and training programs have also been organised by him. His present research area includes differential equations (ordinary, partial and fractional), numerical analysis and computational methods, structural dynamics (FGM, nano) and fluid dynamics, mathematical and uncertainty modeling, soft computing and machine intelligence (artificial neural network, fuzzy, interval and affine computations).

Contributors

Amir Ahmad Department of Information Technology, College of Information Technology, UAE University, Al Ain, UAE

A. M. C. H. Attanayake Department of Statistics and Computer Science, Faculty of Science, University of Kelaniya, Kelaniya, Sri Lanka

S. Chakraverty Department of Mathematics, National Institute of Technology Rourkela, Rourkela, Odisha, India

Uddalak Chatterjee Department of Computer Science and Engineering, National Institute of Technology Sikkim, Ravangla, Sikkim, India

Aswani Kumar Cherukuri School of Information Technology and Engineering, VIT University, Vellore, India

Suchismita Chinara Computer Science & Engineering, NIT Rourkela, Rourkela, Odisha, India

Annapurna Jonnalagadda School of Computer Science and Engineering, VIT University, Vellore, India

Moatlhodi Kgosimore Botswana University of Agriculture and Natural Resources, Francistown, Botswana

Gutha Jaya Krishna Center of Excellence in Analytics, Institute for Development and Research in Banking Technology (IDRBT), Masab Tank, Hyderabad, India

Maranya M. Mayengo Nelson Mandela-African Institute of Science and Technology, Arusha, Tanzania

Dhabaleswar Mohapatra Department of Mathematics, National Institute of Technology Rourkela, Odisha, India

Pramod Kumar Parida Department of Systemics, School of Computer Science, University of Petroleum and Energy Studies, Dehradun, UK, India

S. S. N. Perera Research & Development Centre for Mathematical Modelling, Department of Mathematics, Faculty of Science, University of Colombo, Colombo, Sri Lanka

I. T. S. Piyatilake Department of Computational Mathematics, Faculty of Information Technology, University of Moratuwa, Moratuwa, Sri Lanka

Ganghishetti Pradeep Center of Excellence in Analytics, Institute for Development and Research in Banking Technology (IDRBT), Masab Tank, Hyderabad, India

Vadlamani Ravi Center of Excellence in Analytics, Institute for Development and Research in Banking Technology (IDRBT), Masab Tank, Hyderabad, India

Sangram Ray Department of Computer Science and Engineering, National Institute of Technology Sikkim, Ravangla, Sikkim, India

S. Rout Department of Mathematics, National Institute of Technology Rourkela, Rourkela, Odisha, India

Aanshi Rustagi School of Computer Science and Engineering, VIT University, Vellore, India

Arup Kumar Sahoo Department of Mathematics, National Institute of Technology Rourkela, Rourkela, India

Deepti Moyi Sahoo National Institute of Technology Rourkela, Rourkela, Odisha, India

B Venkata Phanikrishna Computer Science & Engineering, NIT Rourkela, Rourkela, Odisha, India

Recent Trends in Interval Regression: Applications in Predicting Dengue Outbreaks

A. M. C. H. Attanayake and S. S. N. Perera

Abstract Dengue disease is a serious threat for the world. The number of infections increase annually forcing implementation of prompt actions in dengue management. Common practice of modelling is associated with point measurements. However, an interval representation for a point measure provides an additional information for the spread, capture uncertainties associated with variables and useful in making more precise decisions. Further, interval predictions are appropriate in the situations of exact predictions are not essential. Interval-valued analysis in the dengue disease is important as actions taking towards controlling the disease do not depend on the exact number but on the magnitude of the values represented by the interval. In the area of regression analysis, there are techniques to handle interval-valued dependent and independent variables. The present chapter discusses theories of interval regression procedures: centre method, centre and range method, constrained centre and range method, interval regression based on interval least squares algorithm and fuzzy regression techniques. The chapter illustrates applications of these methods using interval-valued data in Colombo, Sri Lanka, and Jakarta, Indonesia. Finally, the chapter emphasizes the importance and effectiveness of the interval regressions over traditional linear regression as well as added advantages of soft computing methods.

Keywords Dengue · Interval regression · Soft computing methods

---

A. M. C. H. Attanayake (✉)
Department of Statistics and Computer Science, Faculty of Science, University of Kelaniya, Kelaniya 11600, Sri Lanka
e-mail: succ@kln.ac.lk

S. S. N. Perera
Research & Development Centre for Mathematical Modelling, Department of Mathematics, Faculty of Science, University of Colombo, Colombo 00700, Sri Lanka

S. Chakraverty (ed.), *Soft Computing in Interdisciplinary Sciences*,
Studies in Computational Intelligence 988,
https://doi.org/10.1007/978-981-16-4713-0_1

1 Introduction

Dengue is one of the prominent diseases existing in the world. The disease is becoming a huge treat for the human survival. It is transmitted to humans through the bites of infected mosquitos with the virus. The affecting dengue virus is denoted as DENV, and four serotypes of the virus have been identified. The availability of four serotypes of the virus has complicated not only the control mechanisms that are implemented but also all other phases associated with the disease.

More than 100 countries in the whole wide world have been constantly affected by dengue. In 2019, the majority of dengue cases were reported in Brazil which was 2,225,461 cases [1]. Some countries where the number of dengue cases reported throughout the year 2019 were greater than or near to 100 000 cases are depicted in Fig. 1.

The highest number of deaths due to dengue in 2019 was reported in the Philippines followed by Brazil which were 1 565 and 789 deaths, respectively [1]. The distribution of number of dengue deaths in the same year over some affected countries are shown in Fig. 2.

Not only the people in the depicted countries in Figs. 1 and 2 but also nearly half of the population in the world are now facing problems due to dengue. Globally, an increasing number of mortalities and morbidities was reported annually due to the dengue disease.

Dengue was first identified in Sri Lanka in 1960 [2]. Currently, the disease has grown exponentially from the starting point. A total of 30,802 dengue cases in 2020 and 5 181 dengue cases from January to March 2021 were reported in Sri Lanka [3]. Figure 3 shows the distribution of dengue cases in Sri Lanka from 2010 to 2020. Clearly, in 2017, there was an explosive outbreak of dengue cases as depicted in Fig. 3. The reported number of cases were 186 101. The maximum number of dengue cases is reported in the Western Province of Sri Lanka. The Western Province is one of the nine provinces in Sri Lanka (Fig. 4) and the most crowded province in the country. The Western Province consists of three districts, namely, Colombo,

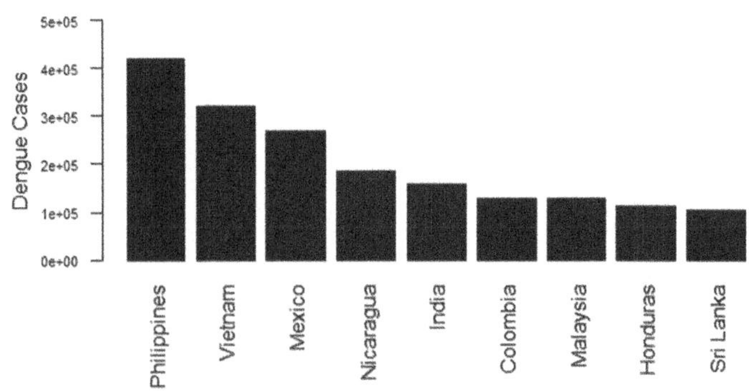

Fig. 1 Distribution of dengue cases in 2019

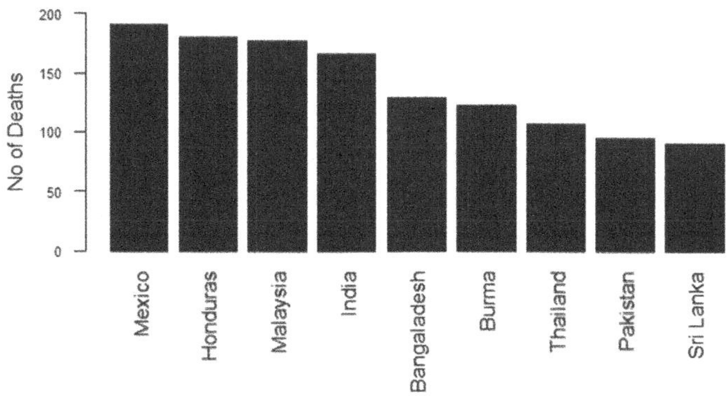

Fig. 2 Distribution of dengue deaths in 2019

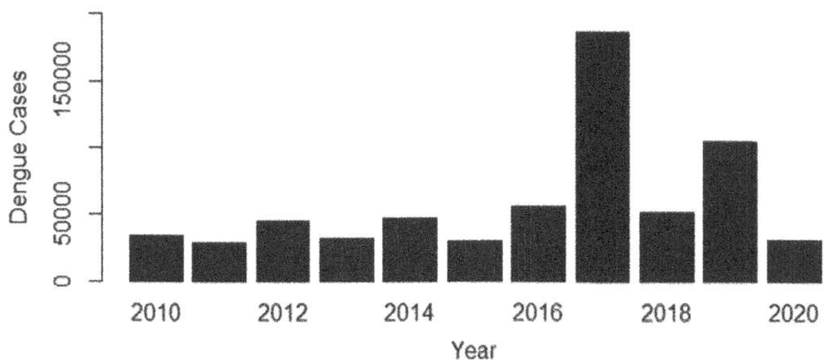

Fig. 3 The distribution of total number of dengue cases in Sri Lanka from 2010 to 2020

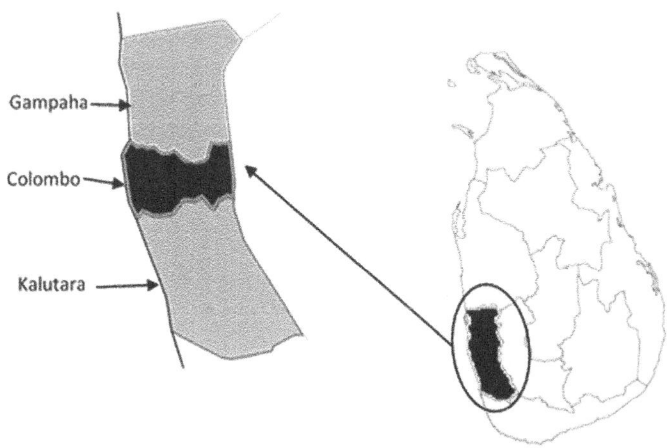

Fig. 4 Gampaha, Colombo and Kalutara districts within the Western Province of Sri Lanka

Table 1 Some characteristics of three districts; Gampaha, Kalutara and Colombo

District	Longitude	Latitude	Population (in 2012)	Land area (Km^2)
Gampaha	7.0840	80.0098	2,324,349	676
Kalutara	6.5854	79.9607	2,304,833	1341
Colombo	6.9271	79.8612	1,221,948	1576

Kalutara and Gampaha. Some characteristics of the three districts are summarized in Table 1. Within the Western Province, the majority of infections were recorded in Colombo district. The highest number of dengue cases was usually reported in Colombo district among all 25 districts in Sri Lanka.

Dengue is one of the leading infections in Indonesia. Dengue was found in Indonesia in 1968 [4]. All provinces of the country are now facing the threat of the disease. Bali, Jakarta and East Java are provinces of Indonesia which have frequently reported dengue cases. The distribution of dengue cases from 2008 to 2016 in Jakarta is shown in Fig. 5. The number of recordings were higher in 2016 than previous years. According to Fig. 5, the second highest was recorded in 2010. Jakarta is the largest city of Indonesia (Fig. 6). Further, it is the capital of Indonesia. Hence, Jakarta is exposed and highly supportive for transmission of dengue. The latitude of Jakarta is 6.2088, whereas longitude is 106.8456.

Controlling dengue and associated hazards is the key requirement of affected countries. Implementation of controlling actions is fuelled by the forecasts produced by the models. Further, models are useful in understanding the underlying structure/pattern of the transmission of the disease. Therefore, modelling and predicting of the dengue are essential constraints in the field of dengue management.

A common practice of modelling and predicting is associated with point measurements or exact data values. However, interval-valued representation for an exact data value provides an additional information for the spread of the value and useful in making more precise decisions. Interval-valued datum represents by a lower bound (limit) and an upper bound (limit) for an exact value. For example, the average male body weight can be represented as 50 kg as a point estimate and it can be represented

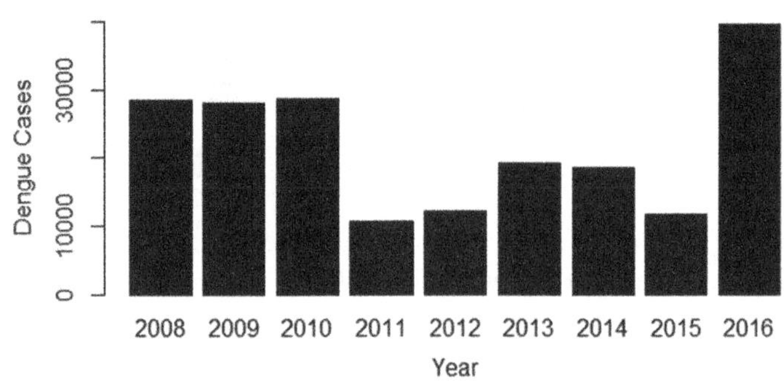

Fig. 5 The distribution of total number of dengue cases in Jakarta, Indonesia

Fig. 6 The position of Jakarta in the map of Indonesia

as [48, 52] as an interval estimate. The interval estimate provides how much variability (possible fluctuations) can be expected in the average male body weight. It depicts approximations for the point value.

There is a growing tendency in the applications of interval-valued data analysis over recent years. Particularly, in the area of regression analysis, various techniques have been developed to handle interval-valued dependent and interval-valued independent variables. These interval-valued representations of data lead to capture uncertainties that are associated with the variables. Further, interval predictions are particularly appropriate in the situations of exact predictions are not essential. Interval-valued analysis in dengue disease is important as actions taking towards controlling the disease do not depend on the exact number but on the magnitude of the values represented by the interval. For an example, the dengue cases in the next month can be predicted as 1000 cases, whereas the interval estimate for the incidence may be [966, 1140] cases. The optimal number of cases are not essential in deciding and implementing the controlling actions but an approximation is sufficient.

Regression analysis is one of the key areas in statistics that can be utilized in modelling and predicting the dengue epidemic. Simple and multiple linear regression are the two main sections of regression analysis. The simple linear regression model develops the relationship of dependent variable with the use of only one independent variable. The multiple linear regression develops the relationship using more than one independent variable. The coefficients of the regression model usually estimate by using the ordinary least squares method which minimizes the error represented by the actual and fitted values of the output variable. By considering reported dengue counts as the output variable (dependent variable) and correlated other factors as input variables, one can model the relationship through multiple regression. The usage of regression models in modelling and predicting dengue case counts can be found in the literature [5–8]. Researchers utilized data to model the relationship of dengue with several climatic and non-climatic factors. The first linear regression model called Centre Method (CM) for interval data prediction was found by Billard and Diday in the year 2000 [9]. In CM, there is no assurance of lower level of predicted value is lower than the upper level. Therefore, Lima Neto and De

Carvalho in 2008 [10] formulated the Centre and Range method (CRM) to capture the additional information provided by the range of the intervals. CRM formulates two separate predictions for centre and range. The final predictions for lower and upper levels of the output variable are found by aggregating the results. Lima Neto and De Carvalho in 2010 [11] invented the Constrained Centre and Range method (CCRM). In the method, it introduces a non-negative constraint on the coefficients of the regression model which is developed on range values. All these mentioned methods are based on the ordinary least squares regression and recently many methods and applications have been developed on interval-valued regressions [12, 13]. Interval regressions based on interval least squares algorithm is one of the newly invented strategies for interval predictions. Applications of interval least squares algorithm in interval regressions rather limited in literature. To the best of our knowledge, no applications can be found in the literature related to the dengue prediction using the interval multiple regression based on interval least squares algorithm approach. Fuzzy linear models deal with vague and imprecise information in order to represent better models [14]. These fuzzy models are appropriate in modelling and predicting dengue disease as the disease associated with various unknown and uncontrollable factors. Particularly, fuzzy regression approaches applicable in modelling uncertainty associated with dengue disease together with several explanatory variables such as rainfall, temperature, wind speed, etc., which are also vague in nature. Numerous fuzzy regression procedures available in theory which would address input/output variables as fuzzy numbers or crisp measures. S. Pavel and M. Jaroslav in 2018 [15] stated that fuzzy regression is an alternative efficient approach for conventional statistical regression and they summarized theories behind various fuzzy regression approaches. Romero et al. in 2019 illustrated the advantages of fuzzy logic-based approaches in modelling vector-borne diseases [16]. No efforts were found in the literature which apply the multiple fuzzy regression approaches in dengue modelling to the best of our knowledge.

The present chapter discusses some of the multiple interval-valued regression procedures, namely, CM, CRM, CCRM, interval regression based on interval least squares algorithm and fuzzy regression techniques (Possibilistic linear regression, fuzzy linear regression using the possibilistic linear regression with least squares method and fuzzy linear regression using the multi-objective fuzzy linear regression method). These powerful techniques accessible through regression analysis can be applied to model the dengue incidence and to manage the disease effectively and efficiently.

The chapter is organized as follows. The first section gives a general overview to the dengue disease, emphasizing the dengue incidence in Colombo, Sri Lanka, and Jakarta, Indonesia. The section outlines the importance of interval representation for a point value and provides a brief literature review. The next section explains some of the interval multiple regression procedures and their respective applications using dengue data from Jakarta and Colombo as test cases. Finally, the chapter continues to explain fuzzy regression techniques based on three fuzzy regression procedures. The conclusion and discussion segment which is the final section summarizes the advantages/drawbacks of regression procedures by highlighting the importance and effectiveness of soft computing methods as a current trend of modelling.

2 Theory and Applications

2.1 The Centre Method (CM)

CM was introduced by Bilard and Diday [9]. Let Y and X are two variables in interval representation, related as in the following relationship:

$$Y^c = X^c\beta + \varepsilon^c$$

$$Y^c = \left(Y_1^c, \ldots Y_n^c\right),$$

$$X^c = \left(\left(x_1^c\right)', \ldots \left(x_n^c\right)'\right)',$$

$$\left(x_i^c\right)' = (1, x_{i1}^c, \ldots x_{ip}^c)$$

$$\beta = \left(\beta_0, \ldots \beta_p\right)'$$

$$\varepsilon^c = (\varepsilon_1^c, \ldots .\varepsilon_n^c)$$

$$x_{ij}^c = \frac{a_{ij} + b_{ij}}{2}$$

$$Y_i^c = (Y_L + Y_U)/2$$

Y_L—lower limit of the interval.
Y_U—upper limit of the interval.
The estimator of β is

$$\widehat{\beta} = (X^{c'}X^c)^{-1}(X^{c'}Y^c)$$

The upper predicted and lower predicted values of the output variable can be denoted as follows:

$$\widehat{Y}_L = X_L{}'\widehat{\beta}$$

$$\widehat{Y}_U = X_U{}'\widehat{\beta}$$

2.2 Centre and Range Method (CRM)

L. Neto with D. Carvalho invented CRM [10]. CRM considers the information of centres as well as ranges/spread of the variables. Let Y and X are two variables in interval representations and related over the centres, then the relationship can be written as

$$Y^c = X^c\beta + \varepsilon^c$$

$$Y^c = \left(Y_1^c, \ldots Y_n^c\right),$$

$$X^c = \left(\left(x_1^c\right)', \ldots \left(x_n^c\right)'\right)',$$

$$\left(x_i^c\right)' = (1, x_{i1}^c, \ldots x_{ip}^c)$$

$$\beta = \left(\beta_0, \ldots \beta_p\right)'$$

$$\varepsilon^c = (\varepsilon_1^c, \ldots .\varepsilon_n^c)$$

$$x_{ij}^c = \frac{a_{ij} + b_{ij}}{2}$$

$$Y_i^c = (Y_L + Y_U)/2$$

$$\widehat{\beta} = (X^{c'}X^c)^{-1}(X^{c'}Y^c)$$

If Y and X are related over the rangers, then the relationship can be written as

$$Y^r = X^r\beta^r + \varepsilon^r$$

$$Y^r = \left(Y_1^r, \ldots Y_n^r\right),$$

$$X^r = \left(\left(x_1^r\right)', \ldots \left(x_n^r\right)'\right)',$$

$$\left(x_i^r\right)' = (1, x_{i1}^r, \ldots x_{ip}^r)$$

$$\beta^r = \left(\beta_0^r, \ldots \beta_p^r\right)'$$

$$\varepsilon^r = (\varepsilon_1^r, \ldots .\varepsilon_n^r)$$

$$x_{ij}^r = \frac{b_{ij} - a_{ij}}{2}$$

$$Y_i^r = (Y_U - Y_L)/2$$

The estimator of β^r is

$$\widehat{\beta}^r = (X^{r'} X^r)^{-1} (X^{r'} Y^r)$$

The upper and lower predicted values of the output variable can be denoted as

$$\widehat{Y}_L = \widehat{Y}^c - \widehat{Y}^r$$

$$\widehat{Y}_U = \widehat{Y}^c + \widehat{Y}^r$$

2.3 *Constrained Centre and Range Method (CCRM)*

L. Neto with Carvalho introduced CCRM [11]. In addition to CRM, there is an additional constraint to ensure the condition $\widehat{Y}_L \leq \widehat{Y}_U$. The relationships over the centres and rangers is as follows:

$$Y^c = X^c \beta + \varepsilon^c$$

$$Y^r = X^r \beta^r + \varepsilon^r$$

with constraint $\beta_j^r \geq 0; \; j = 0, 1, \ldots .p$
The estimators are

$$\widehat{\beta}^r = (X^{r'} X^r)^{-1} (X^{r'} Y^r)$$

$$\widehat{\beta} = (X^{c'} X^c)^{-1} (X^{c'} Y^c)$$

2.4 Applications of CM, CRM and CCRM

Monthly lower limit and upper limit for the dengue count in Colombo, Sri Lanka, were found from the weekly dengue data acquired from the Epidemiology Unit, Ministry of Health, Sri Lanka, for the duration of January 2009–December 2016. Monthly lower limit and upper limit for the rainfalls, monthly lower limit and upper limit for the maximum temperature and monthly lower limit and upper limit for the minimum temperature in Colombo were further identified from the collected weekly data of the Department of Meteorology, Sri Lanka, for the same period. The analysis was done using R software [17].

The distributions of lower and upper boundaries of the dengue counts in Colombo is given in Fig. 7. The reported lower bounds are lower in former months than that of latter months. The gap between lower and upper boundaries are larger in latter months of the data gathering period.

Lower limits and upper limits of monthly maximum temperature depicted in Fig. 8. Some cyclic variances are apparent in Fig. 8.

Figure 9 displays seasonal variations in lower limits as well as in upper limits of monthly minimum temperature in Colombo. Minimum and maximum of the lower

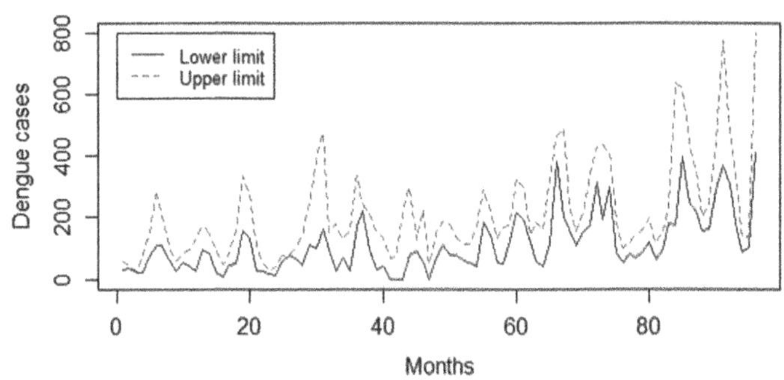

Fig. 7 Distribution of lower limits and upper limits of dengue counts

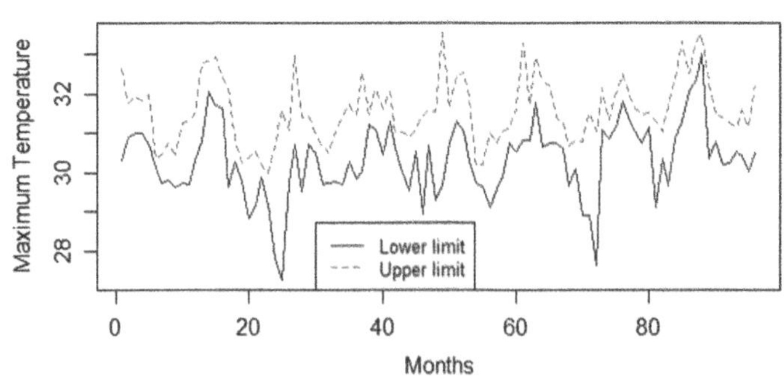

Fig. 8 Distribution of lower limits and upper limits of maximum temperature

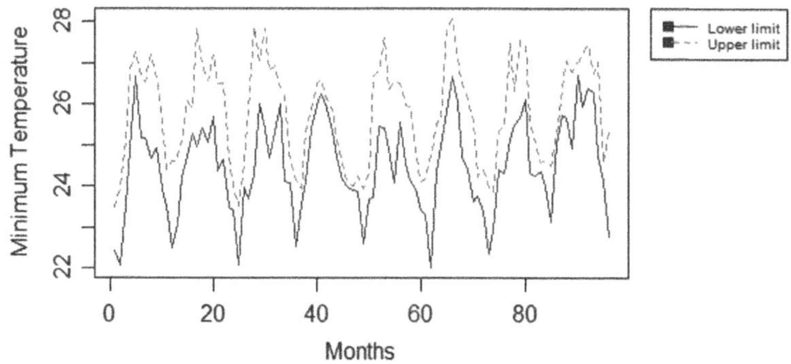

Fig. 9 Distribution of lower limits with upper limits of minimum temperature

bound varies in between 22 and 26.7, whereas that for upper bound varies in between 23.5 and 28.

The distribution of lower limits of monthly rainfall is always fluctuated around zero but variability is high in upper bounds of monthly rainfall creating wider ranges within lower and upper intervals as can be seen in Fig. 10.

Results of cross-correlation analysis are summarized in Table 2. Cross-correlation analysis revealed that 10 weeks leading rainfall and immediate minimum and maximum temperature effect on reported dengue cases. Therefore, 10-week lag period for rainfall with no lags for temperature variables modelled with reported dengue cases in each of the three methods: CM, CRM and CCRM. Data selection for the model development was from January 2009 to December 2015. Data from January to December in 2016 were used for the validation of the three models. The estimated coefficients of regression lines in each of the three models are shown in Table 3 and the summary measures in Table 4.

The highest RMSE (both lower and upper) reported in CM. Although CRM reported the lowest RMSE (both lower and upper), summary results of CRM and CCRM indicate that there is no significance difference between the two methods.

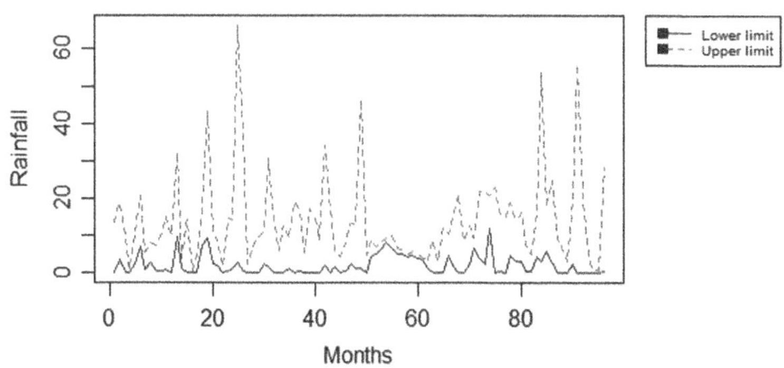

Fig. 10 Distribution of lower limits with upper limits of rainfall

Table 2 Cross-correlations of variables

Lag in weeks	Cross-correlations		
	Rainfall and dengue	Maximum temperature and dengue	Minimum temperature and dengue
12	0.188	0.258	0.120
11	0.217	0.224	0.124
10	0.279	0.172	0.120
9	0.256	0.131	0.131
8	0.219	0.097	0.148
7	0.169	0.057	0.125
6	0.102	0.035	0.127
5	0.030	0.031	0.123
4	0.034	0.042	0.136
3	0.075	0.035	0.129
2	0.126	0.029	0.124
1	0.141	0.080	0.115
0	0.161	0.301	0.146

Table 3 The estimated coefficients of regression lines in CM, CRM and CCRM

	CM	CRM		CCRM	
		Center	Range	Center	Range
Intercept	42.39	42.39	107.67	42.39	77.65
$\widehat{\beta}_0$	−2.75	−2.75	−20.14	−2.75	0.00
$\widehat{\beta}_1$	5.76	5.76	−6.85	5.76	0.00
$\widehat{\beta}_2$	4.79	4.79	2.08	4.79	1.66

Table 4 Summary measures in CM, CRM and CCRM

	RMSE_Lower	RMSE_Upper	R-squared	Standard deviation	
				Center	Range
CM	71.29	118.31	0.40	89.09	–
CRM	69.02	113.73	0.41	89.09	73.62
CCRM	69.33	114.09	0.44	89.09	75.41

Predicted lower and upper limits from each of the models with actual bounds are shown in Figs. 11, 12, 13.

The predicted upper boundaries from CM is lower than the actual lower boundaries of dengue counts. Further, the highest RMSE reported in CM. Hence, it is not suitable for predictions. Validation results from CRM and CCRM were close to each other

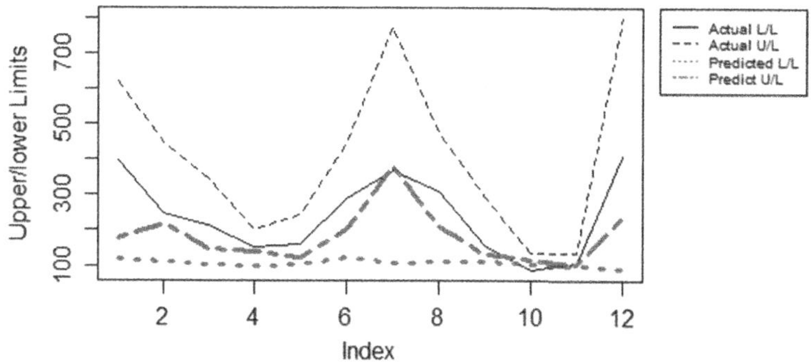

Fig. 11 Actual with forecasted lower/upper boundaries from CM

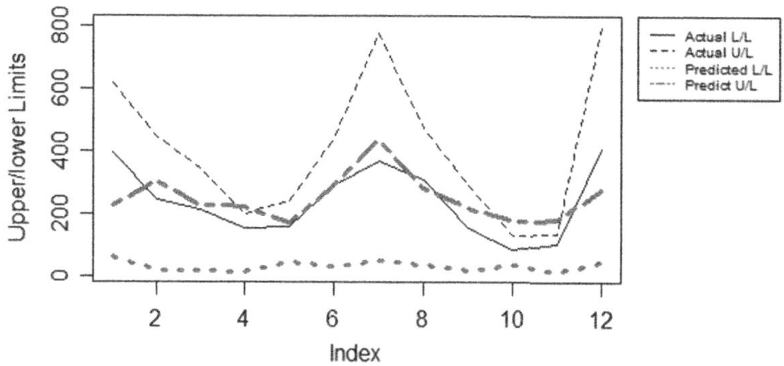

Fig. 12 Actual with forecasted lower/upper boundaries from CRM

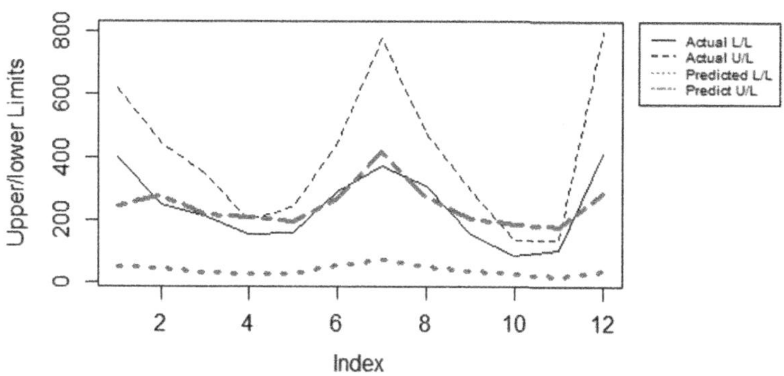

Fig. 13 Actual with forecasted lower/upper boundaries from CCRM

and can use for dengue predictions but more smoothness can be seen in CCRM validation plot.

2.5 Interval Least Squares Algorithm

Interval arithmetic define as follows. Let $c = [\underline{c}, \overline{c}]$ and $d = [\underline{d}, \overline{d}]$ be two non-empty intervals with lower and upper limits and *op* be one of the operators (+, −, *, ÷). If *op* is ÷ then $0 \notin d$. The operations are illustrating below:

$$c + d = [\underline{c} + \underline{d}, \overline{d} + \overline{c}]$$

$$c - d = [\underline{c} - \overline{d}, \overline{c} - \underline{d}]$$

$$c * d = [\min(\underline{c} * \underline{d}, \overline{d} * \overline{c}, \underline{c} * \overline{d}, \overline{c} * \underline{d}), \max(\underline{c} * \underline{d}, \overline{d} * \overline{c}, \underline{c} * \overline{d}, \overline{c} * \underline{d})]$$

$$c/d = [\min(\underline{c}/\underline{d}, \overline{c}/\overline{d}, \underline{c}/\overline{d}, \overline{c}/\underline{d}), \max(\underline{c}/\underline{d}, \overline{c}/\overline{d}, \underline{c}/\overline{d}, \overline{c}/\underline{d})]$$

Consider interval-valued observations for input and output variables. The interval least squares algorithm using interval arithmetic has the following main steps:

- Evaluate the design matrix X as given below with interval arithmetic.

$$X = \begin{pmatrix} 1 & x_{11} & \ldots & x_{p1} \\ 1 & x_{12} & \ldots & x_{p2} \\ \ldots & \ldots & \ldots & \ldots \\ 1 & x_{1n} & \ldots & x_{pn} \end{pmatrix}$$

- Perform the *QR* factorization on the mid-point matrix, X_{mid} of X such that

$$X_{\text{mid}} = Q * R$$

- Calculate $z = Q^T * Y$
- Find an initial β (model parameters) with one or both of the followings:

  (i) Compute an initial point solution of β by solving $z_{mid} = R\beta$, with backward substitution.
  (ii) Compute an initial interval solution of β by solving $z = R\beta$,

with interval arithmetic.

According to Peiris et al. [12], if the centre values of coefficients and boundaries of input and output data are positive, then the boundary estimates for regression coefficients can be found as follows:

Consider the interval multiple regression model as

$$\left[\underline{y_i}, \overline{y_i}\right] = \left[\underline{\beta_0}, \overline{\beta_0}\right] + \left[\underline{\beta_1}, \overline{\beta_1}\right][\underline{x}_{1i}, \overline{x}_{1i}] + [\underline{\beta_2}, \overline{\beta_2}][\underline{x}_{2i}, \overline{x}_{2i}] + \cdots + [\underline{\beta_p}, \overline{\beta_p}][\underline{x}_{pi}, \overline{x}_{pi}]$$

The centre values of interval regression coefficients $\left[\underline{\beta_j}, \overline{\beta_j}\right]$ for $j = 0, 1, 2, \ldots p$ are in vector β'_m.

Consider $\varepsilon_j > 0$ that satisfies

$$\underline{\beta_j} = \beta'_j - \varepsilon_j$$

$$\overline{\beta_j} = \beta'_j + \varepsilon_j$$

Using upper boundary values for dependent variable, y_i can be written by considering interval arithmetic.

$\overline{y_i} = (\beta'_0 + \varepsilon_0) + (\beta'_1 + \varepsilon_1)\overline{x}_{1i} + \ldots + (\beta'_p + \varepsilon_p)\overline{x}_{pi}$

Rearranging the above equation, we have

$$\overline{y_i} = (c_0) + c_1\overline{x}_{1i} + c_2\overline{x}_{2i} + \cdots + c_p\overline{x}_{pi}$$

where $c_j = \beta'_j + \varepsilon_j$ for $j = 0, 1, 2, \ldots p$.

All the variables and coefficients are in the final equation are point values. Fitting an ordinary least squares regression on upper boundary values of independent and dependent variables leads to the estimated values for c_j coefficients in the last equation. Substituting values in the equation, $c_j = \beta'_j + \varepsilon_j$, the only unknown value of ε_j can be estimated. Hence, interval multiple regression equation based on interval least squares algorithm can be formulated.

Let interval $y_{est} = \left[\underline{y}_{iest}, \overline{y}_{iest}\right]$, be an estimation of an interval $y = \left[\underline{y_i}, \overline{y_i}\right]$. The left error is equal to $E_L = \underline{y}_{iest} - \underline{y_i}$ and right error is equal to $E_R = \overline{y}_{iest} - \overline{y_i}$. The interval least squares approximation minimizes the total error of regression which is equal to $\sum_{i=1}^{n} E_L^2 + \sum_{i=1}^{n} E_R^2$.

Accuracy Ratio (AR) and Average Accuracy Ratio (AAR) can be used as a quantitative assessment of quality of the interval approximation. *AR* is given by

$$AR = acc(y, y_{est}) = \begin{cases} 100\% y = y_{est} \\ \frac{w(y \cap y_{est})}{w(y \cup y_{est})} (y \cap y_{est}) \neq \varnothing \\ 0 Otherwise \end{cases}$$

where $w(.)$ denotes width of the interval.

AAR is given by: $AAR = \frac{\sum_{i=1}^{n} acc(y_i, y_{iest})}{n}$

2.6 Applications of Interval Regression Based on Interval Least Squares Algorithm

Monthly lower limit and upper limit for the dengue count as well as monthly lower limit and upper limit for the rainfalls, in Colombo, Sri Lanka, were considered to illustrate the application. Monthly lower and upper limits were identified from the collected weekly data. Data selection for the model development was from January 2009 to December 2015. Data from January to December in 2016 were used to validate the model.

The spread of the dengue highly depends with available mosquito population. Since there is no direct measurement for the mosquito population, 4 weeks leading dengue cases were taken as an estimate. Therefore, 10-week lag period for rainfall with 4 weeks leading dengue cases were modelled with reported dengue cases using interval multiple regression technique. The analysis was done using R software [17].

Following the steps of interval least squares algorithm, an initial point solution for the regression parameters was estimated. Then ordinary least squares regression was fit on the upper boundary values of the variables to estimate the regression parameters. The following two regression lines for lower and upper bounds were generated by adding and subtracting ε_j values from the mid-point solution.

$$Dengue_{upper_{limit}}$$

$$= 34.7 + 2.68701 * Rainfall_{upper\_limit} + 0.6451 * Dengue\_lag\_four_{upper\_limit}$$

$$Dengue_{lower\_limit} = 34.007 + 2.52619 * Rainfall_{lower\_limit} + 0.6521 * Dengue\_lag\_four_{lower\_limit}$$

The plot of actual lower bounds for the dengue cases and the predicted lower bound for the dengue cases are depicted in Fig. 14. That is for upper bounds are shown in Fig. 15. It can be concluded from Figs. 14 and 15, that the predicted values from the interval multiple regression were followed the actual values. The calculated average accuracy ratio was 72%. Further according to validation plot in Fig. 16 it can be seen

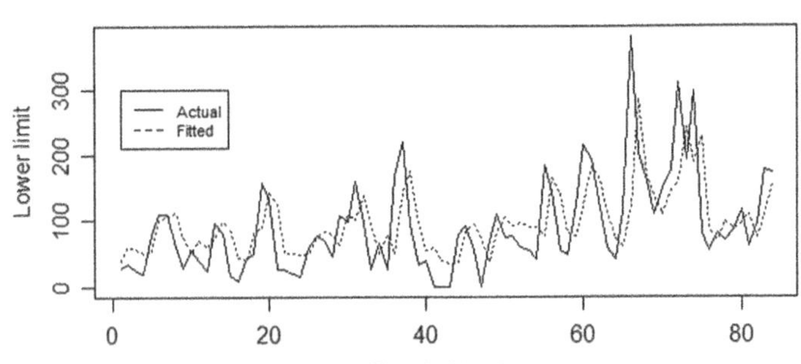

Fig. 14 Actual and predicted lower bounds of dengue cases—Colombo Series

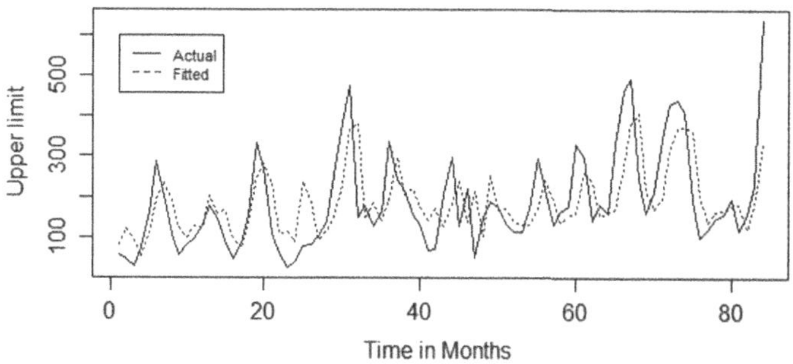

Fig. 15 Actual and predicted upper bounds of dengue cases—Colombo series

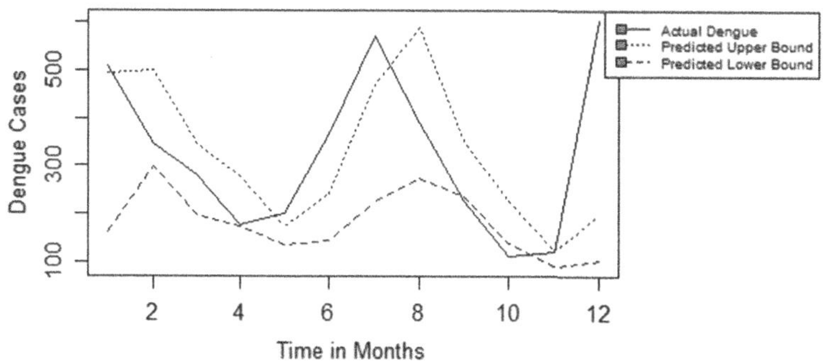

Fig. 16 Plot of validation—Colombo series

that the actual dengue cases were within the predicted lower and upper boundary lines of the interval multiple regression. Exact predictions may not necessary to control the disease while prediction in an interval with expected lower and upper boundaries are sufficient to take decisions on implementing controlling actions. Hence, there is a possibility of predicting the dengue cases in Colombo, Sri Lanka, using the interval regression procedure based on interval least squares algorithm.

Monthly lower limit and upper limit for the dengue count, humidity as well as monthly lower limit and upper limit for the rainfalls, in Jakarta, Indonesia, were further considered to illustrate the application of the interval regression method. Monthly lower and upper limits were identified from the collected weekly data. Climatic data were acquired from the Indonesian Agency for Meteorology Climatology and Geophysics (BMKG) and the dengue data in Jakarta were gathered from the Jakarta health office. Data selection for the model development was from January 2008 to December 2015. Data from January to December in 2016 were used to validate the model.

The distributions of lower and upper limits of the reported dengue cases in Jakarta are depicted in Fig. 17. Some seasonal fluctuations are apparent within every year. Reported lower and upper bounds in 2016 were higher than that of former months indicated the increase in dengue cases in recent years.

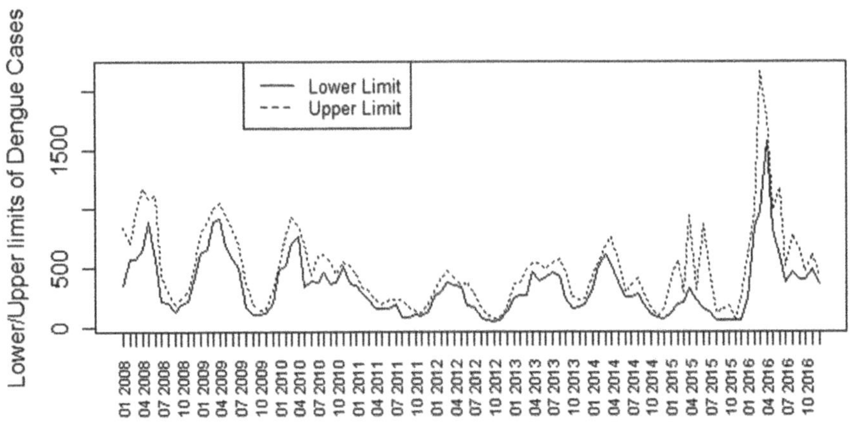

Fig. 17 Distribution of lower limits and upper limits of dengue cases—Jakarta series

According to Fig. 18, the distribution of lower limits of monthly rainfall in Jakarta is always varied around zero. Larger gaps within lower and upper intervals were apparent in Fig. 18. The maximum rainfall was 555 mm during the data gathering period which was occurred on April 2015. Much more variations can be seen in upper limits of rainfall.

The distributions of monthly lower and upper limits of humidity in Jakarta, Indonesia, are depicted in Fig. 19. Some seasonal and cyclic variations are apparent in Fig. 19. The lower limit varies in between 62 and 83, whereas the upper limit in between 69 and 89.

Cross-correlation analysis revealed that two months leading rainfall and 2 month leading humidity effect on reported dengue cases in Jakarta. Two months leading dengue cases were taken as an estimate for the mosquito population. Therefore, considering lag periods for the three variables, the reported dengue cases were

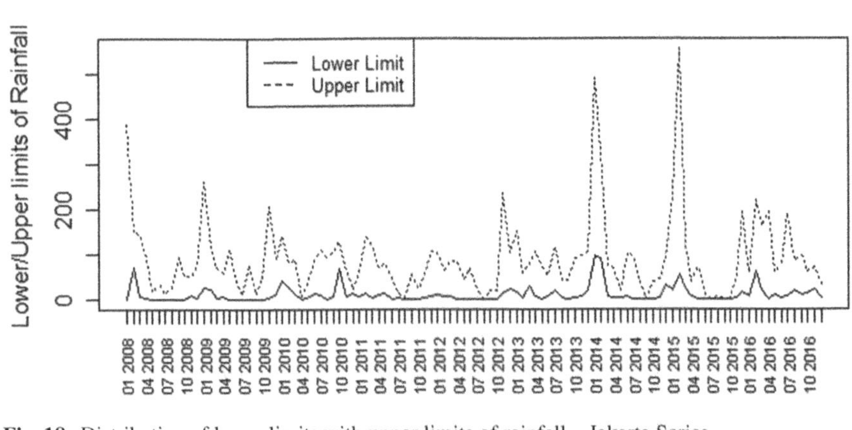

Fig. 18 Distribution of lower limits with upper limits of rainfall—Jakarta Series

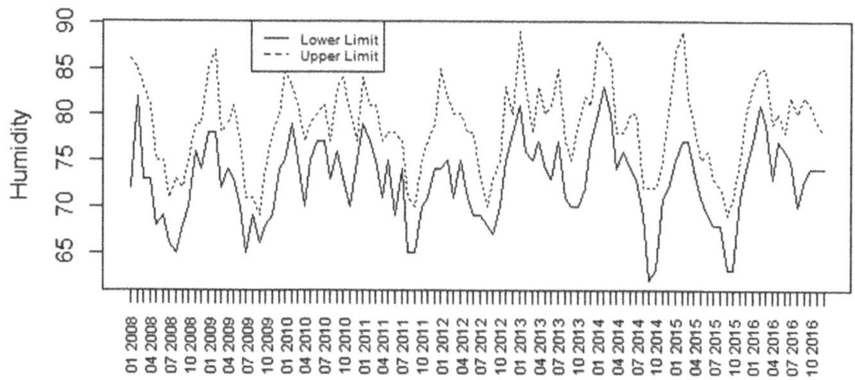

Fig. 19 Distribution of lower limits with upper limits of humidity—Jakarta Series

modelled using interval multiple regression technique. An initial point solution for the regression parameters of the interval regression line was estimated by following the interval least squares algorithm. Then ordinary least squares regression was fit on upper boundary values of the variables to estimate the regression parameters. By equating results $\varepsilon_j;\ j = 1, 2, 3, 4$ values were obtained. The following two regression lines for lower and upper bounds were generated by adding and subtracting ε_j values from the mid-point solution.

$$Dengue_{lower\_limit} = 1262.36 + 0.41 * Rainfall_{lower\_limt} + 2.68 * Humidity_{lower\_limit} +$$

$$0.3514 * Dengue\_lag\_two_{lower\_limit}$$

$$Dengue_{upper\_limit} = 1588 + 0.53 * Rainfall_{upper\_limit} + 3.258 * Humidity_{upper\_limit} +$$

$$0.4314 * Dengue\_lag\_two_{upper\_limit}$$

The plot of actual dengue cases together with predicted dengue cases from the interval regression in the lower bound and upper bound are depicted in Figs. 20 and 21, respectively. It can be concluded from Figs. 20 and 21 that the predicted values generated from the interval multiple regression are following the actual values. The calculated average accuracy ratio was 62%. Further according to validation plot in Fig. 22, it can be seen that the actual dengue cases were within the predicted lower and upper boundary lines of the interval multiple regression. Predictions in intervals with expected lower and upper boundaries are sufficient to take decisions on implementing controlling actions to mitigate the impact of the dengue disease. Hence, there is a possibility of predicting the dengue cases in Jakarta, Indonesia, using the interval regression procedure based on interval least squares algorithm.

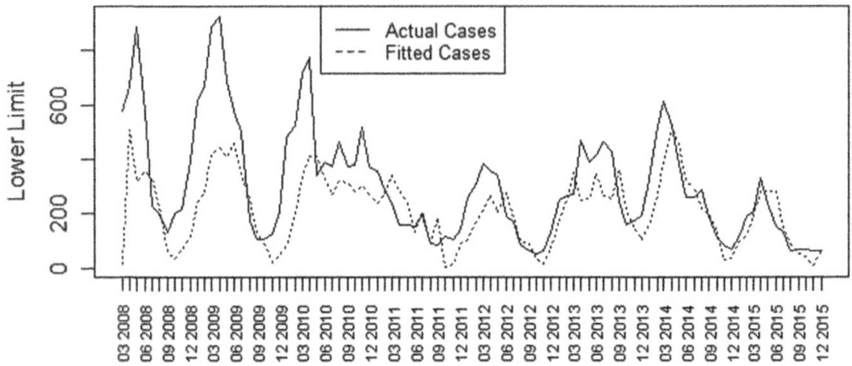

Fig. 20 Actual and predicted lower bounds of dengue cases—Jakarta series

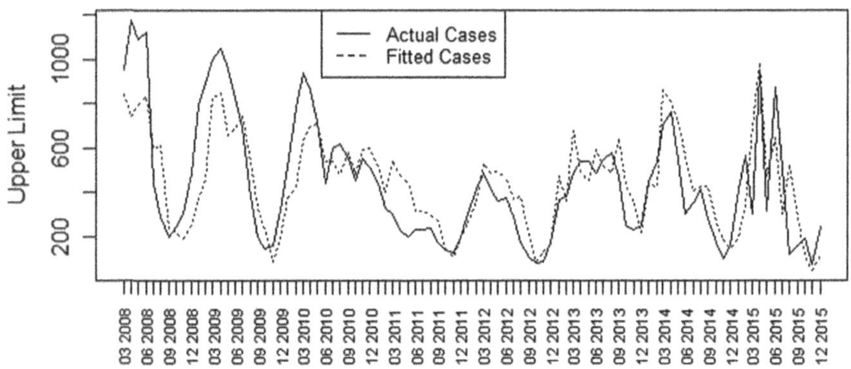

Fig. 21 Actual and predicted upper bounds of dengue cases—Jakarta series

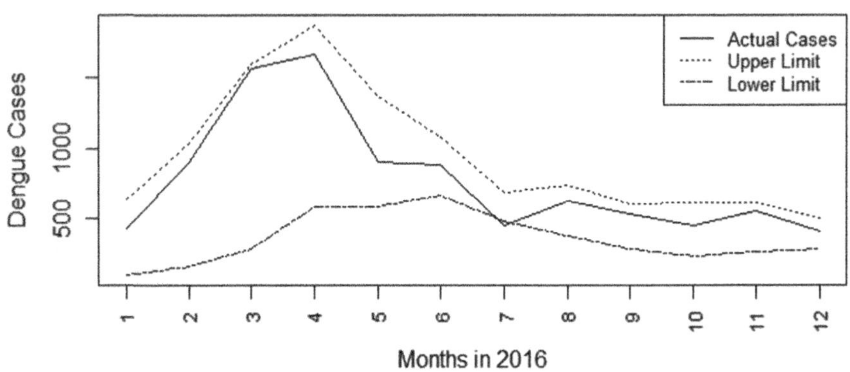

Fig. 22 Plot of validation—Jakarta series

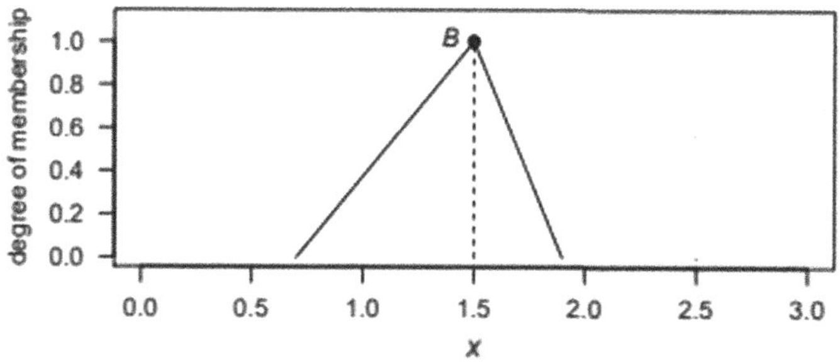

Fig. 23 Triangular fuzzy number

2.7 *Fuzzy Number*

Fuzzy numbers can be defined on uncertainty situations and applicable in any scenario where imprecise information are involved [18]. Suppose a real value number x belongs to the fuzzy set B, with a degree of membership ranges from 0 to 1. The degrees of membership of x are defined by the membership function

$$\mu_B(x) : x \to [0, 1]$$

where $\mu_B(x*) = 0$ means that the value of x^* is not included in the fuzzy number B, while

$$\mu_B(x*) = 1$$

means that x^* is included in B.

2.7.1 Triangular Fuzzy Number (TFN)

TFN is a special class of fuzzy numbers which has three parameters. One of the parameters is for the central value where the degree of membership is equal to 1. The others are for left spread and for the right spread of the data. If the left spread equal to the right spread, then it is defined as a symmetric triangular fuzzy number. A triangular fuzzy number is depicted in the following Fig. 23.

2.8 *Fuzzy Regression*

Fuzzy regression is an application of fuzzy platform for conventional regression analysis. Fuzzy regression analysis gives a fuzzy relationship between dependent and independent variables where vagueness is present in the data [19]. The input data may be crisp values or fuzzy numbers whereas the conventional ordinary least squares regression can handle only crisp measures. When the output is in the form of fuzzy representative then there is a chance to obtain lower and upper approximation models which represent the fuzziness of the output.

2.9 Fuzzy Linear Regression Using the Possibilistic Linear Regression Method

The method was developed by Tanaka et al. in 1989 [21]. The method deals with the response in the form of a symmetric fuzzy number whereas with the predictors as crisp numbers. The fuzzy regression coefficients compromise spread as well as central tendency. That is predictions are in the form of symmetric triangular fuzzy number coefficients. The possibilistic method minimizes the fuzziness of the model through reducing the spread of the fuzzy coefficients. The 'plr' function available in 'fuzzyreg' package of R can be used to fit the possibilistic linear regression.

2.10 Application of Possibilistic Linear Regression (PLR) Method

Monthly reported dengue count, humidity as well as monthly rainfalls, in Jakarta, Indonesia, were considered to illustrate the application of PLR. Considering 2 months' lag period for rainfall and humidity, data were modelled using the possibilistic regression method. Additionally, 2 month leading dengue cases used as an independent variable as an approximation for the available mosquito distribution. Data from 2008 to 2015 were used for the model development and data in 2016 were used for the model validation. All possible combinations of independent variables were tested in the application. The various models that were constructed and the Mean Absolute Percentage Error (MAPE) values of the validation set are summarized in Table 5.

Table 5 Summary of the fitted models and MAPE values—PLR method

Model	Independent variables in the model	MAPE
1	Humidity	0.34
2	Dengue data with lag	0.325
3	Rainfall	0.328
4	Humidity, dengue data with lag	0.49
5	Rainfall, humidity	0.33
6	Rainfall, dengue data with lag	0.23
7	Rainfall, humidity, dengue data with lag	0.46

```
Coefficients in form of symmetric triangular fuzzy numbers:

                     center left.spread right.spread
Intercept        322.299895 279.5075064  279.5075064
Rainfall           1.800611   0.8996712    0.8996712
Dengue with lag    0.234397   0.1987469    0.1987469
```

Fig. 24 Coefficients of Model 6—PLR method

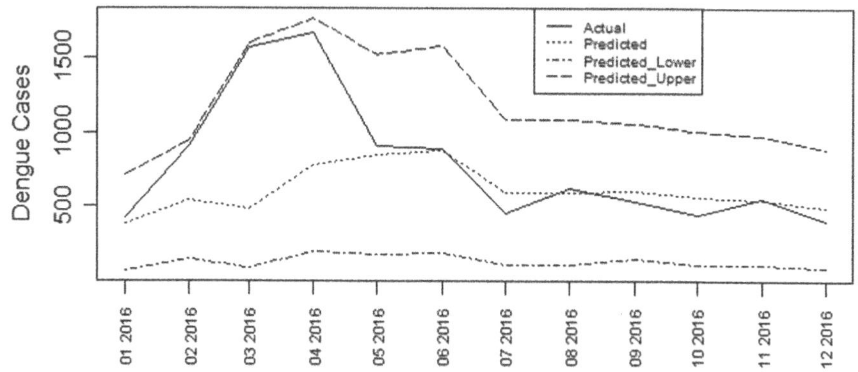

Fig. 25 Dengue cases with lower and upper boundaries—PLR method

Minimum MAPE value is recorded for Model 6 and the value is less than 10%. Therefore, Model 6 can be selected as the most appropriate PLR model to predict monthly dengue cases in Jakarta, Indonesia within the considered independent variables.

The coefficients of Model 6 are given in Fig. 24. The central tendency (Predicted) with left (Predicted_Lower) and right spreads (Predicted_Upper) in Fig. 25 determines the support interval of predictions. Upper approximation model from the right spread and lower approximation model from left spread were obtained to reflect the fuzziness of the dengue count in Jakarta.

The predicted values from the fuzzy regression model (Model 6) and actual reported dengue cases are within the lower and upper approximation models indicates the possibility of prediction through the fitted fuzzy regression model. Therefore, fitted PLR model can be used to predict dengue cases in Jakarta, Indonesia.

2.11 *Possibilistic Linear Regression with Least Squares (PLRLS) Method*

This method is one of the methods available under the fuzzy regression techniques. This method was proposed by H. Lee and H. Tanaka in 1999 to deal with crisp inputs and fuzzy output. This method fits the model which compromises spreads and the central tendency by using the possibility and the least squares approach [20]. The input data represent crisp measures and the model output is in the form of non-symmetrical triangular fuzzy number. The basic idea of the method is to minimize the fuzziness of the model by minimizing the total spread of the fuzzy coefficients subject to including all the given data. The least squares application minimizes the distance between the output and the actual observed output. The general regression model is given as

$$Y = A_0 + A_1 x_1 + \cdots + A_n x_n$$

where Y is the fuzzy output and $A_i^{'}s$, $i = 1, \ldots .n$ are the fuzzy coefficients. x_1 to x_n are non-fuzzy input variables. Fuzzy coefficients assumed as triangular fuzzy numbers. The 'plrls' function available in 'fuzzyreg' package of R can be used to fit the Possibilistic linear regression with least squares method.

2.12 *Application of Possibilistic Linear Regression with Least Squares (PLRLS) Method*

Same data set described in the application of PLR method was used to illustrate the application of PLRLS method. All possible combinations of independent variables were tested under this application. The various PLRLS models that were constructed and the corresponding MAPE values of the validation set are summarized in Table 6.

Minimum MAPE value is recorded for Model 6 and the value is less than 10%. Therefore, Model 6 can be selected as the most appropriate PLRLS model to predict monthly dengue cases in Jakarta, Indonesia, within the considered independent variables.

The coefficients of Model 6 are shown in Fig. 26. The left spread and the right spread determine the lower and upper boundary of the interval where the degree of membership equals to 0. The central tendency (Predicted) with left (Predicted_Lower) and right spreads (Predicted_Upper) in Fig. 27 shows the support interval of predictions. Upper approximation model from the right spread and lower

Table 6 Summary of fitted models and MAPE values—PLRLS method

Model	Independent variables in the model	MAPE
1	Humidity	0.352
2	Dengue data with lag	0.358
3	Rainfall	0.36
4	Humidity, dengue data with lag	0.39
5	Rainfall, humidity	0.33
6	Rainfall, dengue data with lag	0.32
7	Rainfall, humidity, dengue data with lag	0.37

```
Coefficients in form of non-symmetric triangular fuzzy numbers:

                     center  left.spread right.spread
(Intercept)       68.839533   18.2197305     436.2997
Rainfall           1.967385    0.4682200       0.0000
Dengue with lag    0.550309    0.4804088       0.0000
```

Fig. 26 Coefficients of Model 6—PLRLS method

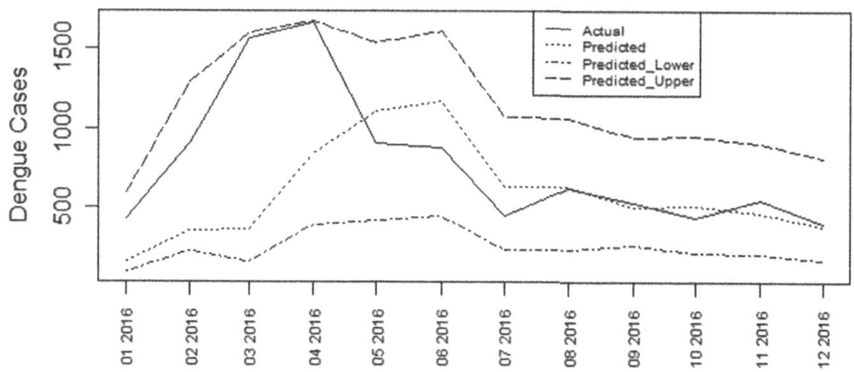

Fig. 27 Dengue cases with lower and upper boundaries—PLRLS method

approximation model from left spread were obtained to reflect the fuzziness of the dengue count in Jakarta.

The predicted values from the fuzzy regression model (Model 6) and actual reported dengue cases are within the lower and upper approximation models (lower and upper boundaries) indicates the possibility of prediction through the fitted PLRLS fuzzy regression model. Therefore, fitted fuzzy regression model can be used to predict dengue cases in Jakarta, Indonesia.

2.13 Fuzzy Linear Regression Using the Multi-objective Fuzzy Linear Regression (MOFLR) Method

The method was developed by Nasrabadi et al. in 2005 [22]. The specialty of the method is that it can handle outliers of the data series if it is present. The method combines the least squares approach with the possibilistic approach in fitting central tendency as well as spreads. The method deals with the response together with the inputs in the form of symmetric triangular fuzzy numbers. The fuzzy regression coefficients of this method compromise spread as well as central tendency. That is, predictions are in the form of symmetric triangular fuzzy number coefficients. The 'moflr' function available in 'fuzzyreg' package of R can be used to fit the MOFLR method.

Table 7 Summary of Fitted Models and MAPE values—MOFLR method

Model	Independent variables in the model	MAPE
1	Humidity	1.25
2	Dengue data with lag	0.37
3	Rainfall	0.36
4	Humidity, dengue data with lag	0.41
5	Rainfall, humidity	0.334
6	Rainfall, dengue data with lag	0.335
7	Rainfall, humidity, dengue data with lag	0.37

```
Coefficients in form of symmetric triangular fuzzy numbers:

                center left.spread right.spread
intercept  -592.813923   -124.0501    -124.0501
Rainfall      1.798511     -0.1419      -0.1419
Humidity     11.716517      3.9917       3.9917
```

Fig. 28 Coefficients of Model 5—MOFLR method

2.14 *Application of MOFLR Method*

The data set used in the Sect. 2.10 was used here to illustrate the application of MOFLR method. All possible combinations of independent variables available in the data set were considered under this method. The MOFLR models that were fitted and the corresponding MAPE values of the validation set are given in Table 7.

Minimum MAPE value is recorded for Model 5. Further, the reported value is less than 10%. Therefore, Model 5 can be selected as the most appropriate MOFLR model to predict monthly dengue cases in Jakarta, Indonesia, within the considered independent variables.

The coefficients of Model 5 are given in Fig. 28. The supported interval of predictions is shown in Fig. 29. The fuzziness of the dengue count in Jakarta is depicted by the upper approximation model from the right spread and lower approximation model from left spread.

The predicted values from the fuzzy regression model (Model 5) as well as upper boundary of Model 5 are lower than the actual reported dengue cases as depicted in Fig. 29. Therefore, the fitted MOFLR model is not appropriate in predicting the dengue cases in Jakarta, Indonesia.

3 Conclusion and Discussion

This chapter deals with multiple interval regression modelling techniques, namely, CM, CRM, CCRM, interval regression based on interval least squares algorithm and fuzzy regression techniques (Possibilistic linear regression, fuzzy linear regression

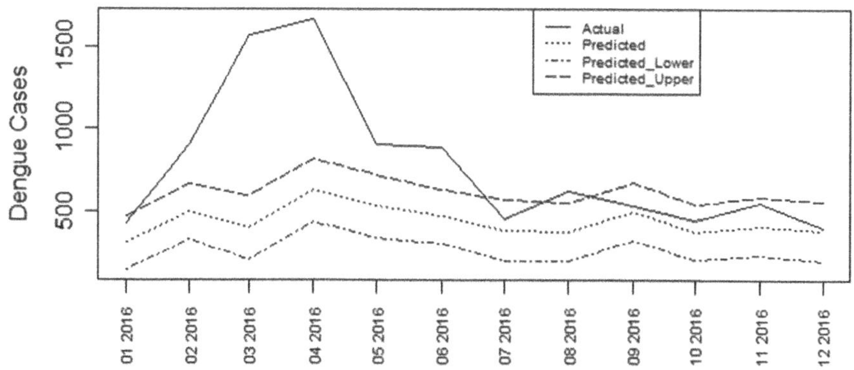

Fig. 29 Dengue cases with lower and upper boundaries—MOFLR method

using the possibilistic linear regression with least squares method and fuzzy linear regression using the multi-objective fuzzy linear regression method). The application of each technique was illustrated using dengue cases reported in Jakarta, Indonesia, and/or dengue cases reported in Colombo, Sri Lanka, as test cases. Monthly rainfall, average minimum temperature and average maximum temperature data in Colombo were used as explanatory variables, whereas monthly rainfall and average humidity data in Jakarta were utilized for the purpose.

The conventional regression analysis is based on the exact data measures of variables. But all of these variables associate with unknown and uncontrollable factors in the real life scenario. Therefore, it is hard to capture these uncertainties through exact measures. Hence, interval representation of exact data point with a lower and an upper boundary is an attempt to capture uncertainties associate with the measure of a variable. There is a growing tendency in the applications of interval-valued data analysis over recent years. Particularly, in the area of regression analysis, there are various techniques to handle interval-valued dependent and interval-valued independent variables. Further, interval predictions are particularly appropriate in the situations of exact predictions are not essential. Interval-valued analysis in dengue disease is important as actions taking towards controlling the disease do not depend on the exact number but on the magnitude of the values represented by the interval. However, interval-valued (lower and upper boundary) representation for an exact data value provides an additional information for the spread of the value and useful in making more precise decisions.

The chapter initiates with the CM, which is the first attempt in interval regressions in the area of statistics. Then the method extended to CRM to capture the spread of the variables and then to CCRM. The CCRM has an advantage of its lower bound cannot be greater than its upper bound. The applications of all these three techniques were illustrated and CCRM was more appropriate for the example given. Generally, the most appropriate interval regression technique depends with the application and the decision-maker should try with the available techniques and invent new techniques if the available ones are not sufficient.

The coefficients of regression parameters usually estimate using ordinary least squares algorithm in classical regression analysis. The interval regression based on interval least squares algorithm is an attempt to capture uncertainties associate with the measures of variables through interval arithmetic. The application of the method was illustrated using test cases in Colombo, Sri Lanka, and Jakarta, Indonesia. The predicted upper and lower boundaries for dengue cases were followed the patterns of the actual distributions of dengue cases suggested that there is a possibility of taking decisions on dengue management based on the technique.

The fuzzy regression is a fuzzy representation for regression analysis. It is a method of soft computing. The method is an attempt to tolerate imprecision and uncertainty associated with exact representation of variables. In this chapter, three fuzzy regression techniques namely, PLR, PLRLS and MOFLR were illustrated using dengue cases reported in Jakarta, Indonesia. Various fuzzy regression procedures are available to deal with fuzzy/non-fuzzy inputs and fuzzy/non-fuzzy outputs. Some characteristics of the three techniques used in the chapter are summarized in Table 8.

Although the MOFLR method showed poor performances in the validation set of the given example, the method deals with symmetric triangular fuzzy representation of variables in both input and output. Actual reported dengue cases in the example were within the lower and upper predictions generated by the PLR and PLRLS methods. Therefore, the methods can be used in Jakarta to implement controlling actions in dengue management. The successfulness of the fuzzy regression approaches is based on the application and the inspiring idea of interval or fuzzy representation of variables is rather important. None of the interval based regression methods works best in all cases. In broad sense, the selection of methods is based on size of the sample, type of variables, number of variables, computational and space complexity etc.

Soft computing is an advanced and innovative methodology to construct intelligent systems, to retain human-like capabilities in the process and has the ability to adjust and study in varying environments. It differs from hard computing which is the conventional computing. Soft computing methods tolerate the uncertainty and imprecision while giving robust and approximate solutions. In real world, many problems cannot be solved logically and some can be solved theoretically but essentially difficult due to its necessity of large number of resources, huge time of computation

Table 8 Some characteristics of the three fuzzy regression procedures

Method	Number of independent variables allowed	Type of the independent variable	Type of the dependent variable
PLR	Infinite	Crisp	Symmetric triangular fuzzy number
PLRLS	Infinite	Crisp	Symmetric triangular fuzzy number
MOFLR	Infinite	Symmetric triangular fuzzy number	Symmetric triangular fuzzy number

etc. Thus, soft computing methodologies can be applied in such scenarios to obtain more realistic solutions. This chapter gives an insight to numerous interval regression techniques that provide predictions in intervals and it is a better approach for conventional regressions that provide point estimates. Although the chapter illustrates applications of interval regression through dengue cases reported in Colombo and Jakarta, the methods can be applied to any area including engineering, actuarial science, agriculture, medical diagnosis, etc. Soft and hard computing methodologies have their inherent pros and cons and therefore by combining these methods to build hybrid models will overcome the drawbacks of each of the methods. This chapter is intended as a guide for implementing interval regressions as well as it will give an inspiration for researchers and decision-makers to develop an extended or new methodologies in the area of interval-valued regressions.

References

1. Wikipedia Contributors (2021) 2019–2020 dengue fever epidemic. In *Wikipedia, the free encyclopedia*. https://en.wikipedia.org/w/index.php?title=2019%E2%80%932020_dengue_fever_epidemic&oldid=1001311223. Accessed 29 March 2021.
2. Sirisena PD, Noordeen F (2014) Evolution of dengue in Sri Lanka: changes in the virus, vector and climate. Int J Infect Dis 19:6–12
3. Epidemiology Unit (2020) Ministry of Healthcare and Nutrition, Sri Lanka, Dengue Update. http://www.epid.gov.lk
4. Setiati TE, Wagenaar JFP, Kruif MD, Mairuhu ATA, Grop ECM, Soemantri A (2006) Changing epidemiology of dengue haemorrhagic fever in Indonesia.
5. Promprou S (2013) Multiple linear regression model to predict dengue haemorrhagic fever (DHF) patients in Kreang Sub-District, Cha-Uat District, Nakhon Si Thammarat, Thailand. J Appl Sci Res 9(12):6193–6197
6. Azman NNB, Karim SABA (2018) Assessing climate factors on dengue spreading in state of Perak. IOP Conf Series: J Phys 1123:012023. https://doi.org/10.1088/1742-6596/1123/1/012026
7. Hii YL, Zhu H, Ng N, Ng LC, Rocklöv J (2012) Forecast of dengue incidence using temperature and rainfall. PLOS Negl Trop Dis. https://doi.org/10.1371/journal.pntd.0001908
8. Cheong YL, Burkart K, Leitão PJ, Lakes T (2013) Assessing weather effects on dengue disease in Malaysia. Int J Environ Res Public Health 10(12):6319–6334. https://doi.org/10.3390/ijerph10126319
9. Billard L, Diday E (2000) Regression analysis for interval-valued data. In: Kiers HAL, Rasson JP, Groenen PJF, Schader M (eds) Data analysis, classification and related methods: proceedings of the seventh conference of the international federation of classification societies, Namur. Springer, Berlin, pp 369–374.
10. Lima ED, Carvalho FDT (2008) Centre and Range method for fitting a linear regression model to symbolic interval data. Comput Stat Data Anal 52(3):1500–1515
11. Lima ED, Carvalho FDT (2010) Constrained linear regression models for symbolic interval-valued variables. Comput Stat Data Anal 54(2):333–347
12. Peiris HOW, Chakraverty S, Perera SSN, Ranwala SMW (2018) Novel interval multiple linear regression model to assess the risk of invasive alien plant species. JSc EUSL 9(1):12–30
13. Billard L, Diday E (2007) Symbolic data analysis: conceptual statistics and data mining. Wiley-Interscience, New York
14. Vilém N, Irina P, Antonín D (2016) Insight into Fuzzy Modeling. Wiley, New York

15. Pavel S., Jaroslav M., (2018), Models used in Fuzzy Linear Regression,17th Conference on Applied Mathematics, Slovak University of Technology, Bratislava, 955–964.
16. Romero D, Olivero J, Real R, Guerrero JC (2019) Applying fuzzy logic to assess the biogeographical risk of dengue in South America. Parasites Vectors 12:428
17. R Core Team (2021) R: a language and environment for statistical computing. R Foundation for Statistical Computing, Vienna, Austria. https://www.R-project.org/
18. Guanrong C, Trung TP (2001) Introduction to fuzzy sets, fuzzy logic, and fuzzy control systems. CRC Press, London.
19. Kahraman C, Beşkese A, Bozbura FT (2006) Fuzzy regression approaches and applications. In: Kahraman C (eds) Fuzzy applications in industrial engineering, studies in fuzziness and soft computing, vol 201. Springer, Berlin.
20. Tanaka H, Watada J (1988) Possibilistic linear systems and their application to the linear regression model. Fuzzy Sets Syst 27:275–289
21. Tanaka H, Hayashi I, Watada J (1989) Possibilistic linear regression analysis for fuzzy data. Eur J Oper Res 40:389–396
22. Nasrabadi MM, Nasrabadi E, Nasrabady AR (2005) Fuzzy linear regression analysis: a multi-objective programming approach. Appl Math Comput 163:245–251

Fuzzy-Affine Approach in Dynamic Analysis of Uncertain Structural Systems

S. Rout and S. Chakraverty

Abstract The dynamic analysis of a structural system with different material and geometric properties leads to linear eigenvalue problem such as generalized (or standard) eigenvalue problem. In general, the material and geometric properties are assumed to be in the form of crisp (or exact). However, due to several errors and insufficient or incomplete information of data, the uncertainties are assumed to be present in the material and geometric properties. These uncertain material and geometric properties may be modeled through convex normalized fuzzy sets. In standard fuzzy arithmetic, all the operands are assumed to be independent of each other. But when they are partially or completely dependent on each other, the standard fuzzy arithmetic results in a wider range. This situation is known as the "dependency problem" or "overestimation problem". In this regard, a fuzzy-affine approach is developed to overcome the dependency problem. This proposed approach may improve the outer enclosures and give tighter bounds to the fuzzy solution set. This chapter deals with the dynamic analysis of various structural systems viz. multi degrees-of-freedom spring-mass structural system, multi-storey frame structural system, etc. by adopting the proposed approach. Several numerical examples related to the dynamic analysis of these structural systems have been worked out to illustrate the reliability and efficiency of the present approach.

Keywords Affine arithmetic · Fuzzy-affine arithmetic · Dynamic analysis of structural system · Fuzzy generalized eigenvalue problem · Fuzzy standard eigenvalue problem

1 Introduction

The linear structural dynamic problems of various structural systems viz. spring-mass structural system with multi degrees of freedom, multi-storeyed shear building structure, multi-storeyed frame structure, etc. having different types of material as well

S. Rout (✉) · S. Chakraverty
Department of Mathematics, National Institute of Technology Rourkela, Rourkela, Odisha, India

S. Chakraverty (ed.), *Soft Computing in Interdisciplinary Sciences*,
Studies in Computational Intelligence 988,
https://doi.org/10.1007/978-981-16-4713-0_2

as geometric properties lead to linear eigenvalue problems (LEPs) [3] such as generalized (or standard) eigenvalue problems (GEPs or SEPs). LEP has a wide variety of applications in several scientific and engineering fields viz. structural mechanics, control theory, fluid dynamics, electrical circuits, etc. For instance, the LEP plays a major role in structural dynamic problems. Hence, generalized (or standard) eigenvalue problem may be considered as the backbone of many science and engineering fields. In this regard, efficient handling of these eigenvalue problems in an uncertain environment is a challenging and important task to deal with.

In general, the material and geometric properties are assumed to be in the form of crisp (or exact). However, due to several errors and insufficient or incomplete information of data, the uncertainties are assumed to be present in the material and geometric properties. Traditionally, these uncertainties are modeled through probabilistic approaches, which are unable to deliver efficient and reliable solutions without a sufficient amount of experimental data. Thus, these uncertain material and geometric properties may be modeled through convex normalized fuzzy sets. Few literature studies related to the basic concepts and properties of fuzzy set theory have been discussed in this section. Zadeh [36] first introduced the concepts of fuzzy sets and fuzzy numbers in 1965. In his work, Zadeh generalized the classical sets with characteristic functions, which vary over [0, 1]. Fuzzy set theory is an important tool to handle uncertain parameters. In this regard, excellent books have been written by different authors viz., [6, 9, 13, 14, 19, 37, 38], etc.

Further, in standard fuzzy arithmetic, all the operands are assumed to be independent of each other. But when they are partially or completely dependent on each other, the standard fuzzy arithmetic results in a wider range. This situation is known as the "dependency problem" or "overestimation problem". Because of this problem, standard fuzzy arithmetic overestimates the width of the resulting interval solutions. In this regard, affine arithmetic proves itself as an efficient tool to handle the overestimation problem and results with comparatively tighter enclosures. The concept of affine arithmetic and its applications in computer graphics are firstly introduced by Comba and Stolfi [11]. After a few years, Stolfi and De Figueiredo [34] illustrated the overestimation problem in the case of standard interval arithmetic and how affine arithmetic can overcome it. Further, the concepts, properties, and several applications of affine arithmetic have also been discussed by De Figueiredo and Stolfi [12]. Miyajima and Kashiwagi [26] proposed a dividing method by utilizing the best multiplication in affine arithmetic. Moreover, Akhmerov [1] developed an interval-affine Gaussian algorithm for constrained systems. A direct method for solving parametric interval linear systems with non-affine dependencies is demonstrated by Skalna [32]. Rump and Kashiwagi [30] discussed the improvements and implementations of affine arithmetic. Skalna and Hladík [33] developed a new algorithm for the Chebyshev minimum-error multiplication of reduced affine forms. Due to the efficacy of affine arithmetic, a fuzzy-affine approach has been proposed in this chapter to improve the fuzzy solution bounds and give tighter ranges of the fuzzy solutions.

As mentioned earlier, the linear structural dynamic problem having fuzzy parameters may lead to a fuzzy generalized (or standard) eigenvalue problem (FGEP or FSEP). Various works are available regarding the solution of the fuzzy generalized (or standard) eigenvalue problem. In this regard, a few of them are included here. Hladík et al. [16] proposed a filtering method for solving interval GEP, which is based on the concept of sufficient conditions for singularity and regularity of interval matrices given by Rex and Rohn [29]. Then, Mahato and Chakraverty [24] extended the same filtering algorithm given by Hladik et al. [16] for fuzzy eigenvalue bounds of fuzzy symmetric matrices to solve FSEP. Also, the filtering algorithm for real eigenvalue bounds of FGEP is illustrated by Mahato and Chakraverty [25]. Xia and Friswell [35] proposed an efficient technique to find the solution of FGEPs in structural dynamics problems. Chakraverty and Behera [4] discussed the parameter identification of multi-storey frame structures with uncertain dynamic data, which may lead to FGEP. Uncertain static and dynamic analysis of the structural systems with imprecise parameters has been described by Chakraverty and Behera [5]. Further [27] computed the eigenvalue bounds for structures with the interval description of uncertain but non-random parameters. Based on the perturbation theory, eigenvalue bounds of structures having uncertain but non-random parameters have been computed by Leng and He [21]. Also, Leng et al. [23] and Leng and He [22] described the computation of the real eigenvalue bounds of the real interval matrices in structural dynamics with interval parameters. The modal analysis of structures and computing the eigenvalue bounds having uncertain but bounded parameters with the help of interval analysis have been studied by Qiu et al. [28] and Sim et al. [31]. Leng [20] solved many application problems of different structural systems having fuzzy uncertain parameters. Hladík and Jaulin [17] discussed the eigenvalue symmetric matrix contractor. Furthermore, Hladík [15] found the eigenvalue solution bounds of the FGEP having both real and complex fuzzy matrices.

The current chapter is organized in the following manner. The present section of the chapter contains the introduction and literature survey. In Sect. 2, necessary preliminaries related to fuzzy numbers and their properties with the basic concepts of affine arithmetic and its operations are discussed. All the discussed concepts are useful for the present investigation. Further, a fuzzy-affine approach to handle the overestimation problems in standard fuzzy arithmetic has been developed in Sect. 3 and the efficacy of the fuzzy-affine arithmetic has also been proven with an example. In Sect. 4, a fuzz-affine approach-based technique has been proposed to handle the fuzzy generalized (or standard) eigenvalue problems whose parameters are taken in the form of triangular or trapezoidal fuzzy numbers. Lastly, five illustrative numerical examples related to different fuzzy linear structural dynamic problems are investigated in Sect. 5, followed by concluding remarks in Sect. 6.

2 Preliminaries

In this section, basic definitions and notations related to the fuzzy number, different types of fuzzy number viz. triangular and trapezoidal fuzzy numbers, and the α-cut technique of fuzzy number are included [2, 6, 8, 9, 13, 14, 19, 36–38]. Further, the basic concepts related to affine arithmetic, the conversion between interval and affine forms, and the affine arithmetic operations have been illustrated [7, 11, 12, 26, 32, 34].

2.1 Fuzzy Number

A particular type of fuzzy set $\tilde{\mathrm{P}}$ may be referred to as a fuzzy number if it satisfies the following properties:

1. $\tilde{\mathrm{P}}$ is normal, (that is $\exists z \in \mathbb{R} | \tilde{\mathrm{P}}(z) = 1$);
2. $\tilde{\mathrm{P}}$ is convex;
3. The membership function $\mu_{\tilde{\mathrm{P}}}(z)$ is piecewise continuous.

2.2 Different Types of Fuzzy Number

Depending upon the geometrical properties, fuzzy numbers are classified into several categories viz. triangular fuzzy number (TFN), Gaussian fuzzy number (GFN), trapezoidal fuzzy number (TrFN), exponential fuzzy number (EFN), etc. For this chapter, the fuzziness has been handled by using the two types of fuzzy numbers such as TFN and TrFN. In this regard, the basic terminologies, behavior, definition, and geometrical representation of both TFN and TrFN are included here.

2.2.1 Triangular Fuzzy Number (TFN)

References [6, 9, 37].

As shown in Fig. 1, generally the TFNs are geometrically represented by a linear graph which is a composition of left-increasing and right-decreasing linear functions. TFNs may be expressed by a triplet $\tilde{\mathrm{P}} = (p_1, p_2, p_3)$ having the membership function $\mu_{\tilde{\mathrm{P}}}(z)$ given as follows:

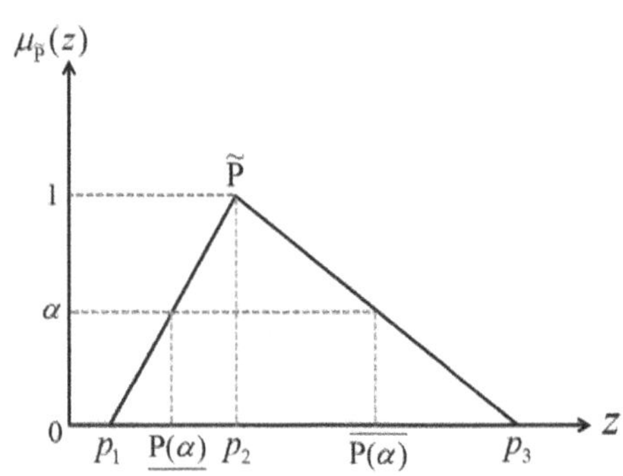

Fig. 1 Geometrical representation of the TFN $\tilde{\mathrm{P}}$

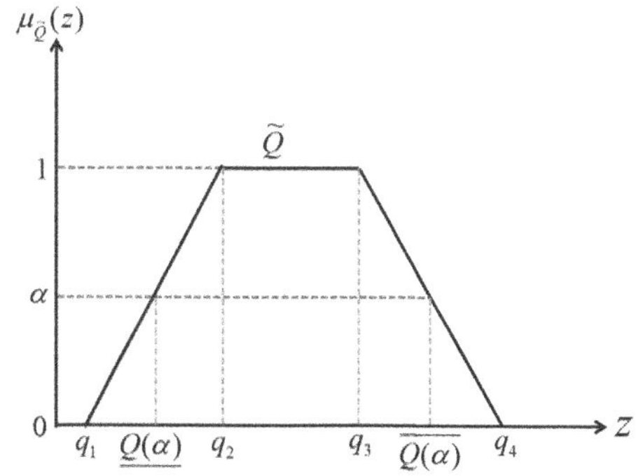

Fig. 2 Geometrical representation of the TrFN $\tilde{Q}$

$$\mu_{\tilde{P}}(z) = \begin{cases} 0, & z < p_1, z > p_3 \\ \frac{z-p_1}{p_2-p_1}, & z \in [p_1, p_2] \\ \frac{p_3-z}{p_3-p_2}, & z \in [p_2, p_3] \end{cases}. \tag{2.1}$$

In the case of TFN, there exists exactly one $z_0 \in \mathbb{R}$ such that the membership function becomes unity (that is $\mu_{\tilde{P}}(z_0) = 1$). Thus, z_0 is called the mean value of the TFN $\tilde{P}$.

2.2.2 Trapezoidal Fuzzy Number (TrFN)

References [2, 8, 14, 37, 38].

In the case of TrFN, there exists an interval $z \in [q_2, q_3]$ such that $\mu_{\tilde{Q}}(z) = 1$ as depicted in Fig. 2. TrFN may be expressed by a quadruplet $\tilde{Q} = (q_1, q_2, q_3, q_4)$ having the membership function $\mu_{\tilde{Q}}(z)$ given as follows:

$$\mu_{\tilde{Q}}(z) = \begin{cases} 0, & z < q_1, z > q_4 \\ \frac{z-q_1}{q_2-q_1}, & z \in [q_1, q_2] \\ 1, & z \in [q_2, q_3] \\ \frac{q_4-z}{q_4-q_3}, & z \in [q_3, q_4] \end{cases}. \tag{2.2}$$

2.3 *α-Cut Technique of Fuzzy Number [2]*

The "α-cut" technique of a fuzzy number $\tilde{P}$ is a crisp set defined as follows:

$$P(\alpha) = \{z \in \mathbb{R} | \tilde{P}(z) \geq \alpha\}. \tag{2.3}$$

Here, each α-cut representation $\mathrm{P}(\alpha)$ of the given fuzzy number $\tilde{\mathrm{P}}$ is a standard closed interval that depends on the value of parameter "α" that varies from 0 to 1 ($\alpha \in [0, 1]$). Thus, the α-cut representation of $\tilde{\mathrm{P}}$ may be expressed as $\mathrm{P}(\alpha) = [\underline{\mathrm{P}(\alpha)}, \overline{\mathrm{P}(\alpha)}]$, where $\underline{\mathrm{P}(\alpha)}$ and $\overline{\mathrm{P}(\alpha)}$ are the respective lower and upper bounds of α-cut representation $\mathrm{P}(\alpha)$.

In this regard, the α-cut representations of the different types of fuzzy numbers are illustrated below.

Let us consider a TFN $\tilde{\mathrm{P}} = (p_1, p_2, p_3)$. By utilizing the α-cut technique, the TFN may be transformed into a fuzzy interval form given as

$$\mathrm{P}(\alpha) = [\underline{\mathrm{P}(\alpha)}, \overline{\mathrm{P}(\alpha)}] = [p_1 + \alpha(p_2 - p_1), p_3 - \alpha(p_3 - p_2)] \text{ for } \alpha \in [0, 1]. \quad (2.4)$$

Similarly, consider a TrFN $\tilde{Q} = (q_1, q_2, q_3, q_4)$. By adopting the α-cut technique, the TrFN may be converted into the fuzzy interval form expressed as

$$Q(\alpha) = [\underline{Q(\alpha)}, \overline{Q(\alpha)}] = [q_1 + \alpha(q_2 - q_1), q_4 - \alpha(q_4 - q_3)] \text{ for } \alpha \in [0, 1]. \quad (2.5)$$

2.4 *Affine Arithmetic*

References [7, 11, 26, 32].

Due to the underlying assumption of the standard interval arithmetic that all the constituting operands during any interval arithmetic operations vary independently over their ranges, it overestimates the range of the resulting interval solution. This "interval dependency problem" or "interval overestimation problem" in the case of standard interval arithmetic is a major hurdle in handling uncertain real-life problems. In this regard, affine arithmetic has been introduced to handle uncertain parameters efficiently.

Affine form representation is a linear polynomial of real variables (ε_l for $l = 1, 2, \ldots, k$). If $\hat{m}$ denotes the affine form representation of an ideal quantity "m", then it may be defined as

$$m \in m_0 + \left[-\sum_{l=1}^{k} |m_l|, \sum_{l=1}^{k} |m_l| \right]. \quad (2.6)$$

Further, $\hat{m}$ is explicitly expressed as

$$m \in \hat{m} = m_0 + \sum_{l=1}^{k} m_l \varepsilon_l = m_0 + m_1 \varepsilon_1 + \cdots + m_{k-1} \varepsilon_{k-1} + m_k \varepsilon_k, \quad (2.7)$$

where these real variables ε_l for $l = 1, 2, \ldots, k$ are known as noise symbols which lie in a particular interval $\Omega = [-1, 1]$. Moreover, each associate coefficient m_l for $l = 1, 2, \ldots, k$ of the respective noise symbol ε_l is known as partial deviation and the initial term m_0 is called as the central value of $\hat{m}$.

2.5 Conversion of Interval to Affine and Vice Versa

References [7, 12, 34].

To convert an interval into its affine form representation, consider an interval $[m] = [\underline{m}, \overline{m}]$. If $\hat{m}$ denotes affine form representation of $[m]$, then $\hat{m}$ may be obtained as follows:

$$\hat{m} = m_0 + m_r \varepsilon_m \text{ for } \varepsilon_m \in \Omega = [-1, 1], \tag{2.8}$$

where $m_0 = \frac{1}{2}(\underline{m} + \overline{m})$ is the center and $m_r = \frac{1}{2}(\overline{m} - \underline{m})$ is the radius (or half-width) of the interval $[m]$.

Conversely, to transform an affine form into its interval bounds, consider an affine form representation (as given in Eq. (2.7)) $\hat{m} = m_0 + \sum_{l=1}^{k} m_l \varepsilon_l = m_0 + m_1 \varepsilon_1 + \cdots + m_{k-1}\varepsilon_{k-1} + m_k \varepsilon_k$. If $[m] = [\underline{m}, \overline{m}]$ is the interval bound of $\hat{m}$, then it is computed as follows:

$$\underline{m} = m_0 - \mathrm{T}_d \text{ and } \overline{m} = m_0 + \mathrm{T}_d, \tag{2.9}$$

where T_d is the total deviation of $\hat{m}$ calculated as the sum of the magnitude of all the partial deviations m_l for $l = 1, 2, \ldots, k$ which may be written as

$$\mathrm{T}_d = \sum_{l=1}^{k} |m_l| = |m_1| + |m_2| + \cdots + |m_k|. \tag{2.10}$$

2.6 Affine Arithmetic Operations

References [7, 12, 26, 32].

The binary operations viz. addition, subtraction, scalar multiplication, multiplication, and division of the affine arithmetic are discussed below.

Let us consider two affine form representations $\hat{m}$ and $\hat{n}$ (as expressed in Eq. (2.7)) given as

$\hat{m} = m_0 + \sum_{l=1}^{k} m_l \varepsilon_l = m_0 + m_1\varepsilon_1 + \cdots + m_k\varepsilon_k$ and $\hat{n} = n_0 + \sum_{l=1}^{k} n_l \varepsilon_l = n_0 + n_1\varepsilon_1 + \cdots + n_k\varepsilon_k$.

Thus, all the corresponding affine arithmetic operations may be defined as follows:

- **Addition**: $\hat{m} + \hat{n} = (m_0 + n_0) + \sum_{l=1}^{k} (m_l + n_l)\varepsilon_l = (m_0 + n_0) + (m_1 + n_1)\varepsilon_1 + \cdots + (m_k + n_k)\varepsilon_k$.
- **Subtraction**: $\hat{m} - \hat{n} = (m_0 - n_0) + \sum_{l=1}^{k} (m_l - n_l)\varepsilon_l = (m_0 - n_0) + (m_1 - n_1)\varepsilon_1 + \cdots + (m_k - n_k)\varepsilon_k$.
- **Scalar multiplication**: $s \cdot \hat{m} = (s \cdot m_0) + \sum_{l=1}^{k} (s \cdot m_l)\varepsilon_l = (s \cdot m_0) + (s \cdot m_1)\varepsilon_1 + \cdots + (s \cdot m_k)\varepsilon_k$, for $s \in \mathbb{R}$.
- **Multiplication**: $\hat{m} \cdot \hat{n} = m_0 n_0 + \sum_{l=1}^{k} (m_0 n_l + m_l n_0)\varepsilon_l + c_t \varepsilon_t$, where $|c_t| \geq \left| \sum_{l=1}^{k} m_l \varepsilon_l \cdot \sum_{l=1}^{k} n_l \varepsilon_l \right|$, $\varepsilon_l \in \Omega = [-1, 1]$ for $l = 1, \ldots, k$, and $\varepsilon_t \in \Omega = [-1, 1]$ is the newly generated noise symbol.
- **Division**: $\frac{\hat{m}}{\hat{n}} = \hat{m} \cdot \frac{1}{\hat{n}} = \frac{m_0}{n_0} + \frac{1}{\hat{n}} \sum_{l=1}^{k} \left(m_l - \frac{m_0}{n_0} n_l \right) \varepsilon_l$, provided $\hat{n} \neq \{0\}$.

3 Fuzzy-Affine Approach

Similar to standard interval arithmetic, standard fuzzy arithmetic also assumes that all the constituting operands during the fuzzy-arithmetic operation are completely independent of each other. But, there may be the case where the operands may partially (or completely) depend upon each other. In that case, the standard fuzzy arithmetic may result in comparatively wider outer enclosures than the actual range. This situation is known as the "dependency problem" (or "overestimation problem"). To overcome the overestimation problem that occurred in standard fuzzy arithmetic, the fuzzy-affine approach has been developed. Firstly, the fuzzy numbers have to transform into their respective fuzzy-affine forms.

In this regard, let us consider a fuzzy number (TFN or TrFN) $\tilde{P}$. To transform it into its fuzzy-affine form, initially $\tilde{P}$ may be converted to a fuzzy interval form by using the α-cut technique as mentioned in Sect. 2.3. Thus, the fuzzy interval form of the fuzzy number $\tilde{P}$ is

$$P(\alpha) = [\underline{P(\alpha)}, \overline{P(\alpha)}] \text{ for } \alpha \in [0, 1], \tag{3.1}$$

where $\underline{P(\alpha)}$ and $\overline{P(\alpha)}$ are the respective lower and upper bounds of the fuzzy interval form $P(\alpha)$. Then, the center $(P_0(\alpha))$ and radius (or half-width) $(P_r(\alpha))$ of $P(\alpha)$ may be computed as follows:

$$\mathrm{P}_0(\alpha) = \frac{1}{2}\left(\underline{\mathrm{P}(\alpha)} + \overline{\mathrm{P}(\alpha)}\right) \text{ and } \mathrm{P}_r(\alpha) = \frac{1}{2}\left(\overline{\mathrm{P}(\alpha)} - \underline{\mathrm{P}(\alpha)}\right). \tag{3.2}$$

Finally, after the conversion of the fuzzy number into its fuzzy interval form, it has been further transformed into the affine form representation by adopting the procedure described in Sect. 2.5 (Eq. (2.8)). Therefore, the fuzzy-affine form ($\mathrm{P}(\alpha, \varepsilon_p)$) of the fuzzy number $\tilde{\mathrm{P}}$ may be obtained as

$$\mathrm{P}(\alpha, \varepsilon_p) = \mathrm{P}_0(\alpha) + \mathrm{P}_r(\alpha)\varepsilon_p = \frac{1}{2}(1 - \varepsilon_p)\underline{\mathrm{P}(\alpha)} + \frac{1}{2}(1 + \varepsilon_p)\overline{\mathrm{P}(\alpha)}, \tag{3.3}$$

for $\alpha \in [0, 1]$ and $\varepsilon_p \in [-1, 1]$. Here, "$\alpha$" is the fuzzy parameter and "ε_p" is the noise symbol of the fuzzy-affine form $\mathrm{P}(\alpha, \varepsilon_p)$.

The fuzzy-affine forms of different types of fuzzy numbers such as TFN and TrFN may be evaluated in the following manner.

3.1 *Fuzzy-Affine form of TFN*

Let us consider a TFN $\tilde{\mathrm{P}} = (p_1, p_2, p_3)$. By utilizing the α-cut technique (as given in Eq. (2.4)), $\tilde{\mathrm{P}}$ has been converted into its respective fuzzy interval form $\mathrm{P}(\alpha)$ given as

$$\mathrm{P}(\alpha) = [\underline{\mathrm{P}(\alpha)}, \overline{\mathrm{P}(\alpha)}] = [p_1 + \alpha(p_2 - p_1), p_3 - \alpha(p_3 - p_2)] \text{ for } \alpha \in [0, 1]. \tag{3.4}$$

Then, the above fuzzy interval form $\mathrm{P}(\alpha)$ further may be transformed into a fuzzy-affine form by utilizing the technique discussed in Sect. 2.5 as follows:

$$\mathrm{P}(\alpha, \varepsilon_p) = \mathrm{P}_0(\alpha) + \mathrm{P}_r(\alpha)\varepsilon_p \text{ for } \varepsilon_p \in [-1, 1], \tag{3.5}$$

where $\mathrm{P}_0(\alpha)$ is the center and $\mathrm{P}_r(\alpha)$ is the radius (or half-width) of the above fuzzy interval form $\mathrm{P}(\alpha)$ which may be computed as

$$\mathrm{P}_0(\alpha) = \frac{1}{2}\left(\underline{\mathrm{P}(\alpha)} + \overline{\mathrm{P}(\alpha)}\right) = \frac{1}{2}[(p_1 + p_3) + \alpha(2p_2 - p_1 - p_3)]; \tag{3.6a}$$

$$\mathrm{P}_r(\alpha) = \frac{1}{2}\left(\overline{\mathrm{P}(\alpha)} - \underline{\mathrm{P}(\alpha)}\right) = \frac{1}{2}[(p_3 - p_1) - \alpha(p_3 - p_1)]. \tag{3.6b}$$

Therefore, the fuzzy-affine form of the given TFN $\tilde{\mathrm{P}} = (p_1, p_2, p_3)$ is

$$\mathrm{P}(\alpha, \varepsilon_p) = \frac{1}{2}[(p_1 + p_3) + \alpha(2p_2 - p_1 - p_3)] + \frac{1}{2}[(p_3 - p_1) - \alpha(p_3 - p_1)]\varepsilon_p, \tag{3.7}$$

for $\alpha \in [0, 1]$ and $\varepsilon_p \in [-1, 1]$ where ε_p is the noise symbol for the fuzzy-affine form $P(\alpha, \varepsilon_p)$.

3.2 Fuzzy-Affine form of TrFN

Let us consider a TrFN $\tilde{Q} = (q_1, q_2, q_3, q_4)$. By adopting the α-cut technique (as given in Eq. (2.5)), $\tilde{Q}$ has been transformed into its fuzzy interval form $Q(\alpha)$ expressed as

$$Q(\alpha) = [\underline{Q(\alpha)}, \overline{Q(\alpha)}] = [q_1 + \alpha(q_2 - q_1), q_4 - \alpha(q_4 - q_3)] \text{ for } \alpha \in [0, 1]. \quad (3.8)$$

Here, the center $Q_0(\alpha)$ and the radius (or half-width) $Q_r(\alpha)$ of the above fuzzy interval form $Q(\alpha)$ may be determined as

$$Q_0(\alpha) = \frac{1}{2}\left(\underline{Q(\alpha)} + \overline{Q(\alpha)}\right) = \frac{1}{2}[(q_1 + q_4) + \alpha\{(q_2 + q_3) - (q_1 - q_4)\}]; \quad (3.9a)$$

$$Q_r(\alpha) = \frac{1}{2}\left(\overline{Q(\alpha)} - \underline{Q(\alpha)}\right) = \frac{1}{2}[(q_4 - q_1) + \alpha\{(q_1 + q_3) - (q_2 + q_4)\}]. \quad (3.9b)$$

Hence, the fuzzy-affine form of the given TrFN $\tilde{Q} = (q_1, q_2, q_3, q_4)$ is obtained as follows:

$$Q(\alpha, \varepsilon_q) = Q_0(\alpha) + Q_r(\alpha)\varepsilon_q, \text{ for } \varepsilon_q \in [-1, 1]. \quad (3.10)$$

That is

$$\begin{aligned} Q(\alpha, \varepsilon_q) = &\frac{1}{2}[(q_1 + q_4) + \alpha\{(q_2 + q_3) - (q_1 - q_4)\}] \\ &+ \frac{1}{2}[(q_4 - q_1) + \alpha\{(q_1 + q_3) - (q_2 + q_4)\}]\varepsilon_q, \end{aligned} \quad (3.11)$$

for $\alpha \in [0, 1]$ and $\varepsilon_q \in [-1, 1]$, where ε_q is the noise symbol for the fuzzy-affine form $Q(\alpha, \varepsilon_q)$.

3.3 Fuzzy-Affine Approach

To perform the fuzzy-affine arithmetic approach, firstly we have to transform all the operands in the form of fuzzy numbers into their fuzzy-affine forms as discussed in the above sections. In this regard, let us consider two fuzzy numbers $\tilde{P}$ and $\tilde{Q}$. Then, these fuzzy numbers have been converted into their respective fuzzy-affine forms given as

$$\tilde{\mathrm{P}} \approx \mathrm{P}(\alpha, \varepsilon_p) \text{ and } \tilde{Q} \approx Q(\alpha, \varepsilon_q), \tag{3.12a}$$

for $\alpha \in [0, 1]$ and $\varepsilon_p, \varepsilon_q \in [-1, 1]$. Here, "$\alpha$" is the fuzzy parameter and "ε_p" and "ε_q" are different noise symbols. Therefore, operations of fuzzy-affine arithmetic approach can be defined as follows:

$$\mathrm{X}(\alpha, \varepsilon_p, \varepsilon_q) = \mathrm{P}(\alpha, \varepsilon_p) * Q(\alpha, \varepsilon_q), \tag{3.12b}$$

where $\mathrm{X}(\alpha, \varepsilon_p, \varepsilon_q)$ is the fuzzy-affine form of the resulting solution having several parameters such as fuzzy parameter (α), all the existing noise symbols (ε_p and ε_q), and with some newly generated noise symbols during the operation.

After the finding of the fuzzy-affine solution, it may be further reconverted into its fuzzy interval form as

$$\mathrm{X}(\alpha) = \mathrm{P}(\alpha) * Q(\alpha) = [\underline{\mathrm{X}(\alpha)}, \overline{\mathrm{X}(\alpha)}] \text{ for } \forall\alpha \in [0, 1], \tag{3.13}$$

$$\text{where } \underline{\mathrm{X}(\alpha)} = \min_{\varepsilon_p, \varepsilon_q \in [-1,1]} \mathrm{X}(\alpha, \varepsilon_p, \varepsilon_q) \text{ and } \overline{\mathrm{X}(\alpha)} = \max_{\varepsilon_p, \varepsilon_q \in [-1,1]} \mathrm{X}(\alpha, \varepsilon_p, \varepsilon_q). \tag{3.14}$$

Finally, the resulting fuzzy solution ($\tilde{\mathrm{X}} = \tilde{\mathrm{P}} * \tilde{Q}$) may be computed by varying the fuzzy parameter (α) from 0 to 1.

3.4 *Efficacy of Fuzzy-Affine Approach*

To show the efficacy and reliability of the fuzzy-affine approach, an example based on a fuzzy nonlinear function with variables in the form of fuzzy numbers (TFN and TrFN) has been worked out.

Example 1: Let us consider a two-dimensional fuzzy nonlinear function $\tilde{\mathrm{N}}(\tilde{\mathrm{P}}, \tilde{Q}) = 9\tilde{\mathrm{P}}^2 - 4\tilde{Q}^2 + 3\tilde{\mathrm{P}}\tilde{Q} - 2\tilde{\mathrm{P}} + \tilde{Q}$. Here, the consistent variables ($\tilde{\mathrm{P}}$ and $\tilde{Q}$) of the fuzzy nonlinear function $\tilde{\mathrm{N}}(\tilde{\mathrm{P}}, \tilde{Q})$ are in the form of fuzzy numbers (TFNs or TrFNs). For the initial case, suppose the fuzzy variables are in the form of TFNs such that

$$\tilde{\mathrm{P}} = (2, 3.5, 5) \text{ and } \tilde{Q} = (1, 2.5, 4).$$

Firstly, by adopting the α-cut technique (as mentioned in Sect. 2.3), the above TFNs are transformed into their fuzzy interval forms ($\mathrm{P}(\alpha)$ and $Q(\alpha)$) as

$$\mathrm{P}(\alpha) = [2 + 1.5\alpha, 5 - 1.5\alpha] \text{ and } Q(\alpha) = [1 + 1.5\alpha, 4 - 1.5\alpha] \text{ for } \alpha \in [0, 1].$$

Thus, by applying standard fuzzy arithmetic [2, 8], the functional value of the given fuzzy nonlinear function $\tilde{N}(\tilde{P}, \tilde{Q})$ in its fuzzy interval form has been computed as follows:

$$N(P(\alpha), Q(\alpha)) = 9P^2(\alpha) - 4Q^2(\alpha) + 3P(\alpha)Q(\alpha) - 2P(\alpha) + Q(\alpha)$$

$$\Rightarrow N(P(\alpha), Q(\alpha)) = [18\alpha^2 + 120\alpha - 31, 18\alpha^2 - 192\alpha + 281] \text{ for } \alpha \in [0, 1].$$

Further to perform the fuzzy-affine approach, the given fuzzy variables $\tilde{P}$ and $\tilde{Q}$ are converted into their respective fuzzy-affine forms (as discussed in Sect. 3.1) as follows:

$$P(\alpha, \varepsilon_p) = 3.5 + (1.5 - 1.5\alpha)\varepsilon_p \text{ and } Q(\alpha, \varepsilon_q) = 2.5 + (1.5 - 1.5\alpha)\varepsilon_q,$$

where $\alpha \in [0, 1]$ and $\varepsilon_p, \varepsilon_q \in [-1, 1]$. Here, ε_p and ε_q are two different noise symbols that occurred during the conversion of the fuzzy-affine forms. Hence, the required fuzzy-affine functional value $(N(P(\alpha, \varepsilon_p), Q(\alpha, \varepsilon_q)))$ of the given fuzzy nonlinear function $(\tilde{N}(\tilde{P}, \tilde{Q}))$ may be determined by adopting the fuzzy-affine approach (given in Sect. 3.3) as follows:

$$\begin{aligned} N(P(\alpha, \varepsilon_p), Q(\alpha, \varepsilon_q)) =& 9P^2(\alpha, \varepsilon_p) - 4Q^2(\alpha, \varepsilon_q) + 3P(\alpha, \varepsilon_p)Q(\alpha, \varepsilon_q) \\ & - 2P(\alpha, \varepsilon_p) + Q(\alpha, \varepsilon_q) \end{aligned}$$

$$\begin{aligned} &= 9(3.5 + (1.5 - 1.5\alpha)\varepsilon_p)^2 - 4(2.5 + (1.5 - 1.5\alpha)\varepsilon_q)^2 \\ &+3(3.5 + (1.5 - 1.5\alpha)\varepsilon_p)(2.5 + (1.5 - 1.5\alpha)\varepsilon_q) \quad , \\ &-2(3.5 + (1.5 - 1.5\alpha)\varepsilon_p) + (2.5 + (1.5 - 1.5\alpha)\varepsilon_q) \end{aligned}$$

where $\alpha \in [0, 1]$ and $\varepsilon_p, \varepsilon_q \in [-1, 1]$. There are also many newly generated noise symbols during the operations of the fuzzy-affine approach.

Now, let us plot a graph of the corresponding TFN solutions computed by using both the fuzzy-affine approach and the standard fuzzy-arithmetic approach. In Fig. 3, the line marked with "o" is used to indicate the TFN solution by using the fuzzy-affine approach, and the dotted line marked with "*" depicts the solution by fuzzy-arithmetic approach.

For the second case, suppose the fuzzy variables are in the form of TrFNs such that

$$\tilde{P} = (2, 3, 4, 5) \text{ and } \tilde{Q} = (1, 1.5, 3.5, 4).$$

Then, their respective fuzzy interval forms and fuzzy-affine forms are calculated in a similar way as given above (for TFN), and these are given as follows:

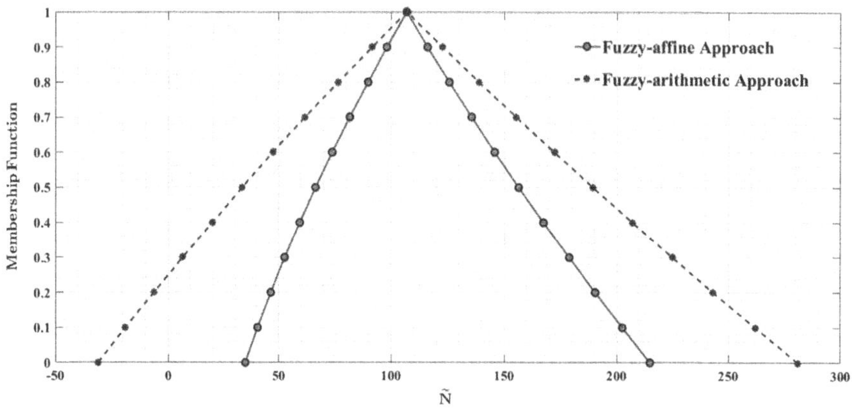

Fig. 3 Comparison plot between fuzzy-affine approach and fuzzy-arithmetic approach for TFN variables of Example 1

$$\mathrm{P}(\alpha) = [2+\alpha, 5-\alpha];\ Q(\alpha) = [1+0.5\alpha, 4-0.5\alpha] \text{ and } \mathrm{P}(\alpha, \varepsilon_p) = 3.5 + (1.5-\alpha)\varepsilon_p;$$
$$Q(\alpha, \varepsilon_q) = 2.5 + (1.5-0.5\alpha)\varepsilon_q \text{ where } \alpha \in [0, 1] \text{ and } \varepsilon_p, \varepsilon_q \in [-1, 1].$$

Similarly, the TrFN functional value of the given fuzzy nonlinear function $\tilde{\mathrm{N}}(\tilde{\mathrm{P}}, \tilde{Q})$ in its fuzzy interval form (by using standard fuzzy arithmetic) and fuzzy-affine form (by using fuzzy-affine approach) has been computed and given below.

$$\mathrm{N}(\mathrm{P}(\alpha), Q(\alpha)) = [9.5\alpha^2 + 60.5\alpha - 31, 9.5\alpha^2 - 116\alpha + 281] \text{ and}$$
$$\begin{aligned}\mathrm{N}(\mathrm{P}(\alpha, \varepsilon_p), Q(\alpha, \varepsilon_q)) &= 9(3.5 + (1.5-\alpha)\varepsilon_p)^2 - 4(2.5 + (1.5-0.5\alpha)\varepsilon_q)^2 \\ &+ 3(3.5 + (1.5-\alpha)\varepsilon_p)(2.5 + (1.5-0.5\alpha)\varepsilon_q) - 2(3.5 + (1.5-\alpha)\varepsilon_p) \\ &+ (2.5 + (1.5-0.5\alpha)\varepsilon_q)\end{aligned},$$

where $\alpha \in [0, 1]$ and $\varepsilon_p, \varepsilon_q \in [-1, 1]$.

Also, the TrFN solution graphs have been plotted for both the cases (that is fuzzy-affine approach and the standard fuzzy-arithmetic approach), and the comparison plots are depicted in Fig. 4.

Lastly, the TFN and TrFN functional values of the given fuzzy nonlinear function $\tilde{\mathrm{N}}(\tilde{\mathrm{P}}, \tilde{Q})$ by both fuzzy-affine approach as well as standard fuzzy-arithmetic approach for several values of the fuzzy parameter ($\alpha \in [0, 1]$) (that is for $\alpha = 0, 0.1, \ldots, 0.9, 1$) are incorporated in Table 1.

It may be noticed, from the plots given in Figs. 3, 4 and from Table 1, that the fuzzy-affine approach gives tighter enclosures than the standard fuzzy arithmetic. Moreover, for $\alpha = 0$, the outer bounds of the functional value (for both the cases that is TFN and TrFN) of the fuzzy nonlinear function $\tilde{\mathrm{N}}(\tilde{\mathrm{P}}, \tilde{Q})$ by fuzzy-affine approach and fuzzy-arithmetic approach are [35.0000, 215.0000] and [−31.0000, 281.0000], respectively. Hence, the fuzzy-affine approach may be more efficient and reliable for the considered example.

Further, it may be noted that in both cases (that is for TFN and TrFN), the fuzzy-affine approach proves its efficacy irrespective of the nature of the fuzzy numbers.

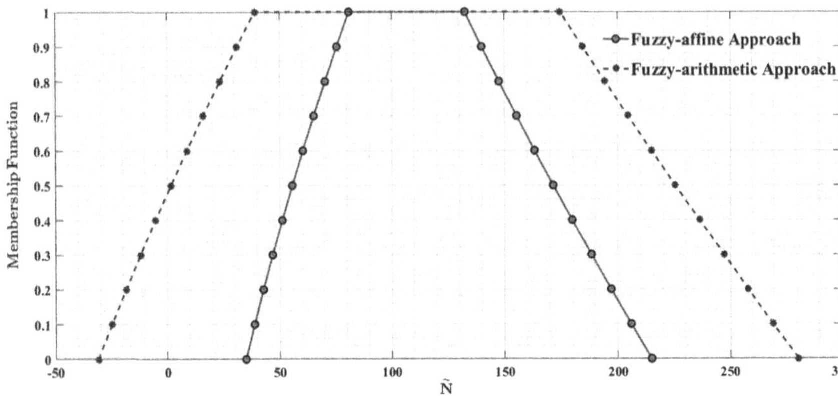

Fig. 4 Comparison plot between fuzzy-affine approach and fuzzy-arithmetic approach for TrFN variables of Example 1

4 Proposed Method

A fuzzy-affine approach-based method has been developed in this section to solve the fuzzy generalized (or standard) eigenvalue problem (FGEP or FSEP) and find its fuzzy eigenvalues.

In this regard, let us consider the FGEP given as follows:

$$\tilde{K}\tilde{x} = \tilde{\lambda}\tilde{M}\tilde{x}, \tag{4.1}$$

where the two coefficient matrices $\tilde{K} = (\tilde{k}_{ij})$ and $\tilde{M} = (\tilde{m}_{ij})$ for $i, j = 1, 2, \ldots, n$ are fuzzy square matrices of order $n \times n$ having elements in the form of fuzzy numbers such that

$$\tilde{K} = \begin{pmatrix} \tilde{k}_{11} & \tilde{k}_{12} & \cdots & \tilde{k}_{1n} \\ \tilde{k}_{21} & \tilde{k}_{22} & \cdots & \tilde{k}_{2n} \\ \vdots & \vdots & \ddots & \vdots \\ \tilde{k}_{n1} & \tilde{k}_{n2} & \cdots & \tilde{k}_{nn} \end{pmatrix} \text{ and } \tilde{M} = \begin{pmatrix} \tilde{m}_{11} & \tilde{m}_{12} & \cdots & \tilde{m}_{1n} \\ \tilde{m}_{21} & \tilde{m}_{22} & \cdots & \tilde{m}_{2n} \\ \vdots & \vdots & \ddots & \vdots \\ \tilde{m}_{n1} & \tilde{m}_{n2} & \cdots & \tilde{m}_{nn} \end{pmatrix}. \tag{4.2}$$

If the elements of the fuzzy coefficient matrices ($\tilde{K}$ and $\tilde{M}$) are in the form of TFNs, then they may be denoted as $\tilde{k}_{ij} = (k_{ij}^1, k_{ij}^2, k_{ij}^3)$ and $\tilde{m}_{ij} = (m_{ij}^1, m_{ij}^2, m_{ij}^3)$ for $i, j = 1, 2, \ldots, n$. Similarly, if the elements are in the form of TrFNs, then may be expressed as $\tilde{k}_{ij} = (k_{ij}^1, k_{ij}^2, k_{ij}^3, k_{ij}^4)$ and $\tilde{m}_{ij} = (m_{ij}^1, m_{ij}^2, m_{ij}^3, m_{ij}^4)$ for $i, j = 1, 2, \ldots, n$. Further, $\tilde{\lambda}$ is the fuzzy eigenvalue and $\tilde{x}$ is the corresponding fuzzy eigenvector of the FGEP.

Firstly, all the constituting elements of the above fuzzy coefficient matrices (given in Eq. (4.2)) are transformed into their respective fuzzy interval forms by adopting

Table 1 Comparison table between fuzzy-affine approach and fuzzy-arithmetic approach for both TFN and TrFN solutions of Example 1 for different values of α

α	Fuzzy-affine approach				Fuzzy-arithmetic approach			
	TFN		TrFN		TFN		TrFN	
	$\underline{N}$	$\overline{N}$	$\underline{N}$	$\overline{N}$	$\underline{N}$	$\overline{N}$	$\underline{N}$	$\overline{N}$
0	35.0000	215.0000	35.0000	215.0000	−31.0000	281.0000	−31.0000	281.0000
0.1	40.5800	202.5800	38.7450	205.8950	−18.8200	261.9800	−24.8550	269.4950
0.2	46.5200	190.5200	42.6800	196.9800	−6.2800	243.3200	−18.5200	258.1800
0.3	52.8200	178.8200	46.8050	188.2550	6.6200	225.0200	−11.9950	247.0550
0.4	59.4800	167.4800	51.1200	179.7200	19.8800	207.0800	−5.2800	236.1200
0.5	66.5000	156.5000	55.6250	171.3750	33.5000	189.5000	1.6250	225.3750
0.6	73.8800	145.8800	60.3200	163.2200	47.4800	172.2800	8.7200	214.8200
0.7	81.6200	135.6200	65.2050	155.2550	61.8200	155.4200	16.0050	204.4550
0.8	89.7200	125.7200	70.2800	147.4800	76.5200	138.9200	23.4800	194.2800
0.9	98.1800	116.1800	75.5450	139.8950	91.5800	122.7800	31.1450	184.2950
1	107.0000	107.0000	81.0000	132.5000	107.0000	107.0000	39.0000	174.5000

the α-cut technique of the fuzzy numbers (TFN or TrFN) discussed in Sect. 2.3. Thus, after the transformation, all the elements in their fuzzy interval forms may be written as

$$k_{ij}(\alpha) = [\underline{k_{ij}}(\alpha), \overline{k_{ij}}(\alpha)] \text{ and } m_{ij}(\alpha) = [\underline{m_{ij}}(\alpha), \overline{m_{ij}}(\alpha)] \text{ for } i, j = 1, 2, \ldots, n. \tag{4.3}$$

Hence, the given FGEP (4.1) is transformed into a fuzzy interval GEP

$$\mathrm{K}(\alpha)x(\alpha) = \lambda(\alpha)\mathrm{M}(\alpha)x(\alpha), \tag{4.4}$$

where the two coefficient matrices are in their fuzzy interval forms $\mathrm{K}(\alpha) = (k_{ij}(\alpha)) = ([\underline{k_{ij}}(\alpha), \overline{k_{ij}}(\alpha)])$ and $\mathrm{M}(\alpha) = (m_{ij}(\alpha)) = ([\underline{m_{ij}}(\alpha), \overline{m_{ij}}(\alpha)])$ for $i, j = 1, 2, \ldots, n$ and may be expressed as

$$\mathrm{K}(\alpha) = \begin{pmatrix} [\underline{k_{11}}(\alpha), \overline{k_{11}}(\alpha)] & [\underline{k_{12}}(\alpha), \overline{k_{12}}(\alpha)] & \cdots & [\underline{k_{1n}}(\alpha), \overline{k_{1n}}(\alpha)] \\ [\underline{k_{21}}(\alpha), \overline{k_{21}}(\alpha)] & [\underline{k_{22}}(\alpha), \overline{k_{22}}(\alpha)] & \cdots & [\underline{k_{2n}}(\alpha), \overline{k_{2n}}(\alpha)] \\ \vdots & \vdots & \ddots & \vdots \\ [\underline{k_{n1}}(\alpha), \overline{k_{n1}}(\alpha)] & [\underline{k_{n2}}(\alpha), \overline{k_{n2}}(\alpha)] & \cdots & [\underline{k_{nn}}(\alpha), \overline{k_{nn}}(\alpha)] \end{pmatrix}; \tag{4.5a}$$

$$\mathrm{M}(\alpha) = \begin{pmatrix} [\underline{m_{11}}(\alpha), \overline{m_{11}}(\alpha)] & [\underline{m_{12}}(\alpha), \overline{m_{12}}(\alpha)] & \cdots & [\underline{m_{1n}}(\alpha), \overline{m_{1n}}(\alpha)] \\ [\underline{m_{21}}(\alpha), \overline{m_{21}}(\alpha)] & [\underline{m_{22}}(\alpha), \overline{m_{22}}(\alpha)] & \cdots & [\underline{m_{2n}}(\alpha), \overline{m_{2n}}(\alpha)] \\ \vdots & \vdots & \ddots & \vdots \\ [\underline{m_{n1}}(\alpha), \overline{m_{n1}}(\alpha)] & [\underline{m_{n2}}(\alpha), \overline{m_{n2}}(\alpha)] & \cdots & [\underline{m_{nn}}(\alpha), \overline{m_{nn}}(\alpha)] \end{pmatrix}, \tag{4.5b}$$

where the fuzzy parameter α varies from 0 to 1.

Further, the fuzzy interval GEP is again converted into its fuzzy-affine form. In this regard, the fuzzy interval coefficient matrices $\mathrm{K}(\alpha)$ and $\mathrm{M}(\alpha)$ (given in Eqs. (4.5a) and (4.5b)) are transformed into the fuzzy-affine matrices by converting each element of the fuzzy interval matrices into their respective fuzzy-affine forms. Therefore, the FGEP (Eq. (4.1)) is finally converted into a fuzzy-affine GEP as

$$\mathrm{K}(\alpha, \varepsilon_{ij})x(\alpha, \varepsilon^{\#}) = \lambda(\alpha, \varepsilon^{\#})\mathrm{M}(\alpha, \varepsilon_{ij}^{*})x(\alpha, \varepsilon^{\#}), \tag{4.6}$$

where $\varepsilon^{\#}$ may be either a newly generated noise symbol or a function existing noise symbols ε_{ij} and ε_{ij}^{*}. Here, $\mathrm{K}(\alpha, \varepsilon_{ij})$ and $\mathrm{M}(\alpha, \varepsilon_{ij}^{*})$ are the fuzzy-affine coefficient matrices expressed as follows:

$$\mathrm{K}(\alpha, \varepsilon_{ij}) = \begin{pmatrix} k_{11}(\alpha, \varepsilon_{11}) & k_{12}(\alpha, \varepsilon_{12}) & \cdots & k_{1n}(\alpha, \varepsilon_{1n}) \\ k_{21}(\alpha, \varepsilon_{21}) & k_{22}(\alpha, \varepsilon_{22}) & \cdots & k_{2n}(\alpha, \varepsilon_{2n}) \\ \vdots & \vdots & \ddots & \vdots \\ k_{n1}(\alpha, \varepsilon_{n1}) & k_{n2}(\alpha, \varepsilon_{n2}) & \cdots & k_{nn}(\alpha, \varepsilon_{nn}) \end{pmatrix}; \tag{4.7a}$$

$$\mathrm{M}(\alpha, \varepsilon^*_{ij}) = \begin{pmatrix} m_{11}(\alpha, \varepsilon^*_{11}) & m_{12}(\alpha, \varepsilon^*_{12}) & \cdots & m_{1n}(\alpha, \varepsilon^*_{1n}) \\ m_{21}(\alpha, \varepsilon^*_{21}) & m_{22}(\alpha, \varepsilon^*_{22}) & \cdots & m_{2n}(\alpha, \varepsilon^*_{2n}) \\ \vdots & \vdots & \ddots & \vdots \\ m_{n1}(\alpha, \varepsilon^*_{n1}) & m_{n2}(\alpha, \varepsilon^*_{n2}) & \cdots & m_{nn}(\alpha, \varepsilon^*_{nn}) \end{pmatrix}, \tag{4.7b}$$

where $\alpha \in [0, 1]$, $\varepsilon_{ij} \in [-1, 1]$, and $\varepsilon^*_{ij} \in [-1, 1]$ for $i, j = 0, 1, \ldots, n$.

Moreover, $\lambda(\alpha, \varepsilon^{\#})$ and $x(\alpha, \varepsilon^{\#})$ are the corresponding fuzzy-affine eigenvalue and the fuzzy-affine eigenvector of the fuzzy-affine GEP (given in Eq. (4.6)), respectively.

The above fuzzy-affine coefficient matrices of the given fuzzy-affine GEP (4.7a, 4.7b) contains various parameters viz. the fuzzy parameter (α) and the noise symbols (ε_{ij} and ε^*_{ij} for $i, j = 0, 1, \ldots, n$). Thus, the required fuzzy-affine eigenvalue solution $\lambda_l(\alpha, \varepsilon^{\#})$ for $l = 1, 2, \ldots, n$ containing these parameters may be obtained by solving the fuzzy-affine GEP (given in Eq. (4.6)) symbolically.

Now, because each noise symbol varies from -1 to 1 (that is $\varepsilon_{ij}, \varepsilon^*_{ij} \in [-1, 1]$ for $i, j = 0, 1, \ldots, n$), the fuzzy interval eigenvalue solutions $\lambda_l(\alpha) = [\underline{\lambda_l}(\alpha), \overline{\lambda_l}(\alpha)]$ for $l = 1, 2, \ldots, n$ may be computed as follows:

$$\underline{\lambda_l}(\alpha) = \min_{-1 \le \varepsilon^{\#} \le 1} \lambda_l(\alpha, \varepsilon^{\#}) \text{ and } \overline{\lambda_l}(\alpha) = \max_{-1 \le \varepsilon^{\#} \le 1} \lambda_l(\alpha, \varepsilon^{\#}), \tag{4.8}$$

for $\alpha \in [0, 1]$ and $l = 1, 2, \ldots, n$.

Lastly, all the fuzzy eigenvalue solutions ($\tilde{\lambda}_l$ for $l = 1, 2, \ldots, n$) of the given FGEP (Eq. (4.1)) can be calculated by varying the fuzzy parameter (α) from 0 to 1, and the fuzzy eigenvalue solution plots have also been constructed by substituting different values of "α" in the lower and upper bounds of the fuzzy interval eigenvalue solutions (4.8).

5 Numerical Examples

To illustrate the applicability of the proposed method, five numerical examples related to fuzzy linear structural dynamic problems (that is FGEP or FSEP) have been solved in the present section. Initially, a FGEP having 4 × 4 fuzzy symmetric coefficient matrices has been solved by adopting the proposed approach. Further, a fuzzy spring-mass structural system with five degrees of freedom is considered in the second problem in which all the stiffness and mass parameters are in the form of TrFNs. In the third problem, a fuzzy multi-storeyed shear building structural system with five degrees of freedom is worked out. Again, a SFEP having 4×4 fuzzy symmetric coefficient matrix has been solved in the fourth problem. Lastly, a fuzzy two-storeyed frame structural problem having fuzzy stiffness and mass parameters is investigated in the fifth problem.

Example 2: For the initial case, consider a FGEP $\tilde{K}\tilde{x} = \tilde{\lambda}\tilde{M}\tilde{x}$ (given in Eq. (4.1)) [16, 21, 24, 28], where one of its coefficient matrices is a fuzzy symmetric matrix of dimension 4×4 in which all of its elements are considered in the form of TFNs and the other coefficient matrix is a 4×4 identity matrix given as follows:

$$\tilde{K} = \begin{bmatrix} (2975, 3000, 3025) & (-2015, -2000, -1985) & 0 & 0 \\ (-2015, -2000, -1985) & (4965, 5000, 5035) & (-3020, -3000, -2980) & 0 \\ 0 & (-3020, -3000, -2980) & (6955, 7000, 7045) & (-4025, -4000, -3975) \\ 0 & 0 & (-4025, -4000, -3975) & (8945, 9000, 9055) \end{bmatrix}$$

$$\text{and } \tilde{M} = I_{4\times 4} = \begin{pmatrix} 1\,0\,0\,0 \\ 0\,1\,0\,0 \\ 0\,0\,1\,0 \\ 0\,0\,0\,1 \end{pmatrix}.$$

To solve the given FGEP $\tilde{K}\tilde{x} = \tilde{\lambda}\tilde{M}\tilde{x}$, we have to transform the fuzzy coefficient matrix ($\tilde{K}$) into its corresponding fuzzy-affine form. For this, firstly convert $\tilde{K}$ into its fuzzy interval form ($K(\alpha)$) by adopting the α-cut technique of TFNs as discussed in Sect. 2.3. Thus, the fuzzy interval form of $\tilde{K}$ is obtained as

$$K(\alpha) = \begin{pmatrix} [2975 + 25\alpha, 3025 - 25\alpha] & [-2015 + 15\alpha, -1985 - 15\alpha] & 0 & 0 \\ [-2015 + 15\alpha, -1985 - 15\alpha] & [4965 + 35\alpha, 5035 - 35\alpha] & [-3020 + 20\alpha, -2980 - 20\alpha] & 0 \\ 0 & [-3020 + 20\alpha, -2980 - 20\alpha] & [6955 + 45\alpha, 7045 - 45\alpha] & [-4025 + 25\alpha, -3975 - 25\alpha] \\ 0 & 0 & [-4025 + 25\alpha, -3975 - 25\alpha] & [8945 + 55\alpha, 9055 - 55\alpha] \end{pmatrix},$$

where $\alpha \in [0, 1]$.

Then, the fuzzy interval matrix $K(\alpha)$ has been further converted into a fuzzy-affine matrix by transforming each element of $K(\alpha)$ to their respective fuzzy-affine forms of TFNs as described in Sect. 3.1. Hence, the fuzzy-affine coefficient matrix $(K(\alpha, \varepsilon_i)_{i=1,\ldots,10})$ is computed as

$$K(\alpha, \varepsilon_i)_{i=1,\ldots,10} = \begin{pmatrix} 3000 + (25 - 25\alpha)\varepsilon_1 & -2000 + (15 - 15\alpha)\varepsilon_2 & 0 & 0 \\ -2000 + (15 - 15\alpha)\varepsilon_3 & 5000 + (35 - 35\alpha)\varepsilon_4 & -3000 + (20 - 20\alpha)\varepsilon_5 & 0 \\ 0 & -3000 + (20 - 20\alpha)\varepsilon_6 & 7000 + (45 - 45\alpha)\varepsilon_7 & -4000 + (25 - 25\alpha)\varepsilon_8 \\ 0 & 0 & -4000 + (25 - 25\alpha)\varepsilon_9 & 9000 + (55 - 55\alpha)\varepsilon_{10} \end{pmatrix},$$

where $\alpha \in [0, 1]$ and $\varepsilon_i \in [-1, 1]$ for $i = 1, \ldots, 10$.

Now after the conversions, the given FGEP is transformed into a fuzzy-affine GEP $K(\alpha, \varepsilon_i)_{i=1,\ldots,10}x(\alpha, \varepsilon^{\#}) = \lambda(\alpha, \varepsilon^{\#})x(\alpha, \varepsilon^{\#})$, with different parameters in the form of noise symbols (ε_i for $i = 1, \ldots, 10$) and fuzzy parameter (α). Hence, all the fuzzy-affine eigenvalue solutions ($\lambda_l(\alpha, \varepsilon^{\#})$ for $l = 1, 2, 3, 4$) are evaluated by adopting the fuzzy-affine approach described in Sect. 4. Finally, the fuzzy-affine eigenvalue solutions ($\lambda_l(\alpha, \varepsilon^{\#})$) are reconverted into their respective fuzzy eigenvalue solutions ($\tilde{\lambda}_l$ for $l = 1, 2, 3, 4$).

All the fuzzy eigenvalue plots are depicted in Figs. 5, 6, 7, 8. Also for comparison, the fuzzy eigenvalue solution plots of Mahato and Chakraverty [24] are incorporated in these figures. In Figs. 5, 6, 7, 8, the line marked with "○" is used to indicate the fuzzy eigenvalues by using the fuzzy-affine approach, and the dotted line marked with "*" depicts the fuzzy eigenvalues by Mahato and Chakraverty [24].

Moreover, for some particular values of the fuzzy parameter viz. $\alpha = 0, 0.1, 0.5, 0.8, 1$, the fuzzy eigenvalue solutions ($\tilde{\lambda}_l$ for $l = 1, 2, 3, 4$) obtained by adopting the fuzzy-affine approach and the solutions by Mahato and Chakraverty [24] are listed in Table 2.

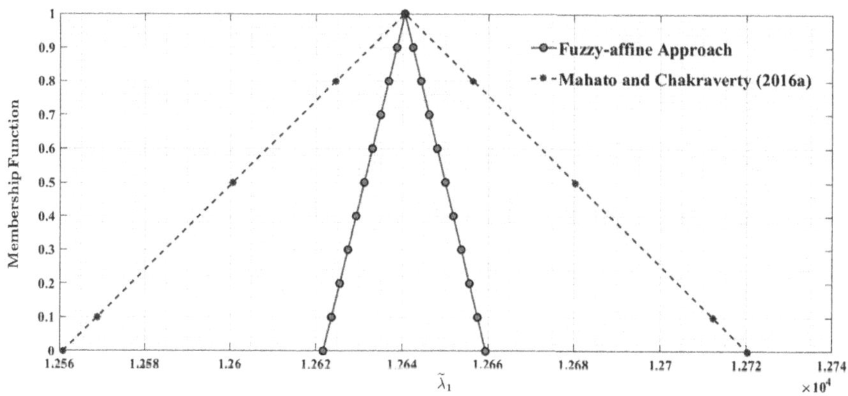

Fig. 5 Comparison plot of first fuzzy eigenvalue between fuzzy-affine approach and Mahato and Chakraverty [24] for Example 2

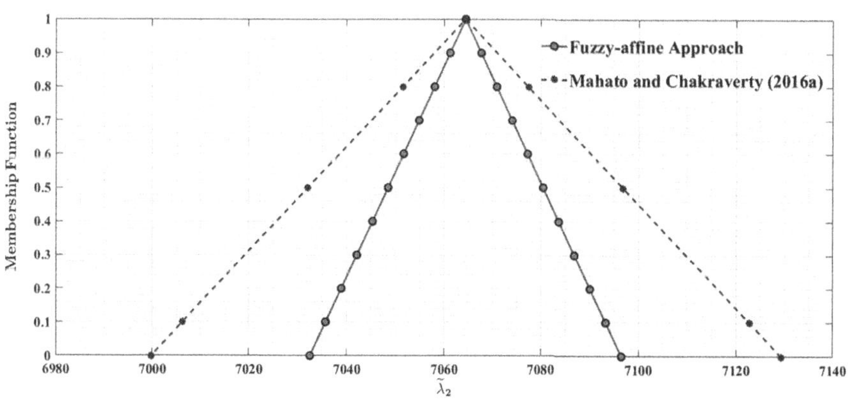

Fig. 6 Comparison plot of second fuzzy eigenvalue between fuzzy-affine approach and Mahato and Chakraverty [24] for Example 2

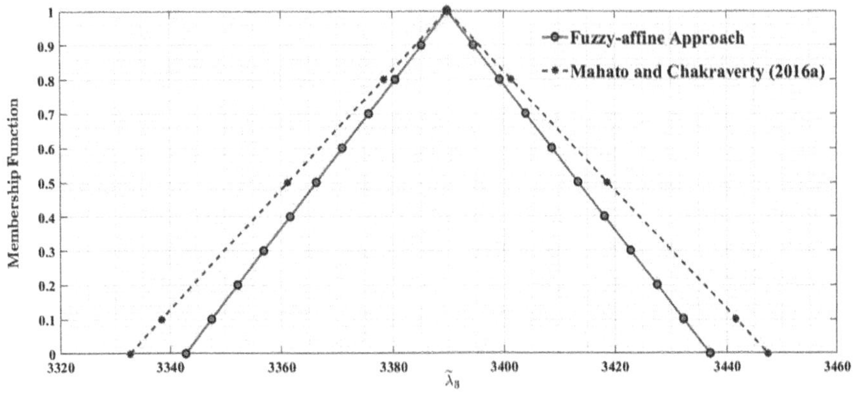

Fig. 7 Comparison plot of third fuzzy eigenvalue between fuzzy-affine approach and Mahato and Chakraverty [24] for Example 2

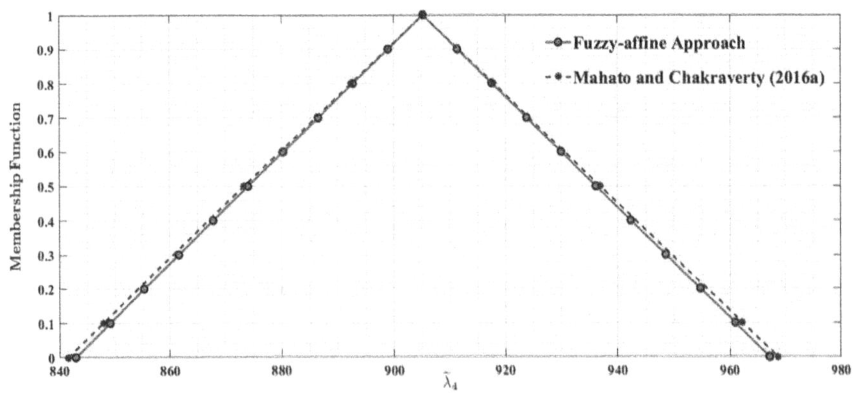

Fig. 8 Comparison plot of fourth fuzzy eigenvalue between fuzzy-affine approach and Mahato and Chakraverty [24] for Example 2

Lastly, a comparison of the lower and upper bounds of the fuzzy eigenvalue solutions by fuzzy-affine approach with other methods by different authors [16, 21, 24, 28], is incorporated in Table 3.

It may be seen from the above tables (Tables 2 and 3) that the proposed fuzzy-affine approach computes better and tighter outer bounds of the fuzzy eigenvalue solutions of the given FGEP in Example 2. Therefore, we may state that the fuzzy-affine approach-based technique results in tighter outer bounds of the eigenvalue solutions.

Example 3: Let us consider a fuzzy spring-mass structural system with five degrees of freedom [5, 10] as shown in Fig. 9. Here, the stiffness parameters ($\tilde{k}_i$ (N/m) for $i = 1, \ldots, 6$) and mass parameters ($\tilde{m}_i(kg)$ for $i = 1, \ldots, 5$) of the fuzzy spring-mass structural system are in the form of TrFNs given as follows:

Table 2 Comparison table of fuzzy eigenvalues between fuzzy-affine approach and Mahato and Chakraverty [24] for Example 2 for different values of α

α	Lower and upper bounds	Fuzzy-affine approach				Mahato and Chakraverty [24]			
		$\tilde{\lambda}_1$	$\tilde{\lambda}_2$	$\tilde{\lambda}_3$	$\tilde{\lambda}_4$	$\tilde{\lambda}_1$	$\tilde{\lambda}_2$	$\tilde{\lambda}_3$	$\tilde{\lambda}_4$
0	$\underline{\lambda_l}$	12,621.7205	7032.5430	3342.8114	842.9251	12,560.8129	6999.7924	3332.7559	841.5328
	$\overline{\lambda_l}$	12,659.4237	7096.5043	3436.9638	967.1082	12,720.2472	7129.2716	3447.4627	968.5661
0.1	$\underline{\lambda_l}$	12,623.5979	7035.7288	3347.5116	849.1616	12,568.7853	7006.2518	3338.4377	847.9047
	$\overline{\lambda_l}$	12,657.5308	7093.2939	3432.2486	960.9267	12,712.2756	7122.7844	3441.6816	962.2368
0.5	$\underline{\lambda_l}$	12,631.1158	7048.4848	3366.3202	874.0792	12,600.6735	7032.1055	3361.2260	873.3721
	$\overline{\lambda_l}$	12,649.9673	7080.4652	3413.3962	936.1713	12,680.3891	7096.8487	3418.6010	936.8949
0.8	$\underline{\lambda_l}$	12,636.7627	7058.0653	3380.4348	892.7372	12,624.5883	7051.5126	3378.3812	892.4517
	$\overline{\lambda_l}$	12,644.3033	7070.8575	3399.2651	917.5741	12,656.4744	7077.4108	3401.3364	917.8623
1	$\underline{\lambda_l}$	12,640.5313	7064.4588	3389.8484	905.1615	12,640.5313	7064.4588	3389.8484	905.1615
	$\overline{\lambda_l}$	12,640.5313	7064.4588	3389.8484	905.1615	12,640.5313	7064.4588	3389.8484	905.1615

Table 3 Comparison of outer bounds of fuzzy eigenvalues for Example 2

Comparisons	Outer bounds	Fuzzy eigenvalues			
		$\tilde{\lambda}_1$	$\tilde{\lambda}_2$	$\tilde{\lambda}_3$	$\tilde{\lambda}_4$
Fuzzy-affine Approach	Lower ($\underline{\lambda_l}$)	12,621.7205	7032.5430	3342.8114	842.9251
	Upper ($\overline{\lambda_l}$)	12,659.4237	7096.5043	3436.9638	967.1082
Mahato and Chakraverty [24]	Lower ($\underline{\lambda_l}$)	12,560.81292	6999.79244	3332.75588	841.53281
	Upper ($\overline{\lambda_l}$)	12,720.24723	7129.27159	3447.46269	968.56612
Hladik et al. [16]	Lower ($\underline{\lambda_l}$)	12,560.8129	6999.8026	3332.7944	841.5328
	Upper ($\overline{\lambda_l}$)	12,720.2472	7129.2716	3447.4628	968.5505
Leng and He [21]	Lower ($\underline{\lambda_l}$)	12,550.53133	6974.45882	3299.84838	815.16148
	Upper ($\overline{\lambda_l}$)	12,730.53133	7154.45882	3479.84838	995.16148
Qiu et al. [28]	Lower ($\underline{\lambda_l}$)	12,588.29000	7000.19500	3331.16200	826.73720
	Upper ($\overline{\lambda_l}$)	12,692.77000	7128.72300	3448.53500	983.58580

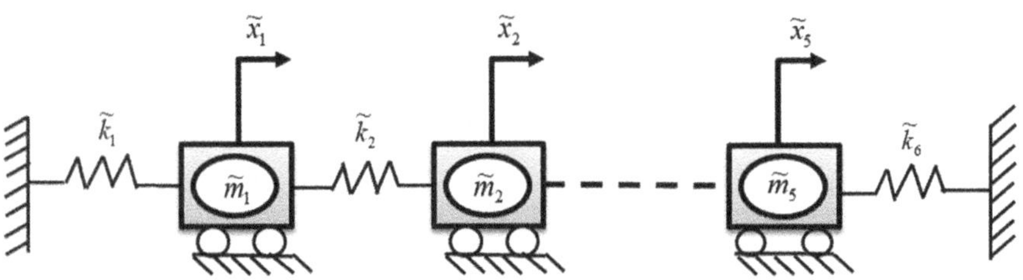

Fig. 9 Fuzzy spring-mass structural system with five degrees of freedom

$\tilde{k}_1 = (2000, 2025, 2075, 2100)$, $\tilde{k}_2 = (1800, 1820, 1830, 1850)$, $\tilde{k}_3 = (1600, 1610, 1620, 1630)$, $\tilde{k}_4 = (1400, 1405, 1415, 1420)$, $\tilde{k}_5 = (1200, 1202, 1208, 1210)$, and $\tilde{k}_6 = (1000, 1002, 1006, 1008)$.

$\tilde{m}_1 = (10, 10.5, 11.5, 12)$, $\tilde{m}_2 = (12, 12.5, 13.5, 14)$, $\tilde{m}_3 = (14, 14.5, 15.5, 16)$, $\tilde{m}_4 = (16, 16.5, 17.5, 18)$, and $\tilde{m}_5 = (18, 18.5, 19.5, 20)$.

The dynamic analysis of the above fuzzy spring-mass structural system with five degrees of freedom having fuzzy material and geometric properties leads to a FGEP (given in Eq. (4.1)) $\tilde{\mathrm{K}}\tilde{x} = \tilde{\lambda}\tilde{\mathrm{M}}\tilde{x}$, where the fuzzy coefficient matrices ($\tilde{\mathrm{K}}$ and $\tilde{\mathrm{M}}$) are known as fuzzy stiffness and fuzzy mass matrices, respectively. For this case, the corresponding fuzzy stiffness and fuzzy mass matrices are fuzzy symmetric matrices of dimension 5×5 expressed as

$$\tilde{\mathrm{K}} = \begin{pmatrix} \tilde{k}_1 + \tilde{k}_2 & -\tilde{k}_2 & 0 & 0 & 0 \\ -\tilde{k}_2 & \tilde{k}_2 + \tilde{k}_3 & -\tilde{k}_3 & 0 & 0 \\ 0 & -\tilde{k}_3 & \tilde{k}_3 + \tilde{k}_4 & -\tilde{k}_4 & 0 \\ 0 & 0 & -\tilde{k}_4 & \tilde{k}_4 + \tilde{k}_5 & -\tilde{k}_5 \\ 0 & 0 & 0 & -\tilde{k}_5 & \tilde{k}_5 + \tilde{k}_6 \end{pmatrix} \text{ and } \tilde{\mathrm{M}} = \begin{pmatrix} \tilde{m}_1 & 0 & 0 & 0 & 0 \\ 0 & \tilde{m}_2 & 0 & 0 & 0 \\ 0 & 0 & \tilde{m}_3 & 0 & 0 \\ 0 & 0 & 0 & \tilde{m}_4 & 0 \\ 0 & 0 & 0 & 0 & \tilde{m}_5 \end{pmatrix}.$$

As given in Example 2, firstly we have to transform the FGEP $\tilde{K}\tilde{x} = \tilde{\lambda}\tilde{M}\tilde{x}$ into a fuzzy-affine GEP. For this, the fuzzy coefficient matrices viz. fuzzy stiffness matrix ($\tilde{K}$) and fuzzy mass matrix ($\tilde{M}$) are transformed into their respective fuzzy-affine forms by converting all the fuzzy stiffness and mass parameters into their corresponding form of fuzzy-affine parameters by adopting the procedure described in Sect. 3.2. Thus, the fuzzy-affine stiffness parameters are calculated as follows:

$k_1(\alpha, \varepsilon_1) = 2050 + (50 - 25\alpha)\varepsilon_1$, $k_2(\alpha, \varepsilon_2) = 1825 + (25 - 20\alpha)\varepsilon_2$, $k_3(\alpha, \varepsilon_3) = 1615 + (15 - 10\alpha)\varepsilon_3$, $k_4(\alpha, \varepsilon_4) = 1410 + (10 - 5\alpha)\varepsilon_4$, $k_5(\alpha, \varepsilon_5) = 1205 + (5 - 2\alpha)\varepsilon_5$, and $k_6(\alpha, \varepsilon_6) = 1004 + (4 - 2\alpha)\varepsilon_6$, where $\alpha \in [0, 1]$ and $\varepsilon_i \in [-1, 1]$ for $i = 1, 2, 3, 4, 5, 6$.

Hence, the fuzzy-affine stiffness matrix may be obtained as

$$\mathrm{K}(\alpha, \varepsilon_i)_{i=1,\ldots,6} = \begin{pmatrix} k_1(\alpha,\varepsilon_1)+k_2(\alpha,\varepsilon_2) & -k_2(\alpha,\varepsilon_2) & 0 & 0 & 0 \\ -k_2(\alpha,\varepsilon_2) & k_2(\alpha,\varepsilon_2)+k_3(\alpha,\varepsilon_3) & -k_3(\alpha,\varepsilon_3) & 0 & 0 \\ 0 & -k_3(\alpha,\varepsilon_3) & k_3(\alpha,\varepsilon_3)+k_4(\alpha,\varepsilon_4) & -k_4(\alpha,\varepsilon_4) & 0 \\ 0 & 0 & -k_4(\alpha,\varepsilon_4) & k_4(\alpha,\varepsilon_4)+k_5(\alpha,\varepsilon_5) & -k_5(\alpha,\varepsilon_5) \\ 0 & 0 & 0 & -k_5(\alpha,\varepsilon_5) & k_5(\alpha,\varepsilon_5)+k_6(\alpha,\varepsilon_6) \end{pmatrix}.$$

That is

$$\mathrm{K}(\alpha, \varepsilon_i)_{i=1,\ldots,6} = \begin{pmatrix} 3875+(50-25\alpha)\varepsilon_1+(25-20\alpha)\varepsilon_2 & -1825-(25-20\alpha)\varepsilon_2 & 0 & 0 & 0 \\ -1825-(25-20\alpha)\varepsilon_2 & 3440+(25-20\alpha)\varepsilon_2+(15-10\alpha)\varepsilon_3 & -1615-(15-10\alpha)\varepsilon_3 & 0 & 0 \\ 0 & -1615-(15-10\alpha)\varepsilon_3 & 3025+(15-10\alpha)\varepsilon_3+(10-5\alpha)\varepsilon_4 & -1410-(10-5\alpha)\varepsilon_4 & 0 \\ 0 & 0 & -1410-(10-5\alpha)\varepsilon_4 & 2615+(10-5\alpha)\varepsilon_4+(5-2\alpha)\varepsilon_5 & -1205-(5-2\alpha)\varepsilon_5 \\ 0 & 0 & 0 & -1205-(5-2\alpha)\varepsilon_5 & 2209+(5-2\alpha)\varepsilon_5+(4-2\alpha)\varepsilon_6 \end{pmatrix},$$

where $\alpha \in [0, 1]$ and $\varepsilon_i \in [-1, 1]$ for $i = 1, 2, 3, 4, 5, 6$.

Similarly, the fuzzy-affine mass matrix may be computed by converting each fuzzy mass parameter into its respective fuzzy-affine forms. That is

$m_1(\alpha, \varepsilon_1^*) = 11 + (1 - 0.5\alpha)\varepsilon_1^*$, $m_2(\alpha, \varepsilon_2^*) = 13 + (1 - 0.5\alpha)\varepsilon_2^*$, $m_3(\alpha, \varepsilon_3^*) = 15 + (1 - 0.5\alpha)\varepsilon_3^*$, $m_4(\alpha, \varepsilon_4^*) = 17 + (1 - 0.5\alpha)\varepsilon_4^*$, and $m_5(\alpha, \varepsilon_5^*) = 19 + (1 - 0.5\alpha)\varepsilon_5^*$, where $\alpha \in [0, 1]$ and $\varepsilon_i^* \in [-1, 1]$ for $i = 1, 2, 3, 4, 5$.

Thus, the fuzzy-affine mass matrix is

$$\mathrm{M}(\alpha, \varepsilon_i^*)_{i=1,\ldots,5} = \begin{pmatrix} m_1(\alpha,\varepsilon_1^*) & 0 & 0 & 0 & 0 \\ 0 & m_2(\alpha,\varepsilon_2^*) & 0 & 0 & 0 \\ 0 & 0 & m_3(\alpha,\varepsilon_3^*) & 0 & 0 \\ 0 & 0 & 0 & m_4(\alpha,\varepsilon_4^*) & 0 \\ 0 & 0 & 0 & 0 & m_5(\alpha,\varepsilon_5^*) \end{pmatrix}.$$

Hence, we may have

$$\mathrm{M}(\alpha,\varepsilon_i^*)_{i=1,\ldots,5}=\begin{pmatrix} 11+(1-0.5\alpha)\varepsilon_1^* & 0 & 0 & 0 & 0\\ 0 & 13+(1-0.5\alpha)\varepsilon_2^* & 0 & 0 & 0\\ 0 & 0 & 15+(1-0.5\alpha)\varepsilon_3^* & 0 & 0\\ 0 & 0 & 0 & 17+(1-0.5\alpha)\varepsilon_4^* & 0\\ 0 & 0 & 0 & 0 & 19+(1-0.5\alpha)\varepsilon_5^* \end{pmatrix},$$

where for $\alpha \in [0, 1]$ and $\varepsilon_i^* \in [-1, 1]$ for $i = 1, \ldots, 5$.

After the transformation of the given FGEP into a fuzzy-affine GEP $\mathrm{K}(\alpha, \varepsilon_i)_{i=1,\ldots,6}x(\alpha, \varepsilon^\#) = \lambda(\alpha, \varepsilon^\#)\mathrm{M}(\alpha, \varepsilon_i^*)_{i=1,\ldots,5}x(\alpha, \varepsilon^\#)$, the fuzzy-affine eigenvalue solutions ($\lambda_l(\alpha, \varepsilon^\#)$ for $l = 1, \ldots, 5$) have been computed by utilizing the fuzzy-affine approach proposed in Sect. 4. Therefore, the fuzzy eigenvalue solutions ($\tilde{\lambda}_l$ for $l = 1, \ldots, 5$) are evaluated by reconverting $\lambda_l(\alpha, \varepsilon^\#)$, and the fuzzy solution plots are incorporated in Figs. 10, 11, 12, 13, 14 with the fuzzy eigenvalue solution plots of Chakraverty and Behera [5].

Further, the fuzzy eigenvalue solutions ($\tilde{\lambda}_l$ for $l = 1, \ldots, 5$) calculated by the fuzzy-affine approach and the solutions by Chakraverty and Behera [5] are listed in Table 4 for different values of the fuzzy parameter (α) such as $\alpha = 0, 0.2, 0.4, 0.5, 0.6, 0.8, 1$. Also, a comparison of the lower and upper bounds of the fuzzy eigenvalue solutions by the fuzzy-affine approach with [5, 10] is included in Table 5.

It may be clearly noticed from Tables 4 and 5 that the fuzzy-affine approach-based technique results in tighter outer bounds as compared to the results of [5, 10].

Example 4: In this example, consider a fuzzy multi-storey shear structural system [20, 25, 31] with five degrees of freedom as shown in Fig. 15 having fuzzy material and geometric properties. Thus, the stiffness and mass parameters are in the form of fuzzy numbers. For this example, suppose the fuzzy parameters are in the form of

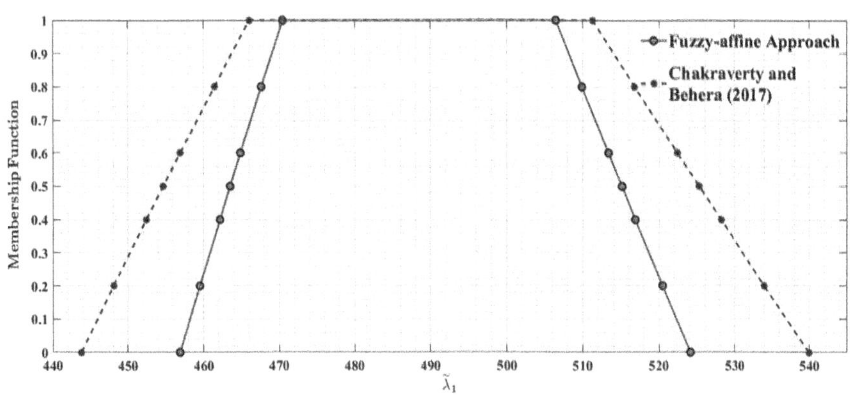

Fig. 10 Comparison plot of first fuzzy eigenvalue between fuzzy-affine approach and Chakraverty and Behera [5] for Example 3

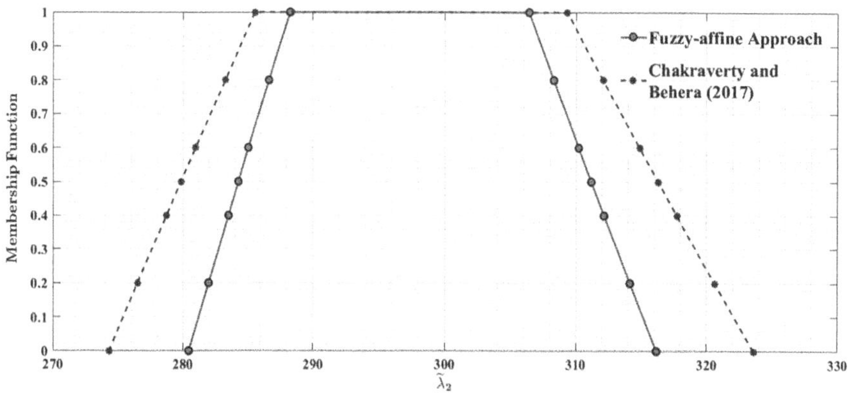

Fig. 11 Comparison plot of second fuzzy eigenvalue between fuzzy-affine approach and Chakraverty and Behera [5] for Example 3

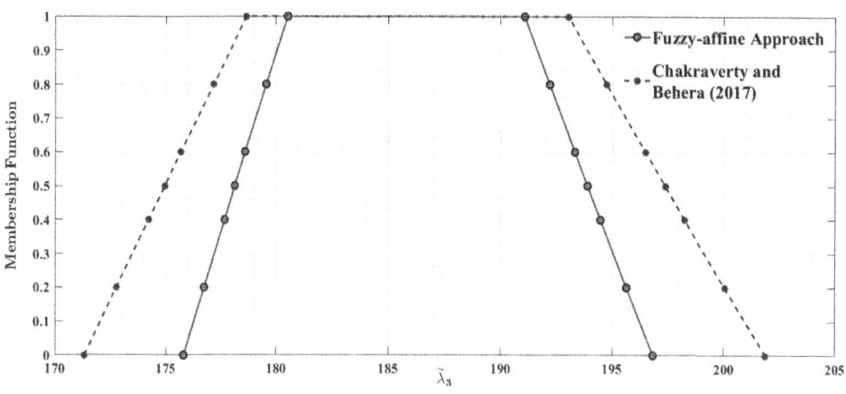

Fig. 12 Comparison plot of third fuzzy eigenvalue between fuzzy-affine approach and Chakraverty and Behera [5] for Example 3

TFNs. Then, the fuzzy stiffness parameters ($\tilde{k}_i$ (N/m) for $i = 1, 2, \ldots, 5$) and the fuzzy mass parameters ($\tilde{m}_i$ (kg) for $i = 1, 2, \ldots, 5$) at each floor level are taken as follows:

$\tilde{k}_1 = (2000, 2010, 2020)$, $\tilde{k}_2 = (1800, 1825, 1850)$, $\tilde{k}_3 = (1600, 1615, 1630)$, $\tilde{k}_4 = (1400, 1410, 1420)$, and $\tilde{k}_5 = (1200, 1205, 1210)$.

$\tilde{m}_1 = (29, 30, 31)$, $\tilde{m}_2 = (26, 27, 28)$, $\tilde{m}_3 = (26, 27, 28)$, $\tilde{m}_4 = (24, 25, 26)$, and $\tilde{m}_5 = (17, 18, 19)$.

The equation of motion for the fuzzy multi-storeyed shear building structural system with five degrees of freedom subject to ambient vibration in the fuzzy uncertain environment may lead to a FGEP $\tilde{K}\tilde{x} = \tilde{\lambda}\tilde{M}\tilde{x}$ [3, 5], where the coefficient stiffness and mass matrices are in the form of fuzzy symmetric matrices of dimension 5×5 obtained as follows:

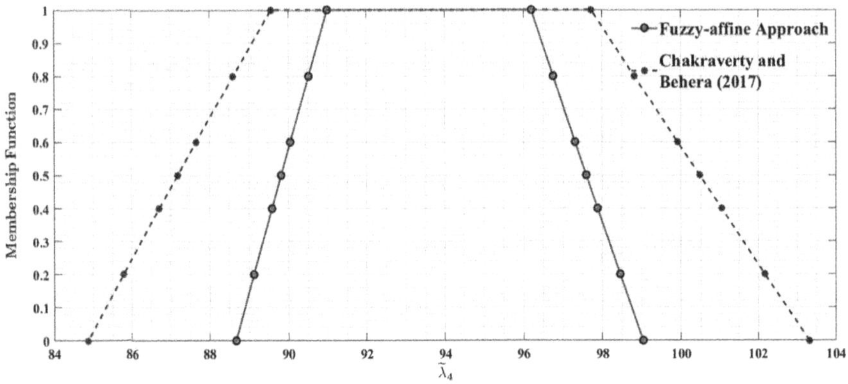

Fig. 13 Comparison plot of fourth fuzzy eigenvalue between fuzzy-affine approach and Chakraverty and Behera [5] for Example 3

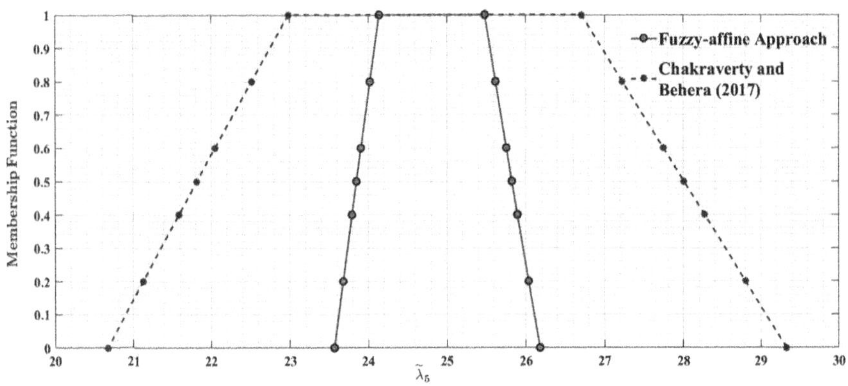

Fig. 14 Comparison plot of fifth fuzzy eigenvalue between fuzzy-affine approach and Chakraverty and Behera [5] for Example 3

$$\tilde{\mathrm{K}} = \begin{pmatrix} \tilde{k}_1 + \tilde{k}_2 & -\tilde{k}_2 & 0 & 0 & 0 \\ -\tilde{k}_2 & \tilde{k}_2 + \tilde{k}_3 & -\tilde{k}_3 & 0 & 0 \\ 0 & -\tilde{k}_3 & \tilde{k}_3 + \tilde{k}_4 & -\tilde{k}_4 & 0 \\ 0 & 0 & -\tilde{k}_4 & \tilde{k}_4 + \tilde{k}_5 & -\tilde{k}_5 \\ 0 & 0 & 0 & -\tilde{k}_5 & \tilde{k}_5 \end{pmatrix} \text{and } \tilde{\mathrm{M}} = \begin{pmatrix} \tilde{m}_1 & 0 & 0 & 0 & 0 \\ 0 & \tilde{m}_2 & 0 & 0 & 0 \\ 0 & 0 & \tilde{m}_3 & 0 & 0 \\ 0 & 0 & 0 & \tilde{m}_4 & 0 \\ 0 & 0 & 0 & 0 & \tilde{m}_5 \end{pmatrix}.$$

Now as given in Example 3, the followings are the respective fuzzy-affine forms of the stiffness parameters $(k_i(\alpha, \varepsilon_i)\ (N/m)$ for $i = 1, 2, \ldots, 5)$.

$k_1(\alpha, \varepsilon_1) = 2010 + (10 - 10\alpha)\varepsilon_1$, $k_2(\alpha, \varepsilon_2) = 1825 + (25 - 25\alpha)\varepsilon_2$, $k_3(\alpha, \varepsilon_3) = 1615 + (15 - 15\alpha)\varepsilon_3$, $k_4(\alpha, \varepsilon_4) = 1410 + (10 - 10\alpha)\varepsilon_4$, and $k_5(\alpha, \varepsilon_5) = 1205 + (5 - 5\alpha)\varepsilon_5$, where $\alpha \in [0, 1]$ and $\varepsilon_i \in [-1, 1]$ for $i = 1, 2, 3, 4, 5$.

Table 4 Comparison table of fuzzy eigenvalues between fuzzy-affine approach and Chakraverty and Behera [5 for Example 3 for different values of α

α	Lower and upper bounds	Fuzzy-affine approach					Chakraverty and Behera [5]				
		$\tilde{\lambda}_1$	$\tilde{\lambda}_2$	$\tilde{\lambda}_3$	$\tilde{\lambda}_4$	$\tilde{\lambda}_5$	$\tilde{\lambda}_1$	$\tilde{\lambda}_2$	$\tilde{\lambda}_3$	$\tilde{\lambda}_4$	$\tilde{\lambda}_5$
0	$\underline{\lambda_l}$	456.8997	280.4391	175.7996	88.6697	23.5660	443.7522	274.3046	171.3316	84.8674	20.6734
	$\overline{\lambda_l}$	524.1465	316.1529	196.8291	99.0332	26.1796	539.9518	323.5612	201.8605	103.3138	29.3269
0.2	$\underline{\lambda_l}$	459.5205	281.9643	176.7202	89.1232	23.6778	448.0719	276.4919	172.7613	85.7835	21.1287
	$\overline{\lambda_l}$	520.4847	314.1461	195.6453	98.4509	26.0368	534.0016	320.6160	200.0486	102.1654	28.7952
0.4	$\underline{\lambda_l}$	462.1824	283.5100	177.6523	89.5824	23.7909	452.4615	278.7131	174.2090	86.7111	21.5870
	$\overline{\lambda_l}$	516.8892	312.1699	194.4779	97.8767	25.8959	528.1621	317.7220	198.2609	101.0317	28.2676
0.5	$\underline{\lambda_l}$	463.5290	284.2906	178.1227	89.8141	23.8479	454.6830	279.8366	174.9397	87.1792	21.8172
	$\overline{\lambda_l}$	515.1157	311.1930	193.9003	97.5926	25.8262	525.2829	316.2938	197.3759	100.4703	28.0053
0.6	$\underline{\lambda_l}$	464.8864	285.0765	178.5960	90.0473	23.9053	456.9225	280.9688	175.6750	87.6502	22.0482
	$\overline{\lambda_l}$	513.3581	310.2235	193.3268	97.3104	25.7569	522.4303	314.8779	196.4968	99.9125	27.7441
0.8	$\underline{\lambda_l}$	467.6335	286.6642	179.5517	90.5180	24.0212	461.4568	283.2598	177.1594	88.6011	22.5124
	$\overline{\lambda_l}$	509.8899	308.3061	192.1914	96.7518	25.6198	516.8030	312.0823	194.7560	98.8076	27.2247
1	$\underline{\lambda_l}$	470.4249	288.2738	180.5196	90.9947	24.1385	466.0662	285.5871	178.6627	89.5639	22.9798
	$\overline{\lambda_l}$	506.4827	306.4172	191.0716	96.2008	25.4844	511.2776	309.3341	193.0381	97.7168	26.7092

Table 5 Comparison of outer bounds of fuzzy eigenvalues for Example 3

Comparisons	Outer bounds	Fuzzy eigenvalues				
		$\tilde{\lambda}_1$	$\tilde{\lambda}_2$	$\tilde{\lambda}_3$	$\tilde{\lambda}_4$	$\tilde{\lambda}_5$
Fuzzy-affine approach	Lower ($\underline{\lambda_I}$)	456.8997	280.4391	175.7996	88.6697	23.5660
	Upper ($\overline{\lambda_I}$)	524.1465	316.1529	196.8291	99.0332	26.1796
Chakraverty and Behera [5]	Lower ($\underline{\lambda_I}$)	443.7522	274.3046	171.3316	84.8674	20.6734
	Upper ($\overline{\lambda_I}$)	539.9518	323.5612	201.8605	103.3138	29.3269
Chen et al. [10]	Lower ($\underline{\lambda_I}$)	443.7521711	274.3045861	171.3316032	84.8673933	20.6733690
	Upper ($\overline{\lambda_I}$)	539.9518289	323.5612202	201.8605276	103.3137982	29.3269368

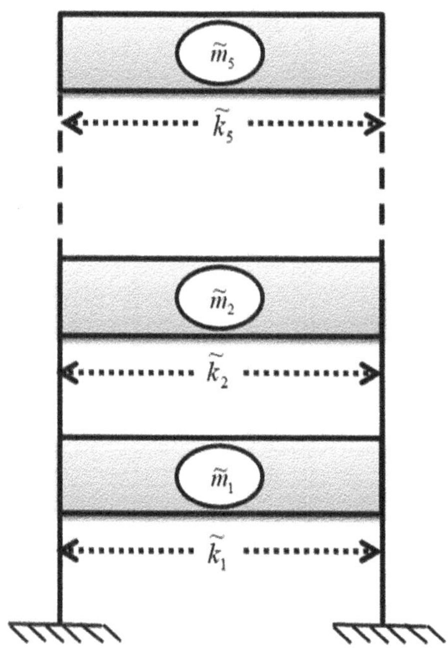

Fig. 15 Fuzzy multi-storeyed shear building structural system with five degrees of freedom

Then, the fuzzy-affine stiffness matrix may be obtained as

$K(\alpha, \varepsilon_i)_{i=1,\ldots,5}$

$$= \begin{pmatrix} k_1(\alpha,\varepsilon_1)+k_2(\alpha,\varepsilon_2) & -k_2(\alpha,\varepsilon_2) & 0 & 0 & 0 \\ -k_2(\alpha,\varepsilon_2) & k_2(\alpha,\varepsilon_2)+k_3(\alpha,\varepsilon_3) & -k_3(\alpha,\varepsilon_3) & 0 & 0 \\ 0 & -k_3(\alpha,\varepsilon_3) & k_3(\alpha,\varepsilon_3)+k_4(\alpha,\varepsilon_4) & -k_4(\alpha,\varepsilon_4) & 0 \\ 0 & 0 & -k_4(\alpha,\varepsilon_4) & k_4(\alpha,\varepsilon_4)+k_5(\alpha,\varepsilon_5) & -k_5(\alpha,\varepsilon_5) \\ 0 & 0 & 0 & -k_5(\alpha,\varepsilon_5) & k_5(\alpha,\varepsilon_5) \end{pmatrix}.$$

That is

$$\mathrm{K}(\alpha,\varepsilon_i)_{i=1,\dots,5}$$
$$= \begin{pmatrix} 3835+(10-10\alpha)\varepsilon_1+(25-25\alpha)\varepsilon_2 & -1825-(25-25\alpha)\varepsilon_2 \\ -1825-(25-25\alpha)\varepsilon_2 & 3440+(25-25\alpha)\varepsilon_2+(15-15\alpha)\varepsilon_3 \\ 0 & -1615-(15-15\alpha)\varepsilon_3 \\ 0 & 0 \\ 0 & 0 \end{pmatrix.}$$
$$\begin{matrix} 0 & 0 & 0 \\ -1615-(15-15\alpha)\varepsilon_3 & 0 & 0 \\ 3025+(15-15\alpha)\varepsilon_3+(10-10\alpha)\varepsilon_4 & -1410-(10-10\alpha)\varepsilon_4 & 0 \\ -1410-(10-10\alpha)\varepsilon_4 & 2615+(10-10\alpha)\varepsilon_4+(5-5\alpha)\varepsilon_5 & -1205-(5-5\alpha)\varepsilon_5 \\ 0 & -1205-(5-5\alpha)\varepsilon_5 & 1205+(5-5\alpha)\varepsilon_5 \end{matrix}\Bigg),$$

where $\alpha \in [0, 1]$ and $\varepsilon_i \in [-1, 1]$ for $i = 1, 2, 3, 4, 5$.

Similarly, the fuzzy-affine forms of the mass parameters ($m_i(\alpha, \varepsilon_i^*)$ (kg) for $i = 1, 2, \dots, 5$) are and the affine mass matrix may be evaluated as follows:

$m_1(\alpha, \varepsilon_1^*) = 30 + (1-\alpha)\varepsilon_1^*$, $m_2(\alpha, \varepsilon_2^*) = 27 + (1-\alpha)\varepsilon_2^*$, $m_3(\alpha, \varepsilon_3^*) = 27 + (1-\alpha)\varepsilon_3^*$, $m_4(\alpha, \varepsilon_4^*) = 25 + (1-\alpha)\varepsilon_4^*$, and $m_5(\alpha, \varepsilon_5^*) = 18 + (1-\alpha)\varepsilon_5^*$, where $\alpha \in [0, 1]$ and $\varepsilon_i^* \in [-1, 1]$ for $i = 1, 2, \dots, 5$.

$$\mathrm{M}(\alpha,\varepsilon_i^*)_{i=1,\dots,5} = \begin{pmatrix} m_1(\alpha,\varepsilon_1^*) & 0 & 0 & 0 & 0 \\ 0 & m_2(\alpha,\varepsilon_2^*) & 0 & 0 & 0 \\ 0 & 0 & m_3(\alpha,\varepsilon_3^*) & 0 & 0 \\ 0 & 0 & 0 & m_4(\alpha,\varepsilon_4^*) & 0 \\ 0 & 0 & 0 & 0 & m_5(\alpha,\varepsilon_5^*) \end{pmatrix}.$$

Thus,

$$\mathrm{M}(\alpha,\varepsilon_i^*)_{i=1,\dots,5}$$
$$= \begin{pmatrix} 30+(1-\alpha)\varepsilon_1^* & 0 & 0 & 0 & 0 \\ 0 & 27+(1-\alpha)\varepsilon_2^* & 0 & 0 & 0 \\ 0 & 0 & 27+(1-\alpha)\varepsilon_3^* & 0 & 0 \\ 0 & 0 & 0 & 25+(1-\alpha)\varepsilon_4^* & 0 \\ 0 & 0 & 0 & 0 & 18+(1-\alpha)\varepsilon_5^* \end{pmatrix},$$

where $\alpha \in [0, 1]$ and $\varepsilon_i^* \in [-1, 1]$ for $i = 1, 2, \dots, 5$.

Therefore, the given FGEP has been transformed into a fuzzy-affine GEP $\mathrm{K}(\alpha, \varepsilon_i)_{i=1,\dots,5}x(\alpha, \varepsilon^\#) = \lambda(\alpha, \varepsilon^\#)\mathrm{M}(\alpha, \varepsilon_i^*)_{i=1,\dots,5}x(\alpha, \varepsilon^\#)$. Then, the fuzzy-affine GEP obtained from the fuzzy multi-storeyed shear building structural system has been solved by using the fuzzy-affine approach proposed in Sect. 4. Further, the fuzzy eigenvalue solutions ($\tilde{\lambda}_l$ for $l = 1, \ldots, 5$) are determined and the fuzzy eigenvalue solutions are plotted and incorporated in Figs. 16, 17, 18, 19, 20 with the fuzzy eigenvalue solution plots of Mahato and Chakraverty [25]

Finally, the resulting fuzzy eigenvalue solutions that are solved by utilizing fuzzy-affine arithmetic and the solutions given in Mahato and Chakraverty [25] are illustrated in Table 6. A comparison of the outer bounds of the fuzzy eigenvalues obtained by using the proposed method with the outer bounds of the solutions of Mahato and Chakraverty [25], Sim et al. [31] and Leng [20] have also been added in Table 7.

From both the comparison tables viz. Tables 6 and 7, it may be seen that all the other fuzzy eigenvalue solutions are comparatively wider in width than the resulting fuzzy eigenvalue solutions by using the fuzzy-affine approach.

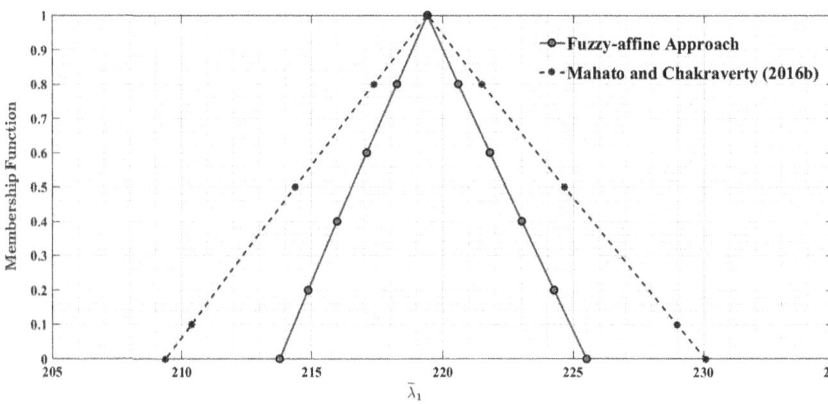

Fig. 16 Comparison plot of first fuzzy eigenvalue between fuzzy-affine approach and Mahato and Chakraverty [25] for Example 4

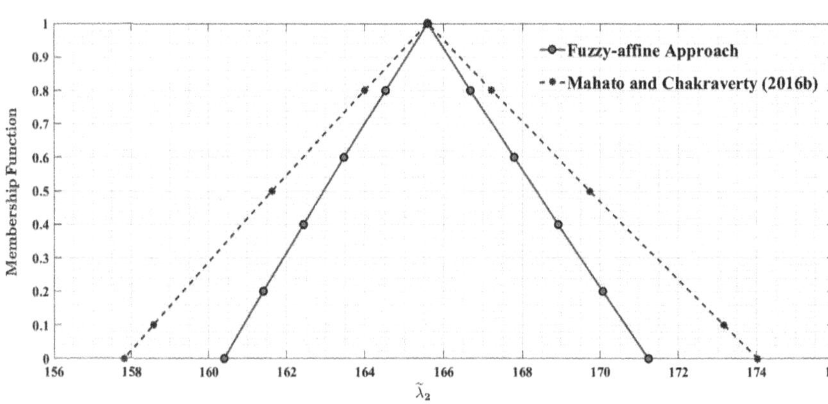

Fig. 17 Comparison plot of second fuzzy eigenvalue between fuzzy-affine approach and Mahato and Chakraverty [25] for Example 4

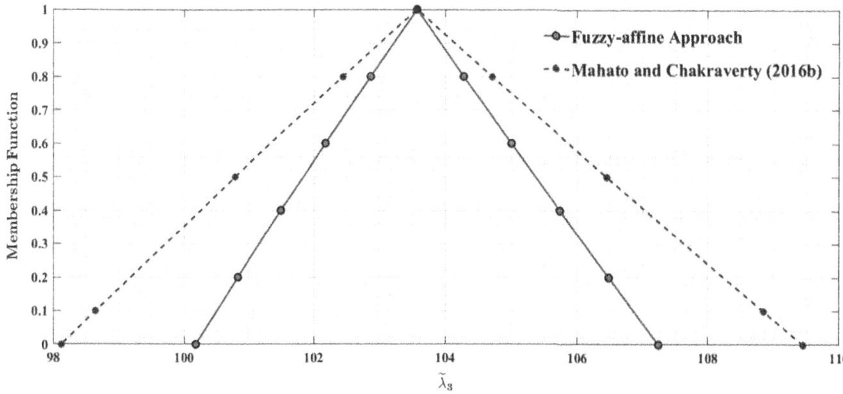

Fig. 18 Comparison plot of third fuzzy eigenvalue between fuzzy-affine approach and Mahato and Chakraverty [25] for Example 4

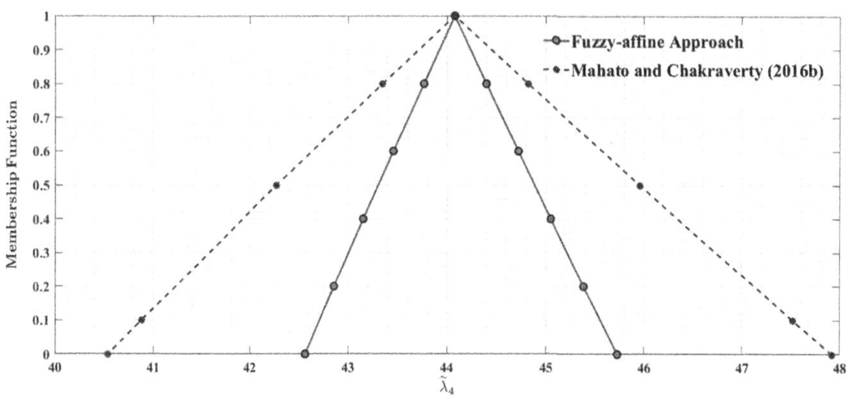

Fig. 19 Comparison plot of fourth fuzzy eigenvalue between fuzzy-affine approach and Mahato and Chakraverty [25] for Example 4

Example 5: Let us consider a FSEP $\tilde{K}\tilde{x} = \tilde{\lambda}\tilde{x}$ (in FSEP the coefficient matrix $\tilde{M}$ of the FGEP is considered to be an identity matrix), whose elements are in the form of TFNs [18]. Here, the coefficient matrix $\tilde{K}$ is considered to be a fuzzy symmetric matrix having elements in the form TFNs given as follows:

$$\tilde{K} = \begin{bmatrix} (1.05, 1.575, 2.1) & (2.9, 3.05, 3.2) & (3.8, 4.0, 4.2) & (4.5, 4.85, 5.2) \\ (2.9, 3.05, 3.2) & (4.9, 5.05, 5.2) & (7.8, 8.0, 8.2) & (4.8, 5.05, 5.3) \\ (3.8, 4.0, 4.2) & (7.8, 8.0, 8.2) & (8.7, 8.95, 9.2) & (7.8, 7.95, 8.1) \\ (4.5, 4.85, 5.2) & (4.8, 5.05, 5.3) & (7.8, 7.95, 8.1) & (5.8, 6.2, 6.6) \end{bmatrix}.$$

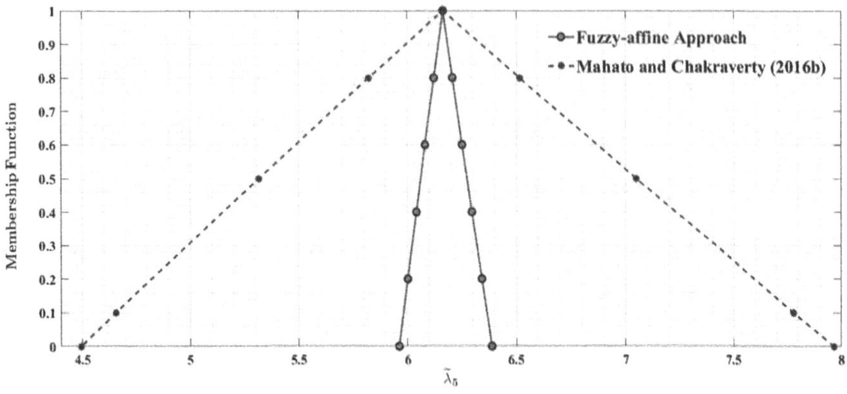

Fig. 20 Comparison plot of fifth fuzzy eigenvalue between fuzzy-affine approach and Mahato and Chakraverty [25] for Example 4

As mentioned earlier, we have to first convert the fuzzy coefficient matrix ($\tilde{K}$) into a fuzzy-affine matrix ($K(\alpha, \varepsilon_i)_{i=1,\ldots,16}$) by transforming each of its fuzzy elements into their respective fuzzy-affine form by following the procedure mentioned in Sect. 3.1. Thus, the fuzzy-affine coefficient matrix may be obtained as

$$K(\alpha, \varepsilon_i)_{i=1,\ldots,16} = \begin{pmatrix} 1.575 + (0.525 - 0.525\alpha)\varepsilon_1 & 3.05 + (0.15 - 0.15\alpha)\varepsilon_2 & 4.0 + (0.2 - 0.2\alpha)\varepsilon_3 & 4.85 + (0.35 - 0.35\alpha)\varepsilon_4 \\ 3.05 + (0.15 - 0.15\alpha)\varepsilon_5 & 5.05 + (0.15 - 0.15\alpha)\varepsilon_6 & 8.0 + (0.2 - 0.2\alpha)\varepsilon_7 & 5.05 + (0.25 - 0.25\alpha)\varepsilon_8 \\ 4.0 + (0.2 - 0.2\alpha)\varepsilon_9 & 8.0 + (0.2 - 0.2\alpha)\varepsilon_{10} & 8.95 + (0.25 - 0.25\alpha)\varepsilon_{11} & 7.95 + (0.15 - 0.15\alpha)\varepsilon_{12} \\ 4.85 + (0.35 - 0.35\alpha)\varepsilon_{13} & 5.05 + (0.25 - 0.25\alpha)\varepsilon_{14} & 7.95 + (0.15 - 0.15\alpha)\varepsilon_{15} & 6.2 + (0.4 - 0.4\alpha)\varepsilon_{16} \end{pmatrix},$$

where $\alpha \in [0, 1]$ and $\varepsilon_i \in [-1, 1]$ for $i = 1, \ldots, 16$.

Finally, by adopting the proposed procedure given in Sect. 4, all the fuzzy eigenvalue solutions ($\tilde{\lambda}_l$ for $l = 1, \ldots, 4$) are evaluated. Further, the fuzzy eigenvalue solution plots by using the fuzzy-affine approach and the solution plots of Jeswal and Chakraverty [18] are depicted in Figs. 21, 22, 23, 24.

Moreover, the fuzzy eigenvalue solutions for different values of the fuzzy parameter (that is $\alpha = 0, 0.3, 0.6, 1$) by utilizing the fuzzy-affine-based technique and the solutions given in Jeswal and Chakraverty [18] are listed in Table 8.

From Table 8, we may observe that the fuzzy eigenvalue solutions of the given FSEP obtained by solving it using the fuzzy-affine-based approach are in good agreement with the fuzzy eigenvalue solutions of Jeswal and Chakraverty [18].

Example 6: Lastly, consider a fuzzy two-storeyed frame structural system [4] as shown in Fig. 25 in an uncertain environment. Suppose, the uncertain parameters for this problem may be considered in the form of TrFNs. Here, the column stiffness parameters ($\tilde{k}_i$ (N/m) for $i = 1, 2, \ldots, 4$) and the floor mass parameters ($\tilde{m}_i$ (kg) for $i = 1, 2$) are considered as follows:

Table 6 Comparison table of fuzzy eigenvalues between fuzzy-affine approach and Mahato and Chakraverty [25] for Example 4 for different values of α

α	Lower and upper bounds	Fuzzy-affine approach					Mahato and Chakraverty [25]				
		$\tilde{\lambda}_1$	$\tilde{\lambda}_2$	$\tilde{\lambda}_3$	$\tilde{\lambda}_4$	$\tilde{\lambda}_5$	$\tilde{\lambda}_1$	$\tilde{\lambda}_2$	$\tilde{\lambda}_3$	$\tilde{\lambda}_4$	$\tilde{\lambda}_5$
0	$\underline{\lambda_l}$	213.7822	160.4049	100.1804	42.5616	5.9620	209.3784	157.8151	98.1225	40.5398	4.4941
	$\overline{\lambda_l}$	225.5186	171.2335	107.2410	45.7281	6.3887	230.0897	174.0310	109.4510	47.9179	7.9657
0.1	$\underline{\lambda_l}$	214.3268	160.9046	100.5071	42.7077	5.9816	210.3901	158.5652	98.6486	40.8811	4.6561
	$\overline{\lambda_l}$	224.8863	170.6469	106.8597	45.5566	6.3656	228.9862	173.1545	108.8414	47.5191	7.7794
0.5	$\underline{\lambda_l}$	216.5470	162.9445	101.8400	43.3042	6.0620	214.3681	161.6253	100.7930	42.2737	5.3153
	$\overline{\lambda_l}$	222.4081	168.3512	105.3659	44.8853	6.2750	224.6548	169.7222	106.4508	45.9577	7.0488
0.8	$\underline{\lambda_l}$	218.2575	164.5190	102.8679	43.7646	6.1240	217.3794	163.9853	102.4446	43.3475	5.8223
	$\overline{\lambda_l}$	220.6011	166.6809	104.2777	44.3968	6.2092	221.4907	167.2229	104.7070	44.8206	6.5153
1	$\underline{\lambda_l}$	219.4201	165.5908	103.5670	44.0780	6.1662	219.4201	165.5908	103.5670	44.0780	6.1662
	$\overline{\lambda_l}$	219.4201	165.5908	103.5670	44.0780	6.1662	219.4201	165.5908	103.5670	44.0780	6.1662

Table 7 Comparison of outer bounds of fuzzy eigenvalues for Example 4

Comparisons	Outer bounds	Fuzzy eigenvalues				
		$\tilde{\lambda}_1$	$\tilde{\lambda}_2$	$\tilde{\lambda}_3$	$\tilde{\lambda}_4$	$\tilde{\lambda}_5$
Fuzzy-affine Approach	Lower ($\underline{\lambda_l}$)	213.7822	160.4049	100.1804	42.5616	5.9620
	Upper ($\overline{\lambda_l}$)	225.5186	171.2335	107.2410	45.7281	6.3887
Mahato and Chakraverty [25]	Lower ($\underline{\lambda_l}$)	209.3784	157.8151	98.1225	40.5398	4.4941
	Upper ($\overline{\lambda_l}$)	230.0897	174.0310	109.4510	47.9179	7.9657
Sim et al. [31]	Lower ($\underline{\lambda_l}$)	211.5100	158.8600	98.5720	40.7540	4.6166
	Upper ($\overline{\lambda_l}$)	227.95	172.89	108.99	47.702	7.8303
Leng [20]	Lower ($\underline{\lambda_l}$)	209.3784	157.8151	98.1225	40.5398	4.4941
	Upper ($\overline{\lambda_l}$)	230.0845	174.0029	109.3994	47.8201	7.8303

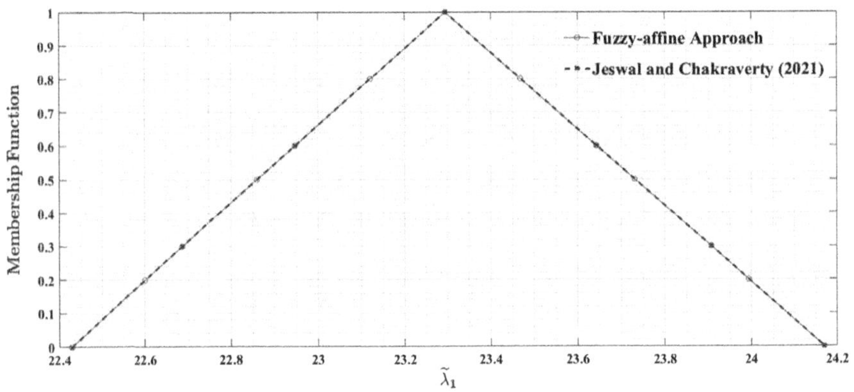

Fig. 21 Comparison plot of first fuzzy eigenvalue between fuzzy-affine approach and Jeswal and Chakraverty [18] for Example 5

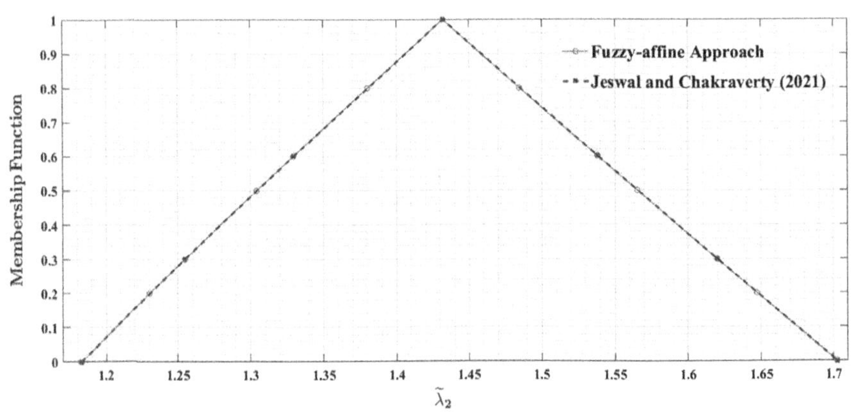

Fig. 22 Comparison plot of second fuzzy eigenvalue between fuzzy-affine approach and Jeswal and Chakraverty [18] for Example 5

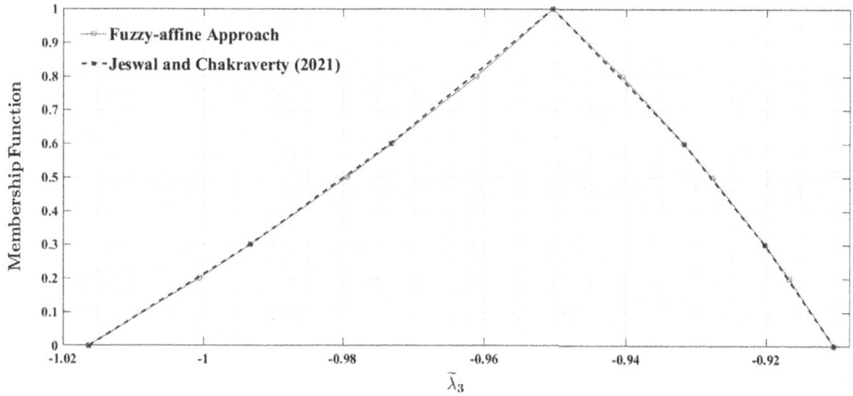

Fig. 23 Comparison plot of third fuzzy eigenvalue between fuzzy-affine approach and Jeswal and Chakraverty [18] for Example 5

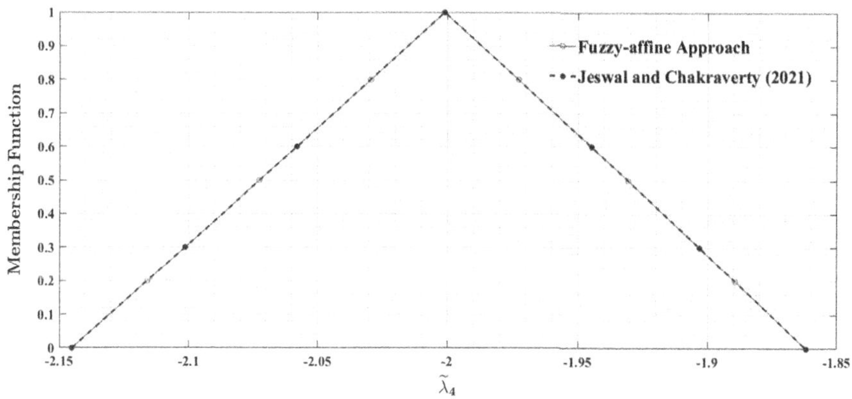

Fig. 24 Comparison plot of fourth fuzzy eigenvalue between fuzzy-affine approach and Jeswal and Chakraverty [18] for Example 5

$$\tilde{k}_1 = \tilde{k}_2 = (5250, 5300, 5500, 5550), \tilde{k}_3 = \tilde{k}_4 = (3425, 3500, 3700, 3775),$$
$$\text{and } \tilde{m}_1 = \tilde{m}_2 = (3200, 3400, 3800, 4000).$$

According to Chakraverty and Behera [4], the dynamic analysis of the given fuzzy two-storeyed frame structural system leads to a FGEP $\tilde{\mathrm{K}}\tilde{x} = \tilde{\lambda}\tilde{\mathrm{M}}\tilde{x}$, where

$$\tilde{\mathrm{K}} = \begin{pmatrix} \tilde{k}_1 + \tilde{k}_2 + \tilde{k}_3 + \tilde{k}_4 & -\tilde{k}_3 - \tilde{k}_4 \\ -\tilde{k}_3 - \tilde{k}_4 & \tilde{k}_3 + \tilde{k}_4 \end{pmatrix} \text{ and } \tilde{\mathrm{M}} = \begin{pmatrix} \tilde{m}_1 & 0 \\ 0 & \tilde{m}_2 \end{pmatrix}$$

are 2×2 fuzzy stiffness and fuzzy mass matrices, respectively.

Table 8 Comparison table of fuzzy eigenvalues between fuzzy-affine approach and Jeswal and Chakraverty [18] for Example 5 for different values of α

α	Lower and upper bounds	Fuzzy-affine approach				Jeswal and Chakraverty [18]			
		$\tilde{\lambda}_1$	$\tilde{\lambda}_2$	$\tilde{\lambda}_3$	$\tilde{\lambda}_4$	$\tilde{\lambda}_1$	$\tilde{\lambda}_2$	$\tilde{\lambda}_3$	$\tilde{\lambda}_4$
0	$\underline{\lambda_l}$	22.428751	1.182763	−1.016353	−2.145161	22.4288	1.1828	−1.0163	−2.1451
	$\overline{\lambda_l}$	24.169729	1.702545	−0.910334	−1.861940	24.1697	1.7025	−0.9103	−1.8619
0.3	$\underline{\lambda_l}$	22.687372	1.254793	−0.993223	−2.101442	22.6874	1.2548	−0.9932	−2.1014
	$\overline{\lambda_l}$	23.906134	1.619662	−0.920164	−1.903132	23.9061	1.6197	−0.9202	−1.9031
0.6	$\underline{\lambda_l}$	22.946912	1.329297	−0.973060	−2.058150	22.9469	1.3294	−0.9732	−2.0580
	$\overline{\lambda_l}$	23.643376	1.538174	−0.931740	−1.944810	23.6434	1.5382	−0.9317	−1.9448
1	$\underline{\lambda_l}$	23.294364	1.432048	−0.950321	−2.001091	23.2944	1.4320	−0.9503	−2.0011
	$\overline{\lambda_l}$	23.294364	1.432048	−0.950321	−2.001091	23.2944	1.4320	−0.9503	−2.0011

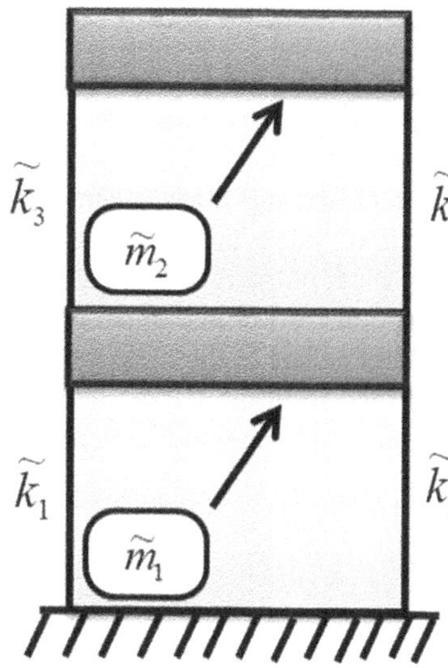

Fig. 25 Fuzzy two-storeyed frame structural system

As mentioned earlier, firstly we have to transform the above fuzzy stiffness and fuzzy mass matrices to their corresponding fuzzy-affine forms. As such the fuzzy-affine column stiffness ($k_i(\alpha, \varepsilon_i)(N/m)$ for $i = 1, \ldots, 4$) and fuzzy-affine floor mass ($m_i(\alpha, \varepsilon_i^*)$ (kg) for $i = 1, 2$) parameters are given below.

$k_1(\alpha, \varepsilon_1) = 5400 + (150 - 50\alpha)\varepsilon_1, k_2(\alpha, \varepsilon_2) = 5400 + (150 - 50\alpha)\varepsilon_2, k_3(\alpha, \varepsilon_3) = 3600 + (175 - 75\alpha)\varepsilon_3$, and $k_4(\alpha, \varepsilon_4) = 3600 + (175 - 75\alpha)\varepsilon_4$, where $\alpha \in [0, 1]$ and $\varepsilon_i \in [-1, 1]$ for $i = 1, \ldots, 4$;

$m_1(\alpha, \varepsilon_1^*) = 3600 + (400 - 200\alpha)\varepsilon_1^*$ and $m_2(\alpha, \varepsilon_2^*) = 3600 + (400 - 200\alpha)\varepsilon_2^*$, where $\alpha \in [0, 1]$ and $\varepsilon_i^* \in [-1, 1]$ for $i = 1, 2$.

Thus, the fuzzy-affine stiffness and mass matrices of the given FGEP are

$$K(\alpha, \varepsilon_i)_{i=1,\ldots,4} = \begin{pmatrix} k_1(\alpha, \varepsilon_1) + k_2(\alpha, \varepsilon_2) + k_3(\alpha, \varepsilon_3) + k_4(\alpha, \varepsilon_4) & -k_3(\alpha, \varepsilon_3) - k_4(\alpha, \varepsilon_4) \\ -k_3(\alpha, \varepsilon_3) - k_4(\alpha, \varepsilon_4) & k_3(\alpha, \varepsilon_3) + k_4(\alpha, \varepsilon_4) \end{pmatrix} \text{ and}$$

$$M(\alpha, \varepsilon_i^*)_{i=1,2} = \begin{pmatrix} m_1(\alpha, \varepsilon_1^*) & 0 \\ 0 & m_2(\alpha, \varepsilon_2^*) \end{pmatrix},$$

where $\alpha \in [0, 1]$ and $\varepsilon_i^* \in [-1, 1]$ for $i = 1, 2$.

That is

$$K(\alpha, \varepsilon_i)_{i=1,\ldots,4} = \begin{pmatrix} 18000 + (150 - 50\alpha)\varepsilon_1 + (150 - 50\alpha)\varepsilon_2 + (175 - 75\alpha)\varepsilon_3 + (175 - 75\alpha)\varepsilon_4 & -7200 - (175 - 75\alpha)\varepsilon_3 - (175 - 75\alpha)\varepsilon_4 \\ -7200 - (175 - 75\alpha)\varepsilon_3 - (175 - 75\alpha)\varepsilon_4 & 7200 + (175 - 75\alpha)\varepsilon_3 + (175 - 75\alpha)\varepsilon_4 \end{pmatrix}$$

and

$$\mathbf{M}(\alpha, \varepsilon_i^*)_{i=1,2} = \begin{pmatrix} 3600 + (400 - 200\alpha)\varepsilon_1^* & 0 \\ 0 & 3600 + (400 - 200\alpha)\varepsilon_2^* \end{pmatrix},$$

where $\alpha \in [0, 1]$, $\varepsilon_i \in [-1, 1]$ for $i = 1, \ldots, 4$, and $\varepsilon_i^* \in [-1, 1]$ for $i = 1, 2$.

Therefore, the fuzzy eigenvalue solutions ($\tilde{\lambda}_l$ for $l = 1, 2$) of the given fuzzy structural system are computed by adopting the procedure given in Sect. 4, and the fuzzy plots of the resulting eigenvalue solutions are depicted in Figs. 26, 27.

Also, the lower and upper bounds of the resulting fuzzy eigenvalue solutions for different values of the fuzzy parameter (α) are listed in Table 9.

All the resulting fuzzy eigenvalue solutions by using the fuzzy-affine approach are found to be in good agreement.

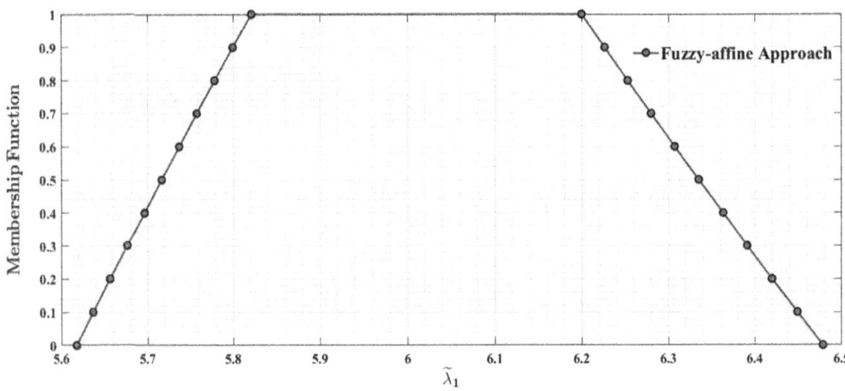

Fig. 26 First fuzzy eigenvalue plot of Example 6

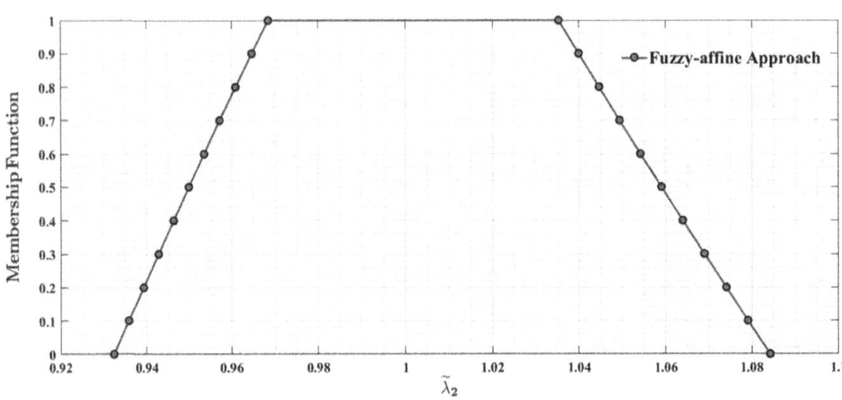

Fig. 27 Second fuzzy eigenvalue plot of Example 6

Table 9 Outer bounds of fuzzy eigenvalues of Example 6 for different values of α

α	First eigenvalue ($\tilde{\lambda}_1$)		Second eigenvalue ($\tilde{\lambda}_2$)	
	Lower bound ($\underline{\lambda_1}$)	Upper bound ($\overline{\lambda_1}$)	Lower bound ($\underline{\lambda_2}$)	Upper bound ($\overline{\lambda_2}$)
0	5.617608	6.478272	0.932392	1.084228
0.1	5.637032	6.448888	0.935833	1.079063
0.2	5.656652	6.419867	0.939307	1.07396
0.3	5.676472	6.391204	0.942817	1.068919
0.4	5.696496	6.362891	0.946361	1.063938
0.5	5.716725	6.334922	0.949942	1.059017
0.6	5.737163	6.307291	0.953558	1.054155
0.7	5.757814	6.279992	0.957212	1.049349
0.8	5.778681	6.253019	0.960902	1.044600
0.9	5.799767	6.226366	0.964631	1.039907
1	5.821075	6.200027	0.968398	1.035267

6 Conclusion

The present chapter proposes a fuzzy-affine approach-based technique for computing the solutions of fuzzy linear structural dynamic problems having parameters in the form of triangular or trapezoidal fuzzy numbers. The fuzzy-affine approach is developed to overcome the overestimation problem that occurred in the case of standard fuzzy arithmetic and it results with comparatively tighter outer bounds of the fuzzy solutions. In this regard, an example of a fuzzy nonlinear function has been worked out to show the efficacy of the fuzzy-affine approach. The fuzzy linear dynamic problems of structural systems lead to FGEP (or FSEP). Thus, a new method is proposed to handle FGEPs having parameters in the form of TFNs or TrFNs. Several numerical examples related to various applications of fuzzy linear structural dynamic problems viz. fuzzy spring-mass structural system with multi degrees of freedom, fuzzy multi-storeyed frame structure, fuzzy multi-storeyed shear building structure, etc. are investigated to illustrate the reliability and efficiency of the present approach.

References

1. Akhmerov RR (2005) Interval-affine Gaussian algorithm for constrained systems. Reliable Comput 11(5):323–341
2. Behera D, Chakraverty S (2015) New approach to solve fully fuzzy system of linear equations using single and double parametric form of fuzzy numbers. Sadhana 40(1):35–49
3. Chakraverty S (2008) Vibration of plates. CRC Press
4. Chakraverty S, Behera D (2014) Parameter identification of multistorey frame structure from uncertain dynamic data. Strojniški Vestn-J Mech Eng 60(5):331–338

5. Chakraverty S, Behera D (2017) Uncertain static and dynamic analysis of imprecisely defined structural systems. In: Fuzzy systems: concepts, methodologies, tools, and applications, pp. 1–30, IGI Global
6. Chakraverty S, Perera S (2018) Recent advances in applications of computational and fuzzy mathematics. Springer Nature Singapore
7. Chakraverty S, Rout S (2020) Affine arithmetic based solution of uncertain static and dynamic problems. Synth Lect Math Stat 12(1):1–170
8. Chakraverty S, Sahoo DM, Mahato NR (2019) Concepts of soft computing: fuzzy and ANN with programming. Springer
9. Chakraverty S, Tapaswini S, Behera D (2016) Fuzzy differential equations and applications for engineers and scientists. CRC Press
10. Chen S, Qiu Z, Song D (1995) A new method for computing the upper and lower bounds on frequecies of structures with interval parameters. Mech Res Commun 22(5):431–439
11. Comba JLD, Stol J (1993) Affine arithmetic and its applications to computer graphics. In: Proceedings of VI SIBGRAPI (Brazilian symposium on computer graphics and image processing), pp 9–18
12. De Figueiredo LH, Stolfi J (2004) Affine arithmetic: concepts and applications. Numer Algorithms 37(1–4):147–158
13. Dubois DJ (1980) Fuzzy sets and systems: theory and applications, vol 144. Academic press
14. Hanss M (2005) Applied fuzzy arithmetic: an introduction with engineering applications/hanss M. Springer 1:100–116
15. Hladík M (2013) Bounds on eigenvalues of real and complex interval matrices. Appl Math Comput 219(10):5584–5591
16. Hladik M, Daney D, Tsigaridas E (2011) A filtering method for the interval eigenvalue problem. Appl Math Comput 217(12):5236–5242
17. Hladík M, Jaulin L (2011) An eigenvalue symmetric matrix contractor. Reliab Comput 27–37
18. Jeswal SK, Chakraverty S (2021) Fuzzy eigenvalue problems of structural dynamics using ANN. In: New paradigms in computational modeling and its applications. Academic Press, pp 145–161
19. Kaufmann A, Gupta MM (1988) Fuzzy mathematical models in engineering and management science. Elsevier Science Inc.
20. Leng H (2014) Real eigenvalue bounds of standard and generalized real interval eigenvalue problems. Appl Math Comput 232:164–171
21. Leng H, He Z (2007) Computing eigenvalue bounds of structures with uncertain-but-non-random parameters by a method based on perturbation theory. Commun Numer Methods Eng 23(11):973–982
22. Leng H, He Z (2010) Computation of bounds for eigenvalues of structures with interval parameters. Appl Math Comput 216(9):2734–2739
23. Leng H, He Z, Yuan Q (2008) Computing bounds to real eigenvalues of real-interval matrices. Int J Numer Meth Eng 74(4):523–530
24. Mahato NR, Chakraverty S (2016a) Filtering algorithm for real eigenvalue bounds of interval and fuzzy generalized eigenvalue problems. ASCE-ASME J Risk Uncertain Eng Syst, Part B: Mech Eng 2(4):044502
25. Mahato NR, Chakraverty S (2016) Filtering algorithm for eigenvalue bounds of fuzzy symmetric matrices. Eng Comput 33(3):855–875
26. Miyajima S, Kashiwagi M (2004) A dividing method utilizing the best multiplication in affine arithmetic. IEICE Electron Express 1(7):176–181
27. Qiu Z, Chen S, Elishakoff I (1996) Bounds of eigenvalues for structures with an interval description of uncertain-but-non-random parameters. Chaos, Solitons Fractals 7(3):425–434
28. Qiu Z, Wang X, Friswell MI (2005) Eigenvalue bounds of structures with uncertain-but-bounded parameters. J Sound Vib 282(1–2):297–312
29. Rex G, Rohn J (1998) Sufficient conditions for regularity and singularity of interval matrices. SIAM J Matrix Anal Appl 20(2):437–445

30. Rump SM, Kashiwagi M (2015) Implementation and improvements of affine arithmetic. Nonlinear Theory Its Appl, IEICE 6(3):341–359
31. Sim J, Qiu Z, Wang X (2007) Modal analysis of structures with uncertain-but-bounded parameters via interval analysis. J Sound Vib 303(1–2):29–45
32. Skalna I (2009) Direct method for solving parametric interval linear systems with non-affine dependencies. In: International conference on parallel processing and applied mathematics. Springer, Berlin, Heidelberg, pp 485–494
33. Skalna I, Hladík M (2017) A new algorithm for Chebyshev minimum-error multiplication of reduced affine forms. Numer Algorithms 76(4):1131–1152
34. Stolfi J, De Figueiredo LH (2003) An introduction to affine arithmetic. Trends Appl Comput Math 4(3):297–312
35. Xia Y, Friswell M (2014) Efficient solution of the fuzzy eigenvalue problem in structural dynamics. Eng Comput 31(5):864–878
36. Zadeh LA (1965) Fuzzy sets. Inf Control 8(3):338–353
37. Zadeh LA, Fu KS, Tanaka K (eds) (2014) Fuzzy sets and their applications to cognitive and decision processes. In: Proceedings of the us–japan seminar on fuzzy sets and their applications, held at the University of California, Berkeley, California. Academic press, July 1–4, 1974
38. Zimmermann HJ (2011) Fuzzy set theory—and its applications. Springer Science & Business Media

Fuzzy Application: Develop a Weather Index

I. T. S. Piyatilake and S. S. N. Perera

Abstract The most visible consequences in the world in recent past decades is an increase in the frequency and intensity of unexpected weather conditions. These unexpected weather-related situations kill approximately 60,000 people per year, globally and most of them were due to the flood and drought. In 2018, the world economy lost about 225 billion US dollars due to the weather-related disasters. According to the global records, most of weather-related disasters were reported from developing countries and this is a critical burden for their economy. These circumstance alarming the relevant authorities to identify the highly vulnerable areas in context of extreme weather conditions. Therefore, developing a weather index which gives the information regarding the regions which have been affected by the impact of extreme weather conditions is a timely required task. This also provides opportunities to implement predetermine actions against the extreme weather-related disasters and minimize damage due to it.

Developing a weather index is a quite sophisticated task due to the uncertainty of the factors associated with the phenomena. Fuzzy theory is a modeling tool which can be used to handle the uncertainty of the real-world problems. Therefore, aim of the present work is to utilize fuzzy theories to handle the uncertainty of the factors and develop an index to identify the weather-related risk in different regions. This chapter describes essential fuzzy theories, modeling tasks, and procedures used to describe extreme weather-related risk in regions. By using literature review necessary factors/parameters for the weather index are recognized. Fuzzy Analytic Hierarchy Process (FAHP) is used to model the risk of regions. The obtained results are compared with the available records to check the validity of the proposed model. Sensitivity analysis is carried out to get a clear picture about the contribution of the

---

I. T. S. Piyatilake (✉)
Department of Computational Mathematics, Faculty of Information Technology, University of Moratuwa, Moratuwa, Sri Lanka
e-mail: thilinisp@uom.lk

S. S. N. Perera
Research & Development Centre for Mathematical Modelling, Department of Mathematics, Faculty of Science, University of Colombo, Colombo, Sri Lanka
e-mail: ssnp@maths.cmb.ac.lk

S. Chakraverty (ed.), *Soft Computing in Interdisciplinary Sciences*,
Studies in Computational Intelligence 988,
https://doi.org/10.1007/978-981-16-4713-0_3

risk factors towards the risk index measurements. Finally, a case study is carried out to find the extreme weather-related risk in 25 regional areas in Sri Lanka.

Keywords Weather · Index · Fuzzy AHP · Sensitivity

1 Introduction

A *disaster* is an unexpected situation in the nature that interrupt day-to-day operations in the society. Natural reasons and human activities are the main causes for disasters. Earthquakes, droughts, floods, volcanoes, wildfires, hurricane, and tsunamis are examples for naturally caused disasters, and man-made disasters includes pollution, oil spills, landslides, nuclear explosions, and fires. Among the two types of disasters, natural disasters which caused due to sudden incidents in the environment are very critical, because people do not have ability to control them.

The *International Federation of Red Cross & Red Crescent Societies* defined the incidents which occurred either by sudden or long-term uncertain events in the physical phenomena as the natural disasters [15]. Natural disasters can be divided in to four classes; geophysical, hydrological, meteorological, and biological. Disasters occur because of geological activities in the environment such as change in the tectonic plates of the earth and the volcanic process called as geological disasters. For instance, landslides, earthquakes, and volcanic eruptions can be identified as the geological disasters. The disasters occur as a result of changes in the atmospheric conditions is termed as meteorological disasters. These events include extreme temperatures, hurricanes, droughts, cyclones, and storm. The disasters related to water processes are identified as hydrological disasters. Floods, droughts, and tsunamis falls in to the hydrological class. Biological disasters occur due to the biological activities in the environment such as the spread of infectious diseases and pets. Even though this classification exists, sometimes natural disasters may fall into several classes. For example, the storm can also cause flooding due to heavy rain. This weather situation may also help to transmit vector-borne diseases such as dengue, malaria, and yellow fever. Therefore, this situation would be an example for combination of meteorological, a hydrological and a biological natural disaster because the storm caused the flood and diseases. Unplanned urban settlements, high population density, climate change, poverty, non-engineered constructions, deforestation, and unstable landforms are the aggravating factors of natural disasters. These factors will help to increase the frequency, complexity, and severity of the disasters.

There is a growing trend in losses due to weather-related natural disasters in the last three decades globally [23]. According to the United Nations Development Programme (UNDP) statistics, approximately 216 million people were affected by natural disasters worldwide each year [23]. Three hundred and eighteen weather-related natural disasters from 122 countries was reported worldwide in the year 2017 [7]. The UNDP estimated that approximately 9,503 people died and 96 million people affected in 2017 by natural disasters [23]. In addition to that these natural disasters

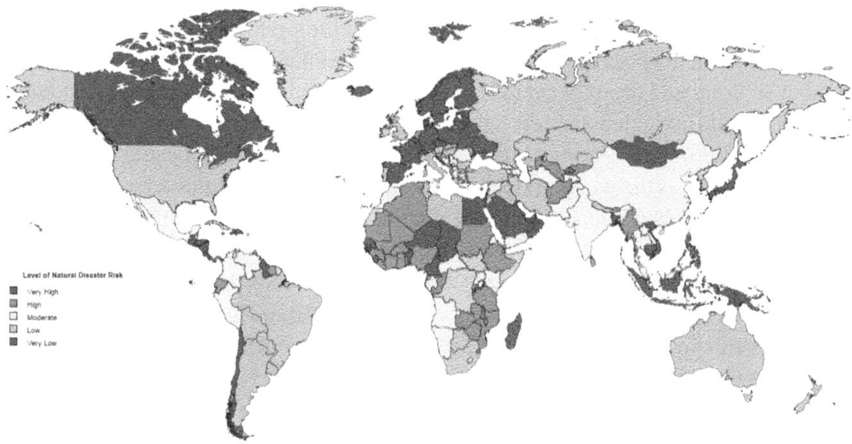

Fig. 1 The hotspot regions for natural disasters in the world [6]

cause many adverse effects to human health and environment. Some of these effects are short term while the others last for many years. These effects include destroy public infrastructure, interrupt basic services in the community, disrupt livelihoods, changes in landscape and natural features, and food scarcity. These sudden effects also tend to increase poverty rates of developing countries and help to cause great financial losses to the world. The estimated global cost due to weather-related natural disaster is about 340 billion dollars [23]. From Fig. 1, it can be seen that the Oceania, South-East Asia, Central America, West Africa, and Central Africa are the hotspots regions for weather-related disasters in the world and most of the countries in these regions are developing countries. Indonesia, India, and the Philippines are identified as the highly vulnerable countries in the world in terms of weather-related disasters. According to the predicted statistics, approximately 325 million people are expected to live in high-risk areas by 2030 [6].

Situation in Sri Lanka is also not different to other developing countries. Sri Lanka is an island in the Indian Ocean which affected seriously by weather-related natural disasters such as floods, strong winds, droughts, cyclones, and landslides. The weather condition of the country is subjected by Southwest and Northeast monsoons. Droughts, floods, and landslides are the commonly experienced disasters in the country due to the variations in two monsoon periods. Compared with countries in the world, people in Sri Lanka do not have experience of terrible earthquakes and volcanoes. However, Sri Lanka was one of the most severely affected countries in the world by the 2004 tsunami in the Indian Ocean. Out of 182 countries in the world, the Germanwatch global climate risk index 2019 ranked Sri Lanka as the 30th country in terms of weather-related loss events and categorized Sri Lanka as a medium risk level country [13]. But, Sri Lanka too has reported an increasing trend in frequency and intensity of extreme weather-related natural disasters in recent past years. According to the statistics of Disaster Management Centre of Sri Lanka, approximately 6.5 million people were affected due to extreme weather events in the year 2015 [11]. Among them, more than 2 million people were affected by the floods and droughts because in the year 2015, Sri Lanka had experienced significant number of flood and drought events [24]. Ampara, Puttalam, Kurunagala, Colombo, Gampaha, and Batticaloa districts have identified as the high-risk areas in the context of floods and droughts. Among them, Ampara district, which is located in the south east of Sri Lanka, is the worst off in 2015, because this district battered by both floods and droughts. Meanwhile, Kurunagala district has reported the highest number of people affected by droughts. In Colombo district, which is the administrative capital of Sri Lanka, individuals are severely threatened by floods due to unplanned urban settlements. The annual estimated economical damages and losses related to extreme weather events in Sri Lanka is about 313 million dollars [24]. These losses severely affected to the economic growth, inflation, and the trade deficit of the country. These circumstance alarming the relevant authorities to identify the highly vulnerable areas in context of extreme weather conditions.

Recently, in Sri Lanka, new development projects were identified to expand power generation, to implement new transportation systems, and to develop tourism industry. Information related to extreme weather events and forecast weather risk play a significant role in locating these tasks in proper regions. Therefore, developing a weather index which gives the information regarding the regions in Sri Lanka which have been affected by impact of extreme weather conditions is a timely required task. This also provides opportunities to implement predetermine actions against the extreme weather-related disasters and minimize damage due to it. In addition, these type of index can help to plan agricultural activities at regional and national levels and help to make awareness among the general public about extreme weather events.

Several organizations and researches in the world developed weather risk indexes to identify the vulnerable countries and regions in terms of extreme weather events. The global climate risk index (CRI) is one such index, which was developed by Germanwatch organization in 2006. In this index, they analyze the impact of weather-related disaster damages in different countries and regions and converted those records in to an risk index which can easily understand by the general public [14]. This index is based on four direct factors: number of deaths occurred due to the disaster, number of deaths occurred per 100,000 people, sum of losses in US dollars, and losses per unit of gross domestic product. These four factors aggregated using weighted average to find the final risk index value. In order to develop this index, they had considered only losses due to the disaster and neglected adaptive and coping capacity of a country to face the disaster. Therefore, this is a limitation of global CRI ranking. The United Nations University Institute for Environment and Human Security was introduced the world risk index (WRI) in 2011 to find the probability a country might be affected by a disaster [25]. This index consist of four main indexes: exposure, vulnerability, coping, and adaptive capacities. These main indexes were further subdivided. The exposure category consists of sub-factors such as population density, affected area, and infrastructure capacity. The vulnerability index is the summation of susceptibility of the society, public infrastructures available in the community, housing facilities, poverty, and economical stability. Finally, these four main indexes were aggregated using weighted average to find the WRI value.

Actuaries climate index (ACI) is another weather risk measurement tool [17]. This index was developed to quantify the frequency of extreme weather events occurred throughout the recent decades in the United States, Canada, and North America. They considered high temperature, low temperature, heavy rainfall, consecutive dry days, high wind, and sea level as the risk factors for their index. Final score of ACI is calculated using arithmetic mean. A suitable research is not observed in Sri Lankan context based on developing weather risk index. Therefore, the aim of this chapter is to develop a weather risk index which consist of different weather events to determine the weather-related risk in different areas in Sri Lanka. Further, such risk index can be utilized to predict the future weather-related risk in different areas.

Weather-related factors in risk indexes are highly uncertain because we cannot determine the exact ranges of these factors and the statistical distribution of these factors are unknown. Uncertainty is a part of fuzzy set theory [27]. Fuzzy set theory was first invented by Zadeh [26]. Fuzzy theory facilitates to model uncertain, vague, and fuzzy information which describe the real-world phenomena. As the complexity of the problem increases, the ability to make precise and significant statements about problem behavior is lessened. Computers cannot satisfactorily solve such problems as machine intellect still employs Boolean logic. The power of human intelligence results from its capacity of treatment for fuzzy statements and decisions by adding logical statements and thinking processes. The human brain has numerous intelligent practices and has superior filtering capacity than the computers. Therefore, fuzzy theory leads vagueness in the decision-making processes that are closer to the human intelligent perform.

Factors in the Risk index is constructed as a hierarchy. Analytical hierarchical process (AHP) is a multistage problem solving tool which was invented by Saaty to handle hierarchical structures [18]. AHP helps to combine expert opinions with the factors in the risk model. However, these expert opinions and factors in the risk index are uncertain. Therefore, Fuzzy AHP is applied to construct the weather risk index.

The remaining sections of this chapter is organized as follows: Sect. 2 provides some essentials preliminaries of fuzzy theory. Preliminaries related to linguistic variables, FAHP, and fuzzy sensitivity theory are also presented in this section. The selected weather events, hierarchical structure, and essential steps of weather index construction process are presented in Sect. 3. Results generated from weather index and validation are presented in Sect. 4. Finally in Sect. 5, the conclusions, remarks, and further possible directions of this work are pointed out.

2 Preliminaries

2.1 Fuzzy Sets

Let x be the elements or objects of the set X. A *fuzzy set* $\tilde{A}$ in X is a collection of ordered pairs and it is denoted by [20, 27]:

$$\tilde{A} = \{(x, \mu_{\tilde{A}}(x)) \mid x \in X\}.$$

Here, the term $\mu_{\tilde{A}}(x)$ is called the membership function or degree of compatibility of x in $\tilde{A}$. Membership function $\mu_{\tilde{A}}(x)$ maps X to membership space [0, 1]. If the value of $\mu_{\tilde{A}}(x)$ is near to unity then there is a higher grade of membership of x in $\tilde{A}$.

The membership function has to be simple, convenient and should be quickly computable. The membership functions are built considering several basic shapes such as piece-wise linear functions, triangular functions, Gaussian functions, and sigmoid functions. Among them, the simplest is the triangular membership function which is designed using straight lines.

2.2 Triangular Membership Function

Let us define the triangular fuzzy membership function as follows [3]:

$$\mu_{\tilde{A}}(x) = \begin{cases} \frac{x}{m-l} - \frac{l}{m-l}, & \text{if } x \in [l, m]; \\ \frac{x}{m-u} - \frac{u}{m-u}, & \text{if } x \in [m, u]; \\ 0, & \text{otherwise,} \end{cases} \quad (1)$$

where $l \le m \le u$. Here, l, m and u represent the smallest possible values, modal value, and largest possible value of the support of $\tilde{A}$, respectively. Support of $\tilde{A}$ is the set of objects defined as $\{x \in X \mid l < x < u\}$.

A number $\tilde{A}$ defined on X is called a triangular fuzzy number if its membership function $\mu_{\tilde{A}}(x)$ is a triangular membership function and it is denoted by a triplet (l, m, u).

2.3 Basic Fuzzy Algebraic Operations Defined on Triangular Fuzzy Number

Let us consider two triangular fuzzy numbers $\tilde{A}_1$ and $\tilde{A}_2$, where $\tilde{A}_1 = (l_1, m_1, u_1)$ and $\tilde{A}_2 = (l_2, m_2, u_2)$. The operational laws based on these two triangular fuzzy numbers can be defined as follows [3, 18]:

1. Addition:

$$\tilde{A}_1 \oplus \tilde{A}_2 = (l_1, m_1, u_1) \oplus (l_2, m_2, u_2) = (l_1 + l_2, m_1 + m_2, u_1 + u_2) \quad (2)$$

2. Multiplication:

$$\tilde{A}_1 \otimes \tilde{A}_2 = (l_1, m_1, u_1) \otimes (l_2, m_2, u_2) = (l_1 l_2, m_1 m_2, u_1 u_2) \tag{3}$$

3. Scalar Multiplication

$$\lambda \otimes \tilde{A}_1 = \lambda \otimes (l_1, m_1, u_1) = (\lambda l_1, \lambda m_1, \lambda u_1), \tag{4}$$

   where $\lambda > 0$ and $\lambda \in X$.
4. Inverse:

$$\tilde{A}_1^{-1} \approx \left(\frac{1}{u_1}, \frac{1}{m_1}, \frac{1}{l_1}\right) \tag{5}$$

2.4 Linguistic Variable

A variable, which use to represent some characteristic of an element is called a linguistic variable [26]. Therefore, the value of this linguistic variable can be a word or phrase. Linguistic variable is also known as fuzzy variable. For instance, consider the variable temperature. We can use words such as cold, warm, and hot to describe the characteristics of temperature. Therefore, temperature is a linguistic variable.

Generally, linguistic variable is denoted by a quintuple $(x, T(x), U, G, \tilde{M})$, where x denotes variable name and $T(x)$ stands for term set of x. That means $T(x)$ represents the name set of linguistic values x which is ranging over universe of discourse U. The terms G and M represent the syntactic and semantic rules associating with each X, respectively. Fuzzy set $\tilde{M}(X)$ is a subset of U. For a given particular set X, a name produced by G is called a "term".

Traditionally, different numerical scales are used to represent linguistic variables. Widely used numerical scales are 1–3, 1– 5, 1–7, and 1–9 [21]. In reality, select the correct characteristic level of the linguistic variable is uncertain. Hence, we can use triangular fuzzy numbers to represent these uncertain judgments. Different characteristic levels and their corresponding triangular fuzzy representation is shown in Table 1 [5].

2.5 Fuzzy Pairwise Comparison Matrix

Using linguistic variables, we can compare different factors with respect to the main goal. Suppose $\boldsymbol{B}$ is a order n fuzzified judgment matrix. This matrix contain all possible pairwise comparisons of different factors with respect to the main goal. Let $\tilde{a}_{ij}$ denotes pairwise comparison between factor i and factor j for all $i, j \in 1, 2, 3, \ldots, n$.

Table 1 Linguistic scales and their importance intensity

Definition	Triangular fuzzy numbers ($\delta = 0.5$)
Absolutely more important	(5/2, 3, 7/2)
Very strongly more important	(2, 5/2, 3)
Strongly more important	(3/2, 2, 5/2)
Weakly more important	(1, 3/2, 2)
Equally important	(1/2, 1, 3/2)
Just equal	(1, 1, 1)

$$\boldsymbol{B} = \begin{array}{c} \\ c_1 \\ c_2 \\ \vdots \\ c_n \end{array} \begin{array}{c} \begin{array}{cccc} c_1 & c_2 & \cdots & c_n \end{array} \\ \begin{bmatrix} (1,1,1) & \tilde{b}_{12} & \ldots & \tilde{b}_{1n} \\ \tilde{b}_{21} & (1,1,1) & \ldots & \tilde{b}_{2n} \\ \vdots & \vdots & \ddots & \vdots \\ \tilde{b}_{n1} & \tilde{b}_{n2} & \ldots & (1,1,1) \end{bmatrix} \end{array}, \tag{6}$$

where $\tilde{b}_{ij} = (1, 1, 1) : \forall i = j$, $\tilde{b}_{ji} = \tilde{b}_{ij}^{-1}$, n stand for number of factors, $c_i = i$th factor, and $\tilde{b}_{ij}$ is the comparison of ith factor with respect to jth factor. All $\tilde{b}_{ij}$ terms can be represent using triangular fuzzy numbers and $\tilde{b}_{ij} = (l_{ij}, m_{ij}, u_{ij})$.

2.6 Analytic Hierarchy Process (AHP)

Analytic Hierarchy Process (AHP) which was invented by Thomas Saaty in 1970 is a decision support tool which we can used to find the ranks of decision alternatives [1, 16]. In this technique, decision-maker can divide the main problem in to small sub-problems and then can arrange these sub-problems into a hierarchical structure. To obtain ranks of different factors in the hierarchical structure, this technique compares each factor with the remaining factors with the aid of linguistic numerical scale. The result of the AHP technique is heavily depend on expert ideas and literature. However, these expert ideas are uncertain and doubtful. This uncertainty and doubtfulness give a negative impact on the results derived from AHP. Due to this reason, Fuzzy Analytic Hierarchy Process (FAHP) was introduced and this technique is a combination of fuzzy logic and traditional AHP [18, 19]. FAHP technique is also similar to traditional AHP. Instead of linguistic numerical scales, FAHP uses triangular fuzzy numbers for the comparisons. Chang's extent analysis method is one such FAHP technique which was introduced by Chang in 1996 to rank factors using triangular fuzzy numbers [3].

2.6.1 Fuzzy Synthetic Extent Values

Let the set X is $\{x_1, x_2, x_3, \ldots, x_n\}$ and set G is $\{g_1, g_2, g_3, \ldots, g_n\}$. The two sets X and G represent the object set and goal set, respectively. Each element of object set is taken and extent analysis [4] is performed with respect to the each element in the goal set. Now, we have m number of extent analysis values for each object and these values are denoted as follows:

$$M^1_{g_i}, M^2_{g_i}, \ldots, M^m_{g_i}, \qquad i = 1, 2, \ldots, n$$

where $M^j_{g_i} (j = 1, 2, \ldots, m)$ are TFNs.

Definition 1 (Fuzzy Synthetic Extent)
Let $M^1_{g_i}, M^2_{g_i}, \ldots, M^m_{g_i}$ be the obtained extent analysis values for ith object for m number of goals. The fuzzy synthetic extent value S_i with respect to ith object can be calculated using the following expression:

$$S_i = \sum_{j=1}^{m} M^j_{g_i} \otimes \left[\sum_{i=1}^{n} \sum_{j=1}^{m} M^j_{g_i} \right]^{-1}. \tag{7}$$

The value of $\sum_{j=1}^{m} M^j_{g_i}$, can be determined using (2) and m number of extent analysis values of the pairwise decision matrix. The value of $\left[\sum_{i=1}^{n} \sum_{j=1}^{m} M^j_{g_i}\right]^{-1}$, can be obtained using $M^j_{g_i} (j = 1, 2, \ldots, m)$ values, and fuzzy operation (2) and (5).

2.6.2 Chang's Extent Analysis Method

The first step of extent analysis method is defining the triangular fuzzy numbers to represent the linguistic scale. Then each factor is compared with the remaining factors with the aid of triangular fuzzy numbers. We can use expert opinions for this purpose. After that, the following steps are considered to find the priority weight of each factor [2, 18, 22].

Step 1: Calculate the value of fuzzy synthetic extent with respect to the ith object using (7).

Step 2: Consider two triangular fuzzy numbers $\tilde{A}_2 = (l_2, m_2, u_2)$ and $\tilde{A}_1 = (l_1, m_1, u_1)$ such that $\tilde{A}_2 \geq \tilde{A}_1$. Then the degree of possibility of $\tilde{A}_1$ and $\tilde{A}_2$ is given by

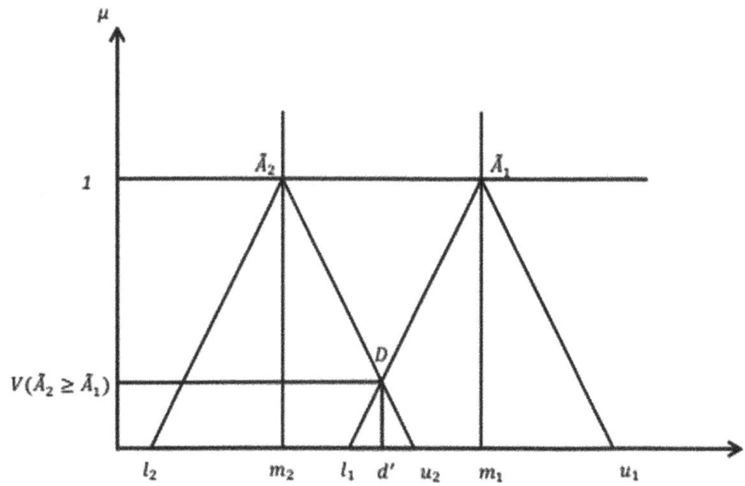

Fig. 2 The ordinate of the highest intersection point D between $\tilde{A}_1$ and $\tilde{A}_2$

$$V(\tilde{A}_2 \geq \tilde{A}_1) = \begin{cases} 1, & \text{if } m_2 \geq m_1; \\ 0, & \text{if } l_1 \geq u_2; \\ \dfrac{l_1 - u_2}{(m_2 - u_2) - (m_1 - l_1)}, & \text{otherwise.} \end{cases} \quad (8)$$

According to the values of $V(\tilde{A}_1 \geq \tilde{A}_2)$ and $V(\tilde{A}_2 \geq \tilde{A}_1)$, two triangular fuzzy numbers can be compared.

Step 3: The ordinate d' of the highest intersection point between $\mu_{\tilde{A}_1}(x)$ and $\mu_{\tilde{A}_2}(x)$ is defined as follows:

$$d'(C_i) = \min V(S_i \geq S_k), \quad (9)$$

for $k = 1, 2, \ldots, n; k \neq i$, This highest distance is shown in Fig. 2 as D.

Step 4: Then the weight vector is given by

$$W' = (d'(A_1), d'(A_2), \ldots, d'(A_n))^T, \quad (10)$$

where A_i, $(i = 1, 2, \ldots, n)$ are n elements.

Step 5: Normalizing (10) can obtain the normalize weight vector

$$W = (d(A_1), d(A_2), \ldots, d(A_n))^T, \quad (11)$$

where $d(A_i) = \dfrac{d'(A_i)}{\sum_{i=1}^{n} d'(A_1)}$. Now W is a non-fuzzy number.

2.7 Value of Degree of Fuzziness

The distance between l_{ij} and u_{ij} of a triangular fuzzy number has a direct influence to the final result. That means element $\tilde{a}_{ij}$ in fuzzy pairwise comparison matrix is influenced by fuzziness. The amount of influence by this fuzziness is called a degree of fuzziness, and it is denoted by δ, where $m_{ij} - l_{ij} = u_{ij} - l_{ij} = \delta$ [18, 22]. The value of δ is a constant and it is considered as an absolute distance from the lower bound value (l_{ij}) to the modal value (m_{ij}) or the modal value (m_{ij}) to the upper bound value (u_{ij}). Given the modal value (m_{ij}), the fuzzy number representing the fuzzy judgment made is defined by $(m_{ij} - \delta, m_{ij}, m_{ij} + \delta)$, with its associated inverse fuzzy number subsequently described by $\left(\frac{1}{m_{ij}+\delta}, \frac{1}{m_{ij}}, \frac{1}{m_{ij}-\delta}\right)$. Here, the effect of the δ value on a fuzzy number (l_{ij}, m_{ij}, u_{ij}) will be elucidated. For example, consider the unbounded scale between 0 to ∞. Around the scale value 1, the domain of the fuzzy scale value measured is between 0 and ∞. With the sub-domain 0 to 1 associated with one direction of preference (e.g., j preferred to i) and 1 to ∞ the reverse preference (e.g., i preferred to j). In the case of fuzzy scale values, there is still a need for the strict partition of the scale value domain. That is, the support of any fuzzy scale value should be in either the 0 to 1 or the 1 to ∞ sub-domains of δ.

To illustrate, consider the fuzzy scale value $m_{ij} = v_k = 2$, where v_k denote the kth scale value. For instance, if $\delta = 2.5$, the associated fuzzy number is $(-0.5, 2, 4.5)$. It follows that the value $l_{ij} = -0.5$ has no meaning as part of a fuzzy judgment. To remove this potential of conflict, a restraint on the l_{ij} value needs to be constructed. Expressed more formally, if m_{ij} is given a fuzzy scale value such that $m_{ij} = v_k \geq 1$, then l_{ij} is bound by $1 \leq l_{ij} \leq m_{ij}$, whose value depends on the value of δ, and is given by

$$l_{ij} = \begin{cases} v_k - \delta(v_k - v_{k-1}), & \delta \leq \dfrac{v_k - 1}{v_k - v_{k-1}}; \\ 1, & \text{otherwise.} \end{cases} \tag{12}$$

This equation for l_{ij} guarantee that irrespective of the value of δ, the support associated with a fuzzy scale value includes no conflicting sub-domain. There is no limit on the upper bound of the fuzzy scale value, and hence, the value of u_{ij} remains $u_{ij} = v_k + \delta(v_{k+1} - v_k)$. Similarly, an expression for the inverse case for $m_{ij} < 1$ can also be developed.

2.8 Convex Combination

Consider the fuzzy sets $\tilde{A}_1, \tilde{A}_2, ..., \tilde{A}_n$. The convex combination $\tilde{B}$ of these fuzzy sets is defined as [18]

Table 2 Risk classification of weather events

Classification	Temperature (°C)	Rainfall (mm)	Number of rainy days
Very low	0–20	0–100	95–105
Low	20–25	100–150	85–95 & 105–115
Moderate	25–30	150–250	75-85 & 115–125
High	30–35	250–350	50–75 & 125–150
Very high	Above 35	Above 350	Above 150 & less than 50

$$\mu_{\tilde{B}}(x) = w_1(x)\mu_{\tilde{A}_1}(x) + \cdots\cdots + w_n(x)\mu_{\tilde{A}_n}(x), \tag{13}$$

where

$$\sum_{i=1}^{n} w_i(x) = 1,$$

and $0 \leq w_i(x)$, for all $x \in X$. A special case of the above occurs when $w_i(x) = w_i$, a constant, $i = 1, \ldots, n$. In this case, $\tilde{B}$ is called a convex linear combination of the $\tilde{A}_i$.

3 Methodology

The risk model is developed considering twenty five main regions in Sri Lankan. The selected regions are Ampara, Anuradhapura, Badulla, Batticaloa, Colombo, Galle, Gampaha, Hambantota, Jaffna, Kalutara, Kandy, Kegalle, Kilinochchi, Kurunegala, Mannar, Matale, Matara, Moneragala, Mullaitivu, Nuwara Eliya, Polonnaruwa, Puttalam, Ratnapura, Trincomalee, and Vavuniya. These regions are shown in Fig. 3. Reviewing the literature available in Sri Lanka three most important weather events were identified. They are maximum average temperature (°C), maximum average rainfall (mm), and number of rainy days per year. The maximum temperature reported in the history of Sri Lanka is 38°C and it is from Vavuniya in 2019 [8]. The mean annual rainfall in Sri Lanka is less than 900 mm in dry regions in the country and it is over 5000 mm in the wet regions [10]. Considering these facts selected weather events are further divided in to five sub-categories: very low, low, moderate, high, and very high. These classifications are shown in Table 2. Using weather events and their sub-categories proposed hierarchical structure is presented in Fig. 4. The first level of this structure includes the aim of this study. That is the weather risk level of different regions. Selected three weather events lie in the second level of the hierarchy and the risk levels include in the third level of this hierarchical structure. Alternatives of this hierarchical structure are the selected 25 regions in Sri Lanka.

Fuzzy pairwise comparisons matrices for the main three factors and five risk level are constructed using expert opinions. Then the weights of factors and risk levels are derived using Chang's extent analysis method. Using these weights factors and

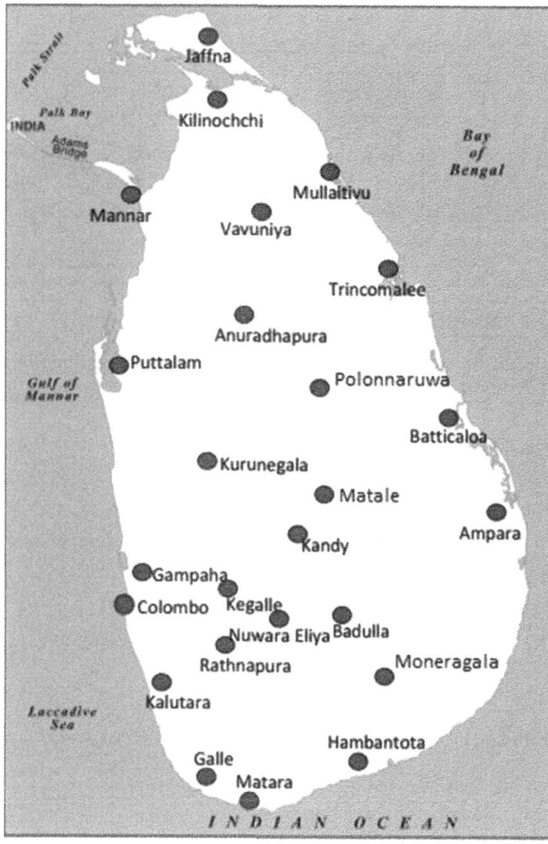

Fig. 3 Selected main regions in Sri Lanka

risk levels are prioritized. Next sensitivity analysis is carried out to find the value of degree of fuzziness. Finally, three weather events aggregated as convex linear combination in order to find the weather-related risk of different regions. Relevant data for this study is gathered from [9]. Simulations are carried out using MATLAB software.

4 Results and Discussion

Step 1: Determine the relative importance of weather events and risk levels using expert opinions

The linguistic scale used to find the relative importance of weather events and risk levels in FAHP process refer to Table 1. Using Table 1, linguistic scale fuzzy pairwise comparison matrices in Tables 3 and 4 are constructed using expert judgments.

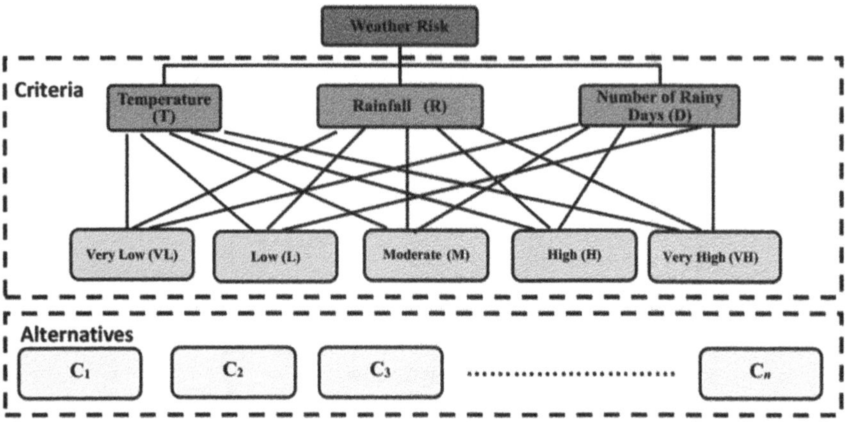

Fig. 4 Proposed hierarchical structure

Table 3 Fuzzy pairwise comparison matrix for the weather events when $\delta = 0.5$

	Rainfall	Temperature	No. of rainy days
Rainfall	(1, 1, 1)	(2, 5/2, 3)	(1/2, 1, 3/2)
Temperature	(1/3, 2/5, 1/2)	(1, 1, 1)	(1, 3/2, 2)
No. of rainy days	(2/3, 1, 2)	(1/2, 2/3, 1)	(1, 1, 1)

Table 4 Fuzzy pairwise comparison matrix for risk levels $\delta = 0.5$

	VH	H	M	L	VL
VH	(1, 1, 1)	(2/3, 1, 2)	(1/2, 2/3, 1)	(2/5, 1/2, 2/3)	(2/7, 1/3, 2/5)
H	(1/2, 1, 3/2)	(1, 1, 1)	(2/3, 1, 2)	(1/2, 2/3, 1)	(2/5, 1/2, 2/3)
M	(1, 3/2, 2)	(1/2, 1, 3/2)	(1, 1, 1)	(2/3, 1, 2)	(1/2, 2/3, 1)
L	(3/2, 2, 5/2)	(1, 3/2, 2)	(1/2, 1, 3/2)	(1, 1, 1)	(2/3, 1, 2)
VL	(5/2, 3, 7/2)	(3/2, 2, 5/2)	(1, 3/2, 2)	(1/2, 2, 3/2)	(1, 1, 1)

Step 2: Determine priority weights using Chang's extent analysis method

Chang's extent analysis method is used to prioritize the weather events and risk levels. Using Tables 3 and (7), the following fuzzy synthetic extent values are derived for the weather events:

$$S_R = (0.2593, 0.447, 0.6875)$$
$$S_T = (0.1728, 0.2881, 0.5)$$
$$S_D = (0.1605, 0.2649, 0.5000)$$

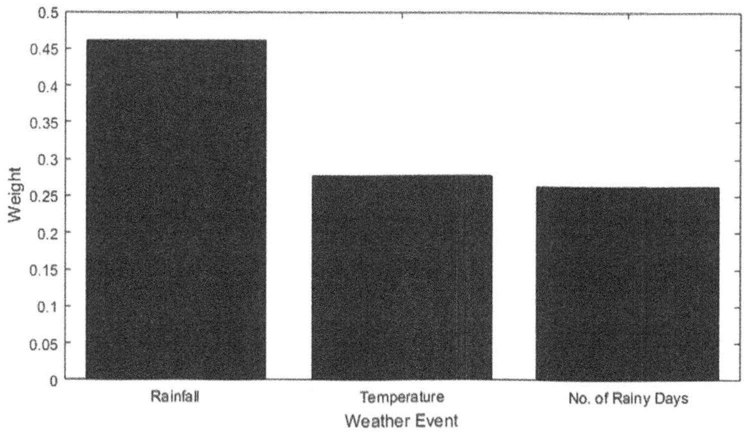

Fig. 5 The weight distribution of weather events

The above three fuzzy synthetic extent values are compared to find the degrees of possibilities. For this purpose, (8) is used.

$$V(S_R \geq S_T) = 1 \quad V(S_T \geq S_R) = 0.6023 \quad V(S_R \geq S_T) = 0.5693$$
$$V(S_R \geq S_D) = 1 \quad V(S_T \geq S_D) = 1 \quad V(S_R \geq S_D) = 0.9338$$

Then the resulted weight vector W' for the three weather events are given by

$$W' = (1.0000, 0.6023, 0.5693).$$

The above W' is normalized as explained in (11) to find the final weight vector W and the resulted weight vector is

$$W = (0.4605, 0.2774, 0.2622).$$

The distribution of the weight among three weather events are given in Fig. 5. According to the results in Fig. 5, we can see that rainfall is the most important factor in context of weather risk. Considering a similar approach generated weights vector for risk levels is

$$W = (0.3182, 0.2501, 0.1957, 0.1430, 0.0930).$$

Distribution of weights of risk level is are shown in Fig. 6. The distribution of weights of risk levels among three weather events are given in Table 5.

Step 3: Determine the value of degree of fuzziness δ

It is important to find out the sensitivity of derived weights in the decision-supportive models. The objective of conducting a sensitivity analysis is to determine whether the derived weights of the weather events changes significantly in case of a small

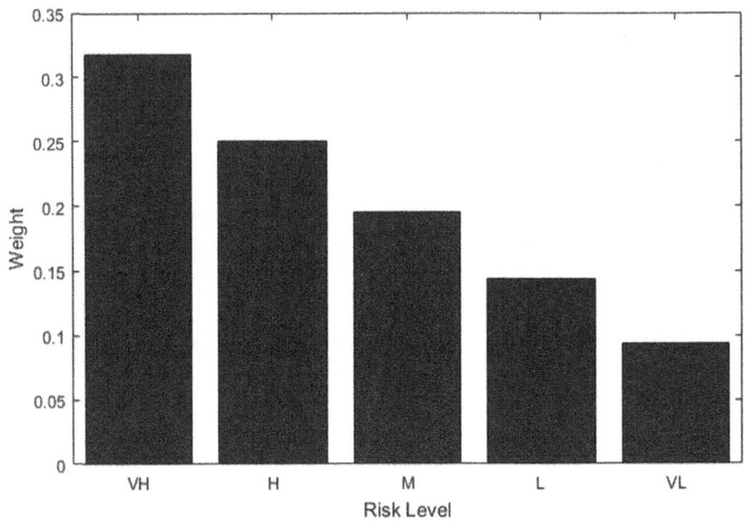

Fig. 6 The weight distribution of risk levels

Table 5 Distribution of risk level weights

Weather event	Weight	Risk level	Weight
Rainfall	0.4605	VH	0.1465
		H	0.1152
		M	0.0901
		L	0.0659
		VL	0.0428
Temperature	0.2774	VH	0.0883
		H	0.0694
		M	0.0543
		L	0.0397
		VL	0.0257
No. of rainy days	0.2622	VH	0.0834
		H	0.0656
		M	0.0513
		L	0.0375
		VL	0.0244

shift of degree of fuzziness δ. Fig. 7 shows how derived weights are varying with respect to the degrees of fuzziness.

From Fig. 7, we can observe that the degrees of fuzziness δ up to 0.2 only one event plays the role. That is rainfall. This means rainfall play a dominant role when making decisions until δ attains 0.2. However, after δ greater than 0.2 the weights

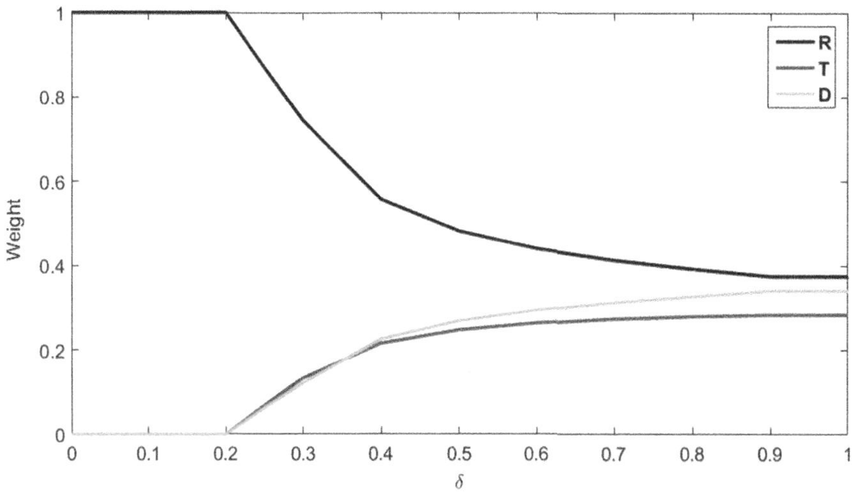

Fig. 7 Weights sensitivity of rainfall, temperature, and rainy days

of the factors, temperature, and number of rainy days start to appear. Therefore, the point 0.2 is called as appearance points. Until δ equal 0.5, these three weight values rapidly changing and attain to steady value after 0.5. This steady value plays a very important role when defining triangular fuzzy number. Hence, minimum workable δ or degree of fuzziness is defined in this study as 0.5.

Step 4: Aggregate the fuzzy membership values to find weather risk

Three weather events are aggregated using convex combination of fuzzy sets to find the weather risk (WR) of different regions and it is given by,

$$WR = 0.4605 \times \mu_{\tilde{A}_R}(x) + 0.2774 \times \mu_{\tilde{A}_T}(x) + 0.2622 \times \mu_{\tilde{A}_D}(x),$$

where $\mu_{\tilde{A}_R}$, $\mu_{\tilde{A}_T}(x)$, $\mu_{\tilde{A}_D}(x)$ denote the membership values of rainfall, temperature, and number of rainy days. The obtained values of WR is compared with threshold values which are defined on the weights of risk level results on FAHP. Comparing these threshold values, proposed weather index (WI) is shown in Table 6. This index contains five weather risk categories: Very low, low, moderate, high, and very high.

Table 6 Weather risk index

Risk level	WR value
Very Low	0–0.1430
Low	0.1430–0.1957
Moderate	0.1957–0.2501
High	0.2501–0.3182
Very High	0.3182–1

Step 5: Identify the weather risk in main regions in Sri Lanka

The proposed weather index is applied to the main regions in Sri Lanka to find the weather-related risk in these regions for the year 2017 and it is shown in Table 7. Finally, these risk levels are compared with the number of affected people from droughts and floods and provide satisfactory results [12].

According to the results generated from proposed weather index, Mannar region can be identified as a very high-risk area in context of weather. This is due to the high temperature in this region. Also noticed, very few number of rainy days in this region annually and it is about 65 days. In addition to that, annual rainfall is also low in this region compared to other regions and it about 59 mm. Therefore, severe droughts reported in this region in year 2017. In addition to that, Moneragala, Mullaitivu, Puttalam, and Ratnapura regions identified as high-risk areas in terms of weather. Model identified Kandy, Kegalle, Kilinochchi, Kurunegala, Matara, Nuwara Eliya, and Polonnaruwa areas as very low-risk region in Sri Lanka in context of weather in the 2017.

5 Conclusion

In this chapter, we developed a weather risk index considering three main weather events to find weather-related risk in main Sri Lankan regions. Selected weather events are rainfall, temperature, and number of rainy days. These weather elements were further sub-divided in five risk levels. Selected weather events were compared with aid of expert opinions. Using these comparisons, fuzzy pairwise comparison matrices were constructed. Chang's extent analysis method was used to derive the weights and to prioritize weather events. Fuzzy membership values of rainfall, temperature, and number of rainy days were aggregated considering their convex combination and developed the risk index. This index contain five risk levels.

For the risk index construction, we considered only three weather-related factors due to the limited accessibility of the resources. Therefore, it is a limitation of this proposed weather index. It is important to consider more weather events such as sun shine hours, coastal sea level variations, and humidity. In addition to that, we can

Table 7 Weather risk in different regions for the year 2017

Region	WR Value	Risk level	Number of affected people
Ampara	0.1421	VL	0
Anuradhapura	0.2145	M	11318
Badulla	0.1325	VL	2593
Batticaloa	0.1342	M	12195
Colombo	0.2387	M	24962
Galle	0.1965	M	12707
Gampaha	0.1993	M	36137
Hambantota	0.1672	L	2000
Jaffna	0.2152	M	45113
Kalutara	0.2301	M	19766
Kandy	0.1435	VL	0
Kegalle	0.1387	VL	125
Kilinochchi	0.1257	VL	0
Kurunegala	0.1624	VL	0
Mannar	0.6782	VH	81918
Matale	0.1287	L	0
Matara	0.1387	VL	0
Moneragala	0.1993	H	38828
Mullaitivu	0.3054	H	43543
Nuwara Eliya	0.1366	VL	0
Polonnaruwa	0.1262	VL	0
Puttalam	0.3182	H	46916
Ratnapura	0.851	H	62455
Trincomalee	0.1982	M	29068
Vavuniya	0.1392	L	0

include intensity of different disasters and disaster management details to the hierarchical structure to generate more accurate results. It is necessary to incorporate time dependency of these weather events because these events are rapidly changing with respect to time. This kind of risk index is very helpful to develop control strategies, and therefore, we can minimize the damage due to weather-related disasters.

References

1. Abedi M, Torabi SA, Norouzi GH (2013) Application of fuzzy AHP method to integrate geophysical data in a prospect scale, a case study: Seridune copper deposit. Bollettino di Geofisica Teorica ed Applicata 54(2):154–164

2. Boender CGE, Grann JGD, Lootsma FA (1989) Multi-criteria decision analysis with fuzzy pairwise comparison. Fuzzy Sets Syst. 29:133–143
3. Chang DY (1996) Applications of the extent analysis method on fuzzy AHP. Eur. J. Oper. Res. 95:649–655
4. Chang DY (1992) Extent analysis and synthetic decision optimization techniques and applications. World Sci. 1:352
5. Chen G, Pham TT (2001) Introduction to fuzzy sets, fuzzy logic, and fuzzy control systems. CRC Press LLC, USA
6. ChildFund (2013) The devastating impact of natural disasters. https://www.childfund.org/Content/NewsDetail/2147489272/. Accessed 20 Dec 2020
7. CRED (2018) Natural disasters in 2017: lower mortality. Higher Cost. https://www.cred.be/publications?field-publication-type-tid=66. Accessed 5 Jan 2021
8. Daily Mirror Online (2021) Sri Lanka experiencing highest temperatures in 140 years, 2019. http://www.dailymirror.lk/breaking-news/SL-experiencing-highest-temperatures-in-140-years. Accessed 2 Feb 2021
9. (2021) Department of Census and Statistics, Sri Lanka. District Statistical HandBook. http://www.statistics.gov.lk/. Accessed 23 Mar 2021
10. (2019) Department of Meteorology, Sri Lanka. Climate of Sri Lanka. http://www.meteo.gov.lk/. Accessed 17 Jan 2021
11. (2009) Disaster Management Centre. Sri Lanka national report on disaster risk, poverty and human development relationship. Accessed 20 Jan 2020
12. (2021) Disaster Management Centre, Sri Lanka. Situation Reports. http://www.dmc.gov.lk
13. Eckstein D, Künzel V, Schüfer L (2021) Global climate risk index 2021. Germanwatch eV t. www.germanwatch.org/en/cri. Accessed 10 Dec 2020
14. (2021). Germanwatch Global Climate Risk Index. https://germanwatch.org/en/cri. Accessed 5 Jan 2021
15. (2021) International federation of red cross and red crescent societies. What is a disaster? https://www.ifrc.org/en/what-we-do/disaster-management/about-disasters/what-is-a-disaster/. Accessed 15 Jan 2021
16. Kabir G, Hasin MAA (2011) Comparative analysis of AHP and Fuzzy AHP models for multi criteria inventory classification. Int. J. Fuzzy Log. Syst. 1(1):1–16
17. (2020) National Association of Insurance Commissioners. Actuaries Climate Index. https://content.naic.org/cipr-topics/topic-actuaries-climate-index.htm. Accessed 8 Feb 2021
18. Piyatilake ITS, Perera SSN (2018) Mathematical model to quantify air pollution in cities. In: Chakraverty S, Perera SSN (eds) Recent advances in applications of computational and fuzzy mathematics. Springer, Singapore, pp 147–178
19. Piyatilake ITS, Perera SSN (2015) Developing a decision support index to quantify air quality: a mathematical modeling approach. J. Basic Appl. Res. Int. 13(2):2395–3438
20. Sakawa M (1993) Fundamentals of fuzzy set theory. Fuzzy sets and interactive multi objective optimization. Applied information technology. Springer, Boston, MA
21. Saaty TL (1980) Analytic hierarchy process. McGraw-Hill, New York
22. Tang YC, Lin TW (2011) Application of the fuzzy analytic hierarchy process to the lead-free equipment selection decision. Int. J. Bus. Syst. Res. 5(1):35–56
23. (2019) United Nations Development Programme. Recovery: Challenges and Lessons. Accessed 6 Feb 2021. https://www.undp.org/content/undp/en/home/librarypage/crisis-prevention-and-recovery/recovery-challenges-and-lessons.html
24. The World Bank. Contingent Liabilities from Natural Disasters: Sri Lanka. https://www.alnap.org/. Accessed 27 Jan 2020
25. (2020) Institute for Environment and Human Security. The World Risk Report. United Nation University. Accessed 5 Feb 2021
26. Zadeh LA (1975) The concept of linguistic variable and its application to approximate Reasoning-I. J. Inf. Sci. 8:199–249
27. Zimmermann HJ (2001) Fuzzy set theory and its applications. Springer Seience + Business Media New York, New York

Type-2 Fuzzy Linear Eigenvalue Problems with Application in Dynamic Structures

Dhabaleswar Mohapatra and S. Chakraverty

Abstract In this chapter, uncertain linear eigenvalue problems are solved, where vagueness is taken as type-2 fuzzy numbers. Triangular perfect quasi type-2 fuzzy numbers (triangular perfect QT2FNs) are used in this discussion. Four parameters are utilized to solve the problem in parametric form. In general, dynamic structural problems lead to eigenvalue problems which may be linear or non-linear eigenvalue problem. The quantities involved in such type of problems are generally taken as crisp for convenience in calculations. But in actual sense there may be some uncertainty in the involved parameters. This is due to the error present in their values that may be because of error in the observation or some other factors. The uncertainty can be handled using probabilistic approach, interval theory, fuzzy theory or any hybridized approach. For probabilistic approach, we need a huge collection of data to deal with the problem, which is not possible in many cases. In such situation, fuzzy theory approach may be advantageous compared to probabilistic case. Numerical examples and application problems are given to check the effectiveness of the proposed method.

1 Introduction

Eigenvalue problems play a great role in many branches of science such as in structural dynamical problems, fluid mechanics, electrical circuits and many more. In general, eigenvalue problems are of two types, viz. linear and non-linear. The present study deals with only linear eigenvalue problems, which is again categorized in to Standard Eigenvalue Problems (SEPs) and Generalized Eigenvalue Problems (GEPs). Eigenvalues, which are to be determined by solving the GEP, are nothing but natural vibration frequencies associated with the corresponding system. For simplicity in calculations, the parameters involved in SEPs and GEPs are taken as crisp by convention. But in actual sense, there may be some uncertainty because of some changes in the environment of the system or some other errors while carrying out the experimental works. Such uncertainty may be handled using probabilistic approach,

D. Mohapatra (✉) · S. Chakraverty
Department of Mathematics, National Institute of Technology Rourkela, Odisha, India

S. Chakraverty (ed.), *Soft Computing in Interdisciplinary Sciences*,
Studies in Computational Intelligence 988,
https://doi.org/10.1007/978-981-16-4713-0_4

if a huge data set is available to us. But in some cases the available sample may not be large enough to implement the probabilistic approach. In such cases, fuzzy theory or interval approaches may be carried out to tackle such situations.

Different methods to handle standard and general eigenvalue problems with crisp parameters are discussed in [1–4]. But there may be possibilities of impreciseness in the involving parameters. So, these problems have been studied by researchers in uncertain environment (interval and type-1 fuzzy). In this regard, Rohn [5], Alefeld and Herzberger [6], and Moore et al. [7] presented different methods to handle interval standard eigenvalue problems (ISEPs) and interval generalized eigenvalue problems (IGEPs). In [8], the bounds enclosing all eigenvalues of interval matrices described in a complex plane over a rectangle. Interval filtering method has been proposed by Hladik et al. [9] which improves the approximation of interval real eigenvalue bounds iteratively. Vibration problems of structural parameters having interval uncertainty have been studied in [10, 11] by implementation of the Rayleigh quotient iteration method and parameter vertex solutions, respectively. By utilizing concepts of modal analysis, properties of continuous functions, matrix perturbation theory and interval perturbation concept different types of methods have been discussed in [12–17]. Behera and Chakravery [18] proposed an algorithm to figure out static and dynamic structural problems assuming the uncertainty in fuzzy environment. Further, filtering algorithm has been implemented to find out the eigenvalue bounds of interval, fuzzy generalized eigenvalue problems and fuzzy symmetric matrices [19]. An procedure to find the column stiffness of multistorey frame structures with uncertain parameters and dynamic data has been given in [20].

Most of the studies in fuzzy eigenvalue problems have been done by dealing the uncertainty in terms of type-1 fuzzy numbers. In type-1 fuzzy numbers, the membership grade gives a crisp value at a particular point of the domain space of the fuzzy set. But, this membership value may vary. The above scenario motivates researchers to implement type-2 fuzzy sets in order to handle uncertainties in such cases, such as in linear eigenvalue problems considered in this chapter.

This chapter is partitioned in to few sections, starting with introduction in the first section, followed by some preliminaries related to fuzzy logic in Sect. 2. In Sects. 3 and 4, type-2 fuzzy linear eigenvalue problems and proposed methodology are discussed. In Sect. 5, three numerical problems and two application problems related to spring–mass systems are discussed. Improper result is obtained in one of the examples, which is written in proper form. Lastly, the conclusion is drawn in Sect. 6.

2 Preliminaries

In this section, a few important basics of fuzzy theory are given for a better understanding of readers.

2.1 Type-1 Fuzzy Numbers

A fuzzy set that is convex and normalized is known as 'type-1 fuzzy number' or simply 'fuzzy number'. A fuzzy set $\tilde{A}$ is an ordered pair [21, 22], given by

$$\tilde{A} = \{(x, \nu_{\tilde{A}}(x))|x \in X \subset \mathbf{R}\} \tag{1}$$

Here, $\nu_{\tilde{A}}(x)$ is the membership function of x.

2.2 Parametric Form of Fuzzy Number

Fuzzy numbers can be expressed as pair of functions using the $\alpha - cut$, where $\alpha \in [0, 1]$, as [21]

$$\tilde{A}(\alpha) = [\underline{A}(\alpha), \overline{A}(\alpha)]. \tag{2}$$

Which is a standard closed interval, where $\underline{A}(\alpha)$ and $\overline{A}(\alpha)$ are lower and upper bounds, with the following criteria

1. $\underline{A}(\alpha)$ is a bounded left continuous non-decreasing function over [0, 1].
2. $\overline{A}(\alpha)$ is a bounded right continuous non-increasing function over [0, 1].
3. $\underline{A}(\alpha) \leq \overline{A}(\alpha), 0 \leq \alpha \leq 1$.

2.3 Type-2 Fuzzy Set

A type-2 fuzzy set $\tilde{W}$ [23] is defined as

$$\tilde{W} = \{((x, v), \nu_{\tilde{W}}(x, v))|x \in X, v \in V_x \subseteq [0, 1]\} \tag{3}$$

where v is the primary membership of x and $\nu_{\tilde{W}}(x, v)$ is the secondary membership function of v such that $0 \leq \nu_{\tilde{W}}(x, u) \leq 1$. Here, x is the primary variable, X is the domain of the fuzzy set and V_x is the domain of the secondary membership function at x.

2.4 Vertical Slice of Type-2 Fuzzy Set

Vertical Slice of a type-2 fuzzy set $\tilde{W}$ [23] is a type-1 fuzzy set, obtained by fixing the primary variable. It is given by

$$\nu_{\tilde{W}}(x = x^*, v) = \nu_{\tilde{W}}(x^*) = \int \frac{f_{x^*}(v)}{v} dv$$

where $v \in V_{x^*} \subseteq [0, 1]$ and $0 \leq f_{x^*}(v) \leq 1$.

2.5 r_1-Plane of Type-2 Fuzzy Set

Let $r_1 \in [0, 1]$ be any real number, the union of all secondary domains of the type-2 fuzzy set $\tilde{A}$ [23] with secondary grade greater or equal to r_1 is called the $r_1 - plane$ of $\tilde{A}$.

2.6 Footprint of Uncertainty

Footprint of uncertainty [23] is obtained from $r_1 - plane$ of the type-2 fuzzy set by substituting $r_1 = 0$.

2.7 Lower Membership Function(LMF) and Upper Membership Function(LMF) of a Type-2 Fuzzy Set

The lower and upper bounds of a type-2 fuzzy set at any level r_1 [23], are known as lower membership function and upper membership function, respectively. That is, $\tilde{A}_{r_1} = (\underline{A}_{r_1}, \overline{A}_{r_1})$.

2.8 Principle Set of $\tilde{A}$ [24]

Principle set is a type-1 fuzzy set, which is achieved for $r_1 = 1$ in the $r_1 - plane$ form of the type-2 fuzzy set.

2.9 r_2- Cut of r_1- Plane [24]

$r_2 - cut$ of a type-2 fuzzy set $\tilde{A}$ at level r_1 is given by $\tilde{A}(r_1, r_2) = (\underline{A}(r_1, r_2), \overline{A}(r_1, r_2))$, i.e. $r_2 - cut$ of LMF and UMF of $\tilde{A}_{r_1}$.

2.10 Triangular Perfect Quasi Type-2 Fuzzy Numbers [25]

A perfect T2FN $\tilde{A}$ [24] is a T2FS , if it holds the conditions below

1. LMF and UMF of footprint of uncertainty of $\tilde{A}$ are type-1 fuzzy numbers.
2. LMF and UMF of principle set of $\tilde{A}$ are also type-1 fuzzy numbers.

Perfect Quasi Type-2 Fuzzy Number [24] is a kind of perfect type-2 fuzzy number, where all vertical slices are type-1 fuzzy numbers and piece-wise functions of the same kind. Triangular perfect QT2FNs was introduced in [25] as

$$\tilde{t} = (\overline{L}_{t_0}, L_{t_1}, \underline{L}_{t_0}, p, \underline{R}_{t_0}, R_{t_1}, \overline{R}_{t_0}). \tag{4}$$

satisfying,

$$\overline{L}_{t_0} \leq L_{t_1} \leq \underline{L}_{t_0} \leq p \leq \underline{R}_{t_0} \leq R_{t_1} \leq \overline{R}_{t_0}.$$

A schematic diagram of the triangular perfect QT2FN $\tilde{t}$ is given in Fig. 1. Now, the r_1, r_2 form is [25]

$$\tilde{t}(r_1, r_2) = (\underline{t}(r_1, r_2), \overline{t}(r_1, r_2))$$

where

$$\underline{t}(r_1, r_2) = (\underline{L}_t(r_1, r_2), \underline{R}_t(r_1, r_2))$$

$$\underline{L}_t(r_1, r_2) = L_{t_1}(r_2) - (1 - r_1)(L_{t_1}(r_2) - \underline{L}_{t_0}(r_2))$$

$$\underline{R}_t(r_1, r_2) = R_{t_1}(r_2) - (1 - r_1)(R_{t_1}(r_2) - \underline{R}_{t_0}(r_2))$$

and

$$\overline{t}(r_1, r_2) = (\overline{L}_t(r_1, r_2), \overline{R}_t(r_1, r_2))$$

$$\overline{L}_t(r_1, r_2) = L_{t_1}(r_2) - (1 - r_1)(L_{t_1}(r_2) - \overline{L}_{t_0}(r_2))$$

$$\overline{R}_t(r_1, r_2) = R_{t_1}(r_2) - (1 - r_1)(R_{t_1}(r_2) - \overline{R}_{t_0}(r_2))$$

Further,

$$\overline{L}_{t_0}(r_2) = p - (1 - r_2)(p - \overline{L}_{t_0})$$

$$L_{t_1}(r_2) = p - (1 - r_2)(p - L_{t_1})$$

$$\underline{L}_{t_0}(r_2) = p - (1 - r_2)(p - \underline{L}_{t_0})$$

$$\overline{R}_{t_0}(r_2) = p - (1 - r_2)(p - \overline{R}_{t_0})$$

$$R_{t_1}(r_2) = p - (1 - r_2)(p - R_{t_1})$$

$$\underline{R}_{t_0}(r_2) = p - (1 - r_2)(p - \underline{R}_{t_0})$$

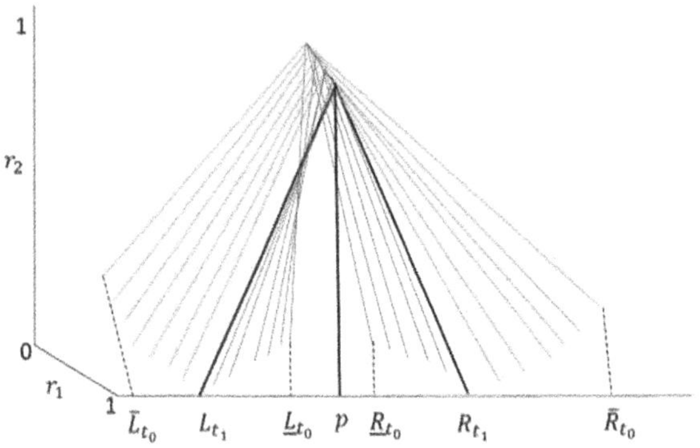

Fig. 1 Schematic diagram of triangular perfect QT2FN $\tilde{t}$

3 Type-2 Fuzzy Linear Eigenvalue Problem

In order to deal with type-2 fuzzy linear eigenvalue problems, let us begin with some basic ideas of linear eigenvalue problem, i.e. crisp linear eigenvalue problems. Crisp linear eigenvalue problems are categorized in to two types based upon their forms,

- Crisp Generalized Eigenvalue Problems(CGEP)
- Crisp Standard Eigenvalue Problems(CSEP)

The general forms of CGEP and CSEP are given as

$$Ax = \lambda Bx \tag{5}$$

$$Cx = \lambda x \tag{6}$$

where A, B are square matrices of same order, i.e. $n \times n$(say), C is also a square matrix, x is the corresponding eigenvector of the eigenvalue λ. It is obvious from Eqs. (5) and (6) that, Eq. (6) is a special case of Eq. (5) when $B = I_n$, i.e. identity matrix of order n. In a very similar way, the type-2 fuzzy linear eigenvalue problem may be categorized as follows:

- Type-2 Fuzzy Generalized Eigenvalue Problem (T2FGEP)
- Type-2 Fuzzy Standard Eigenvalue Problem (T2FSEP)

Suppose $\tilde{A}$ and $\tilde{B}$ be two type-2 fuzzy square matrices of same order (say $n \times n$), then the T2FGEP is

$$\tilde{A}\tilde{x} = \tilde{\lambda}\tilde{B}\tilde{x} \tag{7}$$

Here, $\tilde{x}$ is the type-2 fuzzy eigenvector corresponding to the type-2 fuzzy eigenvalue $\tilde{\lambda}$, matrices $\tilde{A}$ and $\tilde{B}$ are $\tilde{A} = \begin{bmatrix} \tilde{a}_{11} & \tilde{a}_{12} & \dots & \tilde{a}_{1n} \\ \tilde{a}_{21} & \tilde{a}_{22} & \dots & \tilde{a}_{2n} \\ \cdot & \cdot & \cdots & \cdot \\ \cdot & \cdot & \cdots & \cdot \\ \cdot & \cdot & \cdots & \cdot \\ \tilde{a}_{n1} & \tilde{a}_{n2} & \dots & \tilde{a}_{nn} \end{bmatrix}$ and $\tilde{B} = \begin{bmatrix} \tilde{b}_{11} & \tilde{b}_{12} & \dots & \tilde{b}_{1n} \\ \tilde{b}_{21} & \tilde{b}_{22} & \dots & \tilde{b}_{2n} \\ \cdot & \cdot & \cdots & \cdot \\ \cdot & \cdot & \cdots & \cdot \\ \cdot & \cdot & \cdots & \cdot \\ \tilde{b}_{n1} & \tilde{b}_{n2} & \dots & \tilde{b}_{nn} \end{bmatrix}$

Likewise, for type-2 fuzzy square matrix $\tilde{C}$, the T2FSEP is

$$\tilde{C}\tilde{x} = \tilde{\lambda}\tilde{x} \tag{8}$$

where $\tilde{\lambda}$ is the type-2 fuzzy eigenvalue of the type-2 fuzzy matrix

$$\tilde{C} = \begin{bmatrix} \tilde{c}_{11} & \tilde{c}_{12} & \dots & \tilde{c}_{1n} \\ \tilde{c}_{21} & \tilde{c}_{22} & \dots & \tilde{c}_{2n} \\ \cdot & \cdot & \cdots & \cdot \\ \cdot & \cdot & \cdots & \cdot \\ \cdot & \cdot & \cdots & \cdot \\ \tilde{c}_{n1} & \tilde{c}_{n2} & \dots & \tilde{c}_{nn} \end{bmatrix}$$

where $\tilde{a}_{ij}$, $\tilde{b}_{ij}$ and $\tilde{c}_{ij}$, for $i, j = 1, 2, ..., n$, are given by

$$\begin{aligned} \tilde{a}_{ij} &= (\bar{L}_{a_{ij0}}, L_{a_{ij1}}, \underline{L}_{a_{ij0}}, m_{a_{ij}}, \underline{R}_{a_{ij0}}, R_{a_{ij1}}, \bar{R}_{a_{ij0}}) \\ \tilde{b}_{ij} &= (\bar{L}_{b_{ij0}}, L_{b_{ij1}}, \underline{L}_{b_{ij0}}, m_{b_{ij}}, \underline{R}_{b_{ij0}}, R_{b_{ij1}}, \bar{R}_{b_{ij0}}) \\ \tilde{c}_{ij} &= (\bar{L}_{c_{ij0}}, L_{c_{ij1}}, \underline{L}_{c_{ij0}}, m_{c_{ij}}, \underline{R}_{c_{ij0}}, R_{c_{ij1}}, \bar{R}_{c_{ij0}}) \end{aligned}$$

4 Proposed Method to Solve Type-2 Fuzzy Linear Eigenvalue Problem

As mentioned earlier, triangular perfect QT2FNs are considered. For any QT2FN $\tilde{w}$, $r_2 - cut$ of $r_1 - plane$ form is

$$\tilde{w}(r_1, r_2) = (\underline{w}(r_1, r_2), \overline{w}(r_1, r_2)) \tag{9}$$

Another well-known fact in the above expression is that, $\underline{w}(r_1, r_2)$ and $\overline{w}(r_1, r_2)$ are type-1 fuzzy numbers. Further let us introduce two new parameters r_3 and $r_4 \in [0, 1]$ to express the type-2 fuzzy number in quadruple parametric form. By using r_3, we define

$$\tilde{w}(r_1, r_2, r_3) = (\underline{w}(r_1, r_2, r_3), \overline{w}(r_1, r_2, r_3)) \tag{10}$$

where

$$\underline{w}(r_1, r_2, r_3) = r_3(\underline{R}_w(r_1, r_2) - \underline{L}_w(r_1, r_2)) + \underline{L}_w(r_1, r_2) \tag{11}$$

$$\overline{w}(r_1, r_2, r_3) = r_3(\overline{R}_w(r_1, r_2) - \overline{L}_w(r_1, r_2)) + \overline{L}_w(r_1, r_2) \tag{12}$$

As $\underline{w}(r_1, r_2) = (\underline{L}_w((r_1, r_2), \underline{R}_w((r_1, r_2))$ and $\overline{w}(r_1, r_2)=(\overline{L}_w((r_1, r_2), \overline{R}_w((r_1, r_2))$. Now implementing the new parameter r_4 in $\tilde{w}(r_1, r_2, r_3)$ and defining $\tilde{w}(r_1, r_2, r_3, r_4)$, we have

$$\tilde{w}(r_1, r_2, r_3, r_4) = r_4(\overline{w}(r_1, r_2, r_3) - \underline{w}(r_1, r_2, r_3)) + \underline{w}(r_1, r_2, r_3) \tag{13}$$

Further, T2FGEPs and T2FSEPs can be written by using a quadruple parametric form of Triangular Perfect QT2FNs, as

$$\tilde{A}(r_1, r_2, r_3, r_4)\tilde{x}(r_1, r_2, r_3, r_4) = \tilde{\lambda}(r_1, r_2, r_3, r_4)\tilde{B}(r_1, r_2, r_3, r_4)\tilde{x}(r_1, r_2, r_3, r_4) \tag{14}$$

and

$$\tilde{C}(r_1, r_2, r_3, r_4)\tilde{x}(r_1, r_2, r_3, r_4) = \tilde{\lambda}(r_1, r_2, r_3, r_4)\tilde{x}(r_1, r_2, r_3, r_4) \tag{15}$$

Utilizing the above form of the QT2FN, type-2 fuzzy matrices $\tilde{A}$, $\tilde{B}$ and $\tilde{C}$ will become

$$\tilde{A}(r_1, r_2, r_3, r_4) = \begin{bmatrix} \tilde{a}_{11}(r_1, r_2, r_3, r_4) & \tilde{a}_{12}(r_1, r_2, r_3, r_4) & \dots & \tilde{a}_{1n}(r_1, r_2, r_3, r_4) \\ \tilde{a}_{21}(r_1, r_2, r_3, r_4) & \tilde{a}_{22}(r_1, r_2, r_3, r_4) & \dots & \tilde{a}_{2n}(r_1, r_2, r_3, r_4) \\ . & . & \dots & . \\ . & . & \dots & . \\ . & . & \dots & . \\ \tilde{a}_{n1}(r_1, r_2, r_3, r_4) & \tilde{a}_{n2}(r_1, r_2, r_3, r_4) & \dots & \tilde{a}_{nn}(r_1, r_2, r_3, r_4) \end{bmatrix} \tag{16}$$

$$\tilde{B}(r_1, r_2, r_3, r_4) = \begin{bmatrix} \tilde{b}_{11}(r_1, r_2, r_3, r_4) & \tilde{b}_{12}(r_1, r_2, r_3, r_4) & \dots & \tilde{b}_{1n}(r_1, r_2, r_3, r_4) \\ \tilde{b}_{21}(r_1, r_2, r_3, r_4) & \tilde{b}_{22}(r_1, r_2, r_3, r_4) & \dots & \tilde{b}_{2n}(r_1, r_2, r_3, r_4) \\ . & . & \dots & . \\ . & . & \dots & . \\ . & . & \dots & . \\ \tilde{b}_{n1}(r_1, r_2, r_3, r_4) & \tilde{b}_{n2}(r_1, r_2, r_3, r_4) & \dots & \tilde{b}_{nn}(r_1, r_2, r_3, r_4) \end{bmatrix} \tag{17}$$

$$\tilde{C}(r_1, r_2, r_3, r_4) = \begin{bmatrix} \tilde{c}_{11}(r_1, r_2, r_3, r_4) & \tilde{c}_{12}(r_1, r_2, r_3, r_4) & \dots & \tilde{c}_{1n}(r_1, r_2, r_3, r_4) \\ \tilde{c}_{21}(r_1, r_2, r_3, r_4) & \tilde{c}_{22}(r_1, r_2, r_3, r_4) & \dots & \tilde{c}_{2n}(r_1, r_2, r_3, r_4) \\ . & . & \dots & . \\ . & . & \dots & . \\ . & . & \dots & . \\ \tilde{c}_{n1}(r_1, r_2, r_3, r_4) & \tilde{c}_{n2}(r_1, r_2, r_3, r_4) & \dots & \tilde{c}_{nn}(r_1, r_2, r_3, r_4) \end{bmatrix} \tag{18}$$

where $\tilde{\lambda}(r_1, r_2, r_3, r_4)$ and $\tilde{x}(r_1, r_2, r_3, r_4)$ are quadruple parametric forms of type-2 fuzzy eigenvalue and eigenvector, respectively. Now the type-2 fuzzy eigenvalue in quadruple parametric form, of the T2FGEP, can be obtained as

$$\tilde{A}(r_1, r_2, r_3, r_4)\tilde{x}(r_1, r_2, r_3, r_4) = \tilde{\lambda}(r_1, r_2, r_3, r_4)\tilde{B}(r_1, r_2, r_3, r_4)\tilde{x}(r_1, r_2, r_3, r_4) \quad (19)$$

$$\Longrightarrow det\{\tilde{A}(r_1, r_2, r_3, r_4) - \tilde{\lambda}(r_1, r_2, r_3, r_4)\tilde{B}(r_1, r_2, r_3, r_4)\} = 0 \quad (20)$$

Then by assigning different values to the parameters r_1, r_2, r_3 and r_4 the type-2 fuzzy eigenvalue can be achieved. In a very similar way, the T2FSEP can be solved by simplifying

$$det\{\tilde{C}(r_1, r_2, r_3, r_4) - \tilde{\lambda}(r_1, r_2, r_3, r_4)I_n\} = 0 \quad (21)$$

to attain the required type-2 fuzzy eigenvalue.

5 Numerical Examples

Here, a few numerical examples on T2FSEP and T2FGEP are solved along with uncertain dynamic structural application problems.

Example 5.1 Let us start with a simple 2 × 2 T2FSEP,

$$\tilde{A}\tilde{x} = \tilde{\lambda}\tilde{x} \quad (22)$$

where

$$\tilde{A} = \begin{bmatrix} (1, 2, 3, 4, 5, 6, 7) & (2, 3, 4, 5, 6, 7, 8) \\ (2, 3, 4, 5, 6, 7, 8) & (3, 4, 5, 6, 7, 8, 9) \end{bmatrix} \quad (23)$$

By using $r_2 - cut$ of the $r_1 - plane$, each of the type-2 fuzzy entry of $\tilde{A}$

$$\tilde{A}(r_1, r_2) = \begin{bmatrix} \tilde{a}_{11}(r_1, r_2) & \tilde{a}_{12}(r_1, r_2) \\ \tilde{a}_{21}(r_1, r_2) & \tilde{a}_{22}(r_1, r_2) \end{bmatrix} \quad (24)$$

with

$$\tilde{a}_{11}(r_1, r_2) = [(2 + 2r_2) - (1 - r_1)(r_2 - 1), (6 - 2r_2) - (1 - r_1)(1 - r_2)] \quad (25)$$

$$\tilde{a}_{12}(r_1, r_2) = [(3 + 2r_2) - (1 - r_1)(r_2 - 1), (7 - 2r_2) - (1 - r_1)(1 - r_2)] = \tilde{a}_{21}(r_1, r_2) \quad (26)$$

$$\tilde{a}_{22}(r_1, r_2) = [(4 + 2r_2) - (1 - r_1)(r_2 - 1), (8 - 2r_2) - (1 - r_1)(1 - r_2)] \quad (27)$$

Further implementing the parameters r_3 and r_4, one may have

$$\begin{aligned} &\tilde{a}_{11}(r_1, r_2, r_3, r_4) \\ &\quad = r_4[-4r_3(1 - r_1)(r_2 - 1) - 2(1 - r_1)(1 - r_2)] + r_3[(4 - 4r_2) - 2(1 - r_1)(1 - r_2)] \\ &\quad\quad + (2 + 2r_2) - (1 - r_1)(r_2 - 1) \\ &\tilde{a}_{12}(r_1, r_2, r_3, r_4) \\ &\quad = r_4[-4r_3(1 - r_1)(r_2 - 1) - 2(1 - r_1)(1 - r_2)] + r_3[(4 - 4r_2) - 2(1 - r_1)(1 - r_2)] \end{aligned}$$

$$
\begin{aligned}
&+ (3 + r_2) - (1 - r_1)(r_2 - 1)\\
&\tilde{a}_{22}(r_1, r_2, r_3, r_4)\\
&= r_4[-4r_3(1 - r_1)(r_2 - 1) - 2(1 - r_1)(1 - r_2)] + r_3[(4 - 4r_2) - 2(1 - r_1)(1 - r_2)]\\
&+ (4 + 2r_2) - (1 - r_1)(r_2 - 1)
\end{aligned}
\tag{28}
$$

Also $\tilde{a}_{12}(r_1, r_2, r_3, r_4) = \tilde{a}_{21}(r_1, r_2, r_3, r_4)$, then the given T2FSEP can be written as,

$$\tilde{A}(r_1, r_2, r_3, r_4)\tilde{x}(r_1, r_2, r_3, r_4) = \tilde{\lambda}(r_1, r_2, r_3, r_4)\tilde{x}(r_1, r_2, r_3, r_4) \tag{29}$$

$$\implies det\{\tilde{A}(r_1, r_2, r_3, r_4) - \tilde{\lambda}(r_1, r_2, r_3, r_4)I_2\} = 0 \tag{30}$$

By simplifying (30), a quadratic equation will be obtained in $\tilde{\lambda}(r_1, r_2, r_3, r_4)$, which may be solved to get

$$\tilde{\lambda_1} = (4.2360, 6.1623, 8.1231, 10.0990, 12.0827, 14.0710, 16.0622) \tag{31}$$

$$\tilde{\lambda_2} = (-0.2360, -0.1622, -0.1231, -0.0990, -0.0827, -0.0710, -0.0622) \tag{32}$$

Example 5.2 Here, we have taken a T2GEP of 2×2 order, given as

$$\tilde{G}\tilde{x} = \tilde{\lambda}\tilde{H}\tilde{x} \tag{33}$$

where

$$\tilde{G} = \begin{bmatrix} (17700, 17800, 17900, 18000, 18100, 18200, 18300) & (-7350, -7300, -7250, -7200, -7150, -7100, -7050) \\ (-7350, -7300, -7250, -7200, -7150, -7100, -7050) & (7050, 7100, 7150, 7200, 7250, 7300, 7350) \end{bmatrix} \tag{34}$$

$$\tilde{H} = \begin{bmatrix} 3600 & 0 \\ 0 & 3600 \end{bmatrix} \tag{35}$$

Let us define the elements of $\tilde{G}$ as $\tilde{g}_{ij}$ for $i, j = 1, 2$. Then after the implementation of parameters r_1, r_2, r_3 and r_4,

$$
\begin{aligned}
&\tilde{g}_{11}(r_1, r_2, r_3, r_4)\\
&= r_4(r_3(1 - r_1)(400 - 400r_2) + (1 - r_1)(200r_2 - 200)) + r_3((400 - 400r_2) - (1 - r_1)(200 - 200r_2))\\
&+ (17800 + 200r_2) - (1 - r_1)(100r_2 - 100)\\
&\tilde{g}_{12}(r_1, r_2, r_3, r_4)\\
&= r_4(r_3(1 - r_1)(200 - 200r_2) + (1 - r_1)(100r_2 - 100)) + r_3((200 - 200r_2) - (1 - r_1)(100 - 100r_2))+\\
&(-7300 + 100r_2) - (1 - r_1)(50r_2 - 50)\\
&\tilde{g}_{22}(r_1, r_2, r_3, r_4)\\
&= r_4(r_3(1 - r_1)(200 - 200r_2) + (1 - r_1)(100r_2 - 100)) + r_3((200 - 200r_2) - (1 - r_1)(100 - 100r_2))\\
&+ (7100 + 100r_2) - (1 - r_1)(50r_2 - 50)
\end{aligned}
\tag{36}
$$

Then solving

$$\tilde{G}(r_1, r_2, r_3, r_4)\tilde{x}(r_1, r_2, r_3, r_4) = \tilde{\lambda}(r_1, r_2, r_3, r_4)\tilde{H}(r_1, r_2, r_3, r_4)\tilde{x}(r_1, r_2, r_3, r_4) \tag{37}$$

one may get the eigenvalues of this system, in which the parameters r_1, r_2, r_3 and r_4 are involved, then plugging different particular values of the parameters, type-2 fuzzy eigenvalues may be obtained as

$$\tilde{\lambda_1} = (0.9220, 0.9443, 0.9722, 1, 1.0277, 1.0554, 1.0883) \tag{38}$$

$$\tilde{\lambda_2} = (5.9708, 5.9724, 5.9861, 6, 6.0139, 6.0279, 6.0644) \tag{39}$$

Table 1 shows that eigenvalues acquired for the problem 5.2 by the present method using type-2 fuzzy numbers and by [26] using type-1 fuzzy numbers. In the second column of Table 1, type-2 fuzzy solutions are given. In the third column, reduced type-1 fuzzy solutions obtained by the present method are mentioned, whereas in the last column, type-1 fuzzy solutions obtained in [26] are given. From the data of Table-1, it is clear that the obtained solutions by present method is very close to the results in [26].

Example 5.3 Let us solve

$$\begin{bmatrix} \tilde{m}_{11} & \tilde{m}_{12} & 0 \\ \tilde{m}_{12} & \tilde{m}_{11} & \tilde{m}_{12} \\ 0 & \tilde{m}_{12} & \tilde{m}_{11} \end{bmatrix} \tilde{x} = \tilde{\lambda}\tilde{x} \tag{40}$$

where

$$\tilde{m}_{11} = (11.55, 11.7, 11.85, 12, 12.15, 12.3, 12.45) \tag{41}$$

$$\tilde{m}_{12} = (-4.15, -4.1, -4.05, -4, -3.95, -3.9, -3.85) \tag{42}$$

Utilizing the parameters r_1, r_2, r_3 and r_4, one may have

$$\tilde{m}_{11}(r_1, r_2, r_3, r_4) = r_4(r_3(1 - r_1)(0.6 - 0.6r_2) + (1 - r_1)(0.3r_2 - 0.3)) + r_3((0.6 - 0.6r_2)- \tag{43}$$

Table 1 Comparison of type-2 and type-1 eigenvalues

Eigenvalues	Present (type-2)	Present (reduced type-1)	In [26] (type-1)
$\tilde{\lambda}_1$	(0.9220, 0.9443, 0.9722, 1, 1.0277, 1.0554, 1.0883)	(0.9443, 1.0000, 1.0554)	(0.9443, 1.0000, 1.0554)
$\tilde{\lambda}_2$	(5.9708, 5.9724, 5.9861, 6, 6.0139, 6.0279, 6.0644)	(5.9724, 6.0000, 6.0279)	(5.9724, 6.0000, 6.0279)

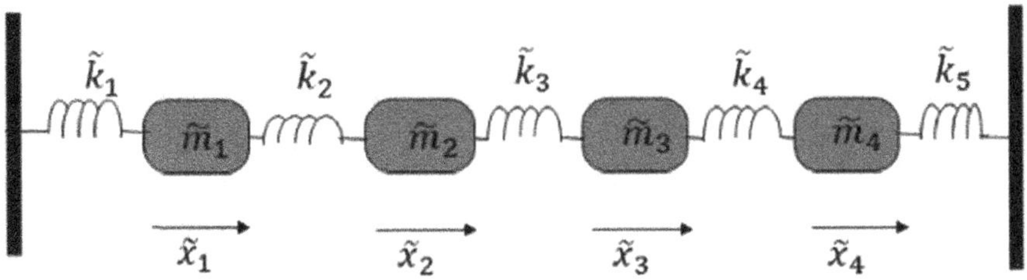

Fig. 2 Spring–mass structure of four degrees of freedom

$$(1-r_1)(0.3-0.3r_2))+(11.7+0.3r_2)-(1-r_1)(0.15r_2-0.15) \tag{44}$$

$$\tilde{m}_{12}(r_1,r_2,r_3,r_4)=r_4(r_3(1-r_1)(0.2-0.2r_2)+(1-r_1)(0.1r_2-0.1))+r_3((0.2-0.2r_2)- \tag{45}$$

$$(1-r_1)(0.1-0.1r_2))+(-4.1+0.1r_2)-(1-r_1)(0.05r_2-0.05) \tag{46}$$

then the given problem may be written as

$$\begin{bmatrix} \tilde{m}_{11}(r_1,r_2,r_3,r_4) & \tilde{m}_{12}(r_1,r_2,r_3,r_4) & 0 \\ \tilde{m}_{12}(r_1,r_2,r_3,r_4) & \tilde{m}_{11}(r_1,r_2,r_3,r_4) & \tilde{m}_{12}(r_1,r_2,r_3,r_4) \\ 0 & \tilde{m}_{12}(r_1,r_2,r_3,r_4) & \tilde{m}_{11}(r_1,r_2,r_3,r_4) \end{bmatrix} \tilde{x}(r_1,r_2,r_3,r_4)=\tilde{\lambda}(r_1,r_2,r_3,r_4)\tilde{x}(r_1,r_2,r_3,r_4) \tag{47}$$

Solving (47), for eigenvalue $\tilde{\lambda}$, one may get

$$\tilde{\lambda}_1=(11.55, 11.7, 11.85, 12, 12.15, 12.3, 12.45) \tag{48}$$

$$\tilde{\lambda}_2=(5.6810, 5.9017, 6.1224, 6.3431, 6.5639, 6.7846, 7.0053) \tag{49}$$

$$\tilde{\lambda}_3=(17.4190, 17.4983, 17.5576, 17.6569, 17.7361, 17.8154, 17.8947) \tag{50}$$

Example 5.4 Here, a spring–mass system with four degrees of freedom is taken as shown in Fig. 2. Where the spring constants are taken as type-2 fuzzy uncertainty and the masses are assumed to be crisp, then the above system can be written in mathematical form as

$$\tilde{K}\tilde{x}=\tilde{\lambda}\tilde{M}\tilde{x} \tag{51}$$

where

$$\tilde{K}=\begin{bmatrix} \tilde{k}_1 & \tilde{k}_2 & 0 & 0 \\ \tilde{k}_2 & \tilde{k}_3 & \tilde{k}_4 & 0 \\ 0 & \tilde{k}_4 & \tilde{k}_5 & \tilde{k}_6 \\ 0 & 0 & \tilde{k}_6 & \tilde{k}_7 \end{bmatrix}, \tilde{M}=\begin{bmatrix} 1 & 0 & 0 & 0 \\ 0 & 1 & 0 & 0 \\ 0 & 0 & 1 & 0 \\ 0 & 0 & 0 & 1 \end{bmatrix}=I_4 \tag{52}$$

Solving in the similar way as above, type-2 fuzzy eigenvalues obtained are

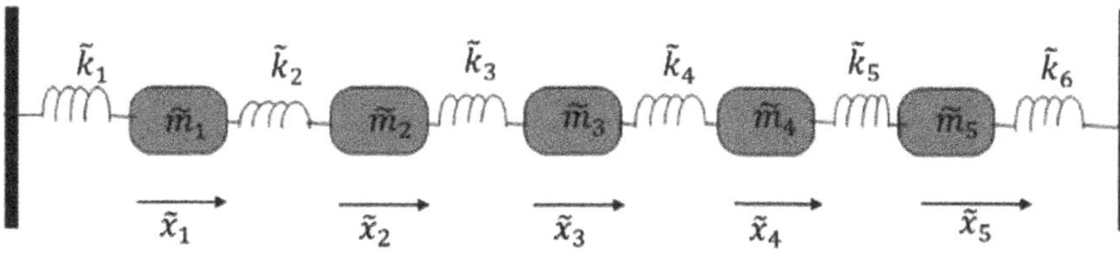

Fig. 3 Spring–mass structure of five degrees of freedom

$$\tilde{\lambda}_1 = (202.2, 229.4, 256.5, 283.6, 310.6, 337.6, 364.6)$$
$$\tilde{\lambda}_2 = (3312.6, 3332.3, 3351.9, 3371.6, 3391.2, 3410.8, 3430.4)$$
$$\tilde{\lambda}_3 = (7237.3, 7259.9, 7282.6, 7305.3, 7328.1, 7350.9, 7373.7)$$
$$\tilde{\lambda}_4 = (9422.9, 9428.4, 9434, 9439.5, 9445.1, 9450.7, 9456.3)$$

Example 5.5 Now let us encounter another application problem, i.e. a five degrees of freedom spring–mass structural system as given in Fig. 3, by the proposed method. Assume that all the structural parameters as type-2 fuzzy numbers, then the above problem leads to a T2FGEP given as

$$\tilde{K}\tilde{x} = \tilde{\lambda}\tilde{M}\tilde{x} \tag{53}$$

Where

$$\tilde{K} = \begin{bmatrix} \tilde{k}_1 + \tilde{k}_2 & -\tilde{k}_2 & 0 & 0 & 0 \\ -\tilde{k}_2 & \tilde{k}_2 + \tilde{k}_3 & -\tilde{k}_3 & 0 & 0 \\ 0 & -\tilde{k}_3 & \tilde{k}_3 + \tilde{k}_4 & -\tilde{k}_4 & 0 \\ 0 & 0 & -\tilde{k}_4 & \tilde{k}_4 + \tilde{k}_5 & -\tilde{k}_5 \\ 0 & 0 & 0 & -\tilde{k}_5 & \tilde{k}_5 + \tilde{k}_6 \end{bmatrix}, \tilde{M} = \begin{bmatrix} c & 0 & 0 & 0 & 0 \\ 0 & \tilde{m}_2 & 0 & 0 & 0 \\ 0 & 0 & \tilde{m}_3 & 0 & 0 \\ 0 & 0 & 0 & \tilde{m}_4 & 0 \\ 0 & 0 & 0 & 0 & \tilde{m}_5 \end{bmatrix} \tag{54}$$

Here,

$$\tilde{k}_1 = (1995, 2000, 2005, 2010, 2015, 2020, 2025)$$
$$\tilde{k}_2 = (1787.5, 1800, 1812.5, 1825, 1837.5, 1850, 1862.5)$$
$$\tilde{k}_3 = (1592.5, 1600, 1607.5, 1615, 1622.5, 1630, 1637.5)$$
$$\tilde{k}_4 = (1395, 1400, 1405, 1410, 1415, 1420, 1425)$$
$$\tilde{k}_5 = (1197.5, 1200, 1202.5, 1205, 1207.5, 1210, 1212.5)$$
$$\tilde{k}_6 = (998, 1000, 1002, 1004, 1006, 1008, 1010)$$
$$\tilde{m}_1 = (9.5, 10, 10.5, 11, 11.5, 12, 12.5)$$
$$\tilde{m}_2 = (11.5, 12, 12.5, 13, 13.5, 14, 14.5)$$
$$\tilde{m}_3 = (13.5, 14, 14.5, 15, 15.5, 16, 16.5)$$
$$\tilde{m}_4 = (15.5, 16, 16.5, 17, 17.5, 18, 18.5)$$
$$\tilde{m}_5 = (17.5, 18, 18.5, 19, 19.5, 20, 20.5)$$

Implementing parameters r_1, r_2, r_3 and r_4, we have

$$\tilde{k}_1(r_1, r_2, r_3, r_4) + \tilde{k}_2(r_1, r_2, r_3, r_4) = r_4(r_3(1-r_1)(70-70r_2) + (1-r_1)(35r_2-35)) + r_3((70-70r_2)- (1-r_1)(35-35r_2)) + (3800+35r_2) - (1-r_1)(17.5r_2-17.5)$$

$$\tilde{k}_2(r_1, r_2, r_3, r_4) + \tilde{k}_3(r_1, r_2, r_3, r_4) = r_4(r_3(1-r_1)(80-80r_2) + (1-r_1)(40r_2-40)) + r_3((80-80r_2)- (1-r_1)(40-40r_2)) + (3400+40r_2) - (1-r_1)(20r_2-20)$$

$$\tilde{k}_3(r_1, r_2, r_3, r_4) + \tilde{k}_4(r_1, r_2, r_3, r_4) = r_4(r_3(1-r_1)(50-50r_2) + (1-r_1)(25r_2-25)) + r_3((50-50r_2)- (1-r_1)(25-25r_2)) + (3000+25r_2) - (1-r_1)(12.5r_2-12.5)$$

$$\tilde{k}_4(r_1, r_2, r_3, r_4) + \tilde{k}_5(r_1, r_2, r_3, r_4) = r_4(r_3(1-r_1)(30-30r_2) + (1-r_1)(15r_2-15)) + r_3((30-30r_2)- (1-r_1)(15-15r_2)) + (2600+15r_2) - (1-r_1)(7.5r_2-7.5)$$

$$\tilde{k}_5(r_1, r_2, r_3, r_4) + \tilde{k}_6(r_1, r_2, r_3, r_4) = r_4(r_3(1-r_1)(18-18r_2) + (1-r_1)(9r_2-9)) + r_3((18-18r_2)- (1-r_1)(9-9r_2)) + (2200+9r_2) - (1-r_1)(4.5r_2-4.5)$$

$$-\tilde{k}_2(r_1, r_2, r_3, r_4) = r_4(r_3(1-r_1)(50-50r_2) + (1-r_1)(25r_2-25)) + r_3((50-50r_2)- (1-r_1)(25-25r_2)) + (-1850+25r_2) - (1-r_1)(12.5r_2-12.5)$$

$$-\tilde{k}_3(r_1, r_2, r_3, r_4) = r_4(r_3(1-r_1)(30-30r_2) + (1-r_1)(15r_2-15)) + r_3((30-30r_2)- (1-r_1)(15-15r_2)) + (-1630+15r_2) - (1-r_1)(7.5r_2-7.5)$$

$$-\tilde{k}_4(r_1, r_2, r_3, r_4) = r_4(r_3(1-r_1)(20-20r_2) + (1-r_1)(10r_2-10)) + r_3((20-20r_2)- (1-r_1)(10-10r_2)) + (-1420+10r_2) - (1-r_1)(5r_2-5)$$

$$-\tilde{k}_5(r_1, r_2, r_3, r_4) = r_4(r_3(1-r_1)(10-10r_2) + (1-r_1)(5r_2-5)) + r_3((10-10r_2)- (1-r_1)(5-5r_2)) + (-1210+5r_2) - (1-r_1)(2.5r_2-2.5)$$

Substituting, above parametric forms of each of the entities involved in this problem and solving for eigenvalues, one may get the eigenvalues in parametric form. Further, using different values of each parameter, type-2 fuzzy eigenvalues are

$$\tilde{\lambda}_1 = (22.8082, 23.4915, 24.1137, 24.6811, 25.1991, 25.6725, 26.1056)$$
$$\tilde{\lambda}_2 = (88.5175, 89.9741, 91.5241, 93.1766, 94.9416, 96.8309, 98.8575)$$
$$\tilde{\lambda}_3 = (171.3082, 175.6269, 180.2168, 185.1040, 190.3183, 195.8933, 201.8677)$$
$$\tilde{\lambda}_4 = (270.0008, 278.1615, 286.8740, 296.1972, 306.1990, 316.9578, 328.5650)$$
$$\tilde{\lambda}_5 = (433.0354, 449.2726, 466.8201, 485.8462, 506.5499, 529.1678, 553.9841)$$

In Table 2, a comparison is given between eigenvalues of spring–mass structure with impreciseness, in triangular perfect QT2FNs(type-2) using present method and in trapezoidal fuzzy numbers(type-1) by [26]. Type-2 fuzzy solutions obtained by present method are given in second column of Table-2. In the third column, reduced type-1(triangular) solutions by present method are given. Type-1(trapezoidal) solutions by [26] are mentioned in last column. It may be observed that, Obtained bounds by present method in type-2 fuzzy case is close to the bounds obtained for type-1 fuzzy in [26].

Table 2 Comparison of type-2 and type-1 eigenvalues

Eigenvalues	Present(type-2)	Present(Reduced type-1)	In [26] (type-1)
$\tilde{\lambda}_1$	(22.8082, 23.4915, 24.1137, 24.6811, 25.1991, 25.6725, 26.1056)	(23.4915, 24.6811, 25.6725)	(23.5660, 24.1774, 25.4399, 26.1796)
$\tilde{\lambda}_2$	(88.5175, 89.9741, 91.5241, 93.1766, 94.9416, 96.8309, 98.8575)	(89.9741, 93.1766, 96.8309)	(88.6697, 90.9004, 96.3431, 99.0332)
$\tilde{\lambda}_3$	(171.3082, 175.6269, 180.2168, 185.1040, 190.3183, 195.8933, 201.8677)	(175.6269, 185.1040, 195.8933)	(175.7996, 181.3439, 190.0671, 196.8291)
$\tilde{\lambda}_4$	(270.0008, 278.1615, 286.8740, 296.1972, 306.1990, 316.9578, 328.5650)	(278.1615, 296.1972, 316.9578)	(280.4391, 286.8788, 308.3185, 316.1529)
$\tilde{\lambda}_5$	(433.0354, 449.2726, 466.8201, 485.8462, 506.5499, 529.1678, 553.9841)	(449.2726, 485.8462, 529.1678)	(456.8997, 469.1946, 508.0278, 524.1465)

6 Conclusion

In this chapter, a method has been discussed to solve T2FGEPs and T2FSEPs using four parameters r_1, r_2, r_3 and r_4. Here, triangular perfect QT2FNs are used to handle type-2 fuzzy uncertainty. Few numerical examples and application problems in spring mass structures have been given along with two comparison tables with type-1 fuzzy results to give a stand to the proposed method.

References

1. Gerald CF (2004) Applied numerical analysis. Pearson Education India
2. Bhat RB, Chakraverty S (2004) Numerical analysis in engineering. Alpha Science Int'l Ltd
3. Humar J (2012) Dynamics of structures. CRC Press
4. Seshu P (2003) Textbook of finite element analysis. PHI Learning Pvt Ltd
5. Rohn J (2005) A handbook of results on interval linear problems
6. Alefeld G, Herzberger J (2012) Introduction to interval computation. Academic Press
7. Moore RE, Kearfott RB, Cloud MJ (2009) Introduction to interval analysis/ramon e. Moore R. Baker Kearfott, Michael J. Cloud, Philadelphia
8. Rohn J (1998) Bounds on eigenvalues of interval matrices. ZAMM-Zeitschrift fur Angewandte Mathematik und Mechanik 78(3):S1049
9. Hladík M, Daney D, Tsigaridas E (2011) A filtering method for the interval eigenvalue problem. Appl. Math. Comput. 217(12):5236–5242

10. Qiu Z, Chen S, Jia H (1995) The rayleigh quotient iteration method for computing eigenvalue bounds of structures with bounded uncertain parameters. Comput. Struct. 55(2):221–227
11. Qiu Z, Wang X, Friswell MI (2005) Eigenvalue bounds of structures with uncertain-but-bounded parameters. J. Sound Vib. 282(1–2):297–312
12. Sim J, Qiu Z, Wang X (2007) Modal analysis of structures with uncertain-but-bounded parameters via interval analysis. J. Sound Vib. 303(1–2):29–45
13. Leng H, He Z, Yuan Q (2008) Computing bounds to real eigenvalues of real-interval matrices. Int. J. Numer. Methods Eng. 74(4):523–530
14. Leng H, He Z (2010) Computation of bounds for eigenvalues of structures with interval parameters. Appl. Math. Comput. 216(9):2734–2739
15. Hladík M (2013) Bounds on eigenvalues of real and complex interval matrices. Appl. Math. Comput. 219(10):5584–5591
16. Leng H, He Z (2007) Computing eigenvalue bounds of structures with uncertain-but-non-random parameters by a method based on perturbation theory. Commun. Numer. Methods Eng. 23(11):973–982
17. Qiu Z, Chen S, Elishakoff I (1996) Bounds of eigenvalues for structures with an interval description of uncertain-but-non-random parameters. Chaos Solitons Fractals 7(3):425–434
18. Chakraverty S, Behera D (2017) Uncertain static and dynamic analysis of imprecisely defined structural systems. In: Fuzzy systems: concepts, methodologies, tools, and applications. IGI Global, pp 1–30
19. Mahato NR, Chakraverty S (2016) Filtering algorithm for real eigenvalue bounds of interval and fuzzy generalized eigenvalue problems. ASCE-ASME J Risk Uncert Eng Sys Part B Mech Eng 2(4) (2016)
20. Chakraverty S, Behera D (2014) Parameter identification of multistorey frame structure from uncertain dynamic data. Strojniski Vestnik/J Mech Eng 60(5)
21. Zimmermann H-J (2001) Introduction to fuzzy sets. Fuzzy set theory—and its applications. Springer, pp 1–8
22. Chakraverty S, Perera S (2018) Recent advances in applications of computational and fuzzy mathematics. Springer
23. Mendel JM, John RB (2002) Type-2 fuzzy sets made simple. IEEE Trans fuzzy Syst 10(2):117–127
24. Hamrawi H (2011) Type-2 fuzzy alpha-cuts, De Montfort University. PhD thesis
25. Mazandarani M, Najariyan M (2014) Differentiability of type-2 fuzzy number-valued functions. Commun Nonlinear Sci Numer Simul 19(3):710–725
26. Chakraverty S, Rout S (2020) Affine arithmetic based solution of uncertain static and dynamic problems. Synth Lect Math Stat 12(1):1–170

Fuzzy Dynamical System in Alcohol-Related Health Risk Behaviors and Beliefs

Maranya M. Mayengo, Moatlhodi Kgosimore, and S. Chakraverty

Abstract In this chapter, we develop a mathematical model for alcohol-related health risks incorporating fuzziness in uncertainties associated with individual risk behavior and induced death rate. In the study, fuzzy numbers (sets) are defined as the degree of peer influence of susceptible individuals into drinking. Using the next generation matrix operator (NGM), we derive the fuzzy reproduction number and characterize the existence of equilibrium states and their stability properties. The study reveals that perceived most influential individuals tend to increase the force of influence. The model has the potential to reveal inherent risk behaviors or cultural beliefs which are critical in the development of alcoholism management strategies.

Keywords Fuzzy models · Health risks · Alcoholism · Cultural beliefs · Fuzzy risk reproduction number

1 Introduction

Alcoholic beverages are an integral part of cultures around the world [1, 2] because of their wide use in rituals and societal artefacts (festivals). Hence, culture plays an important role in promoting drinking practices that are positive while discouraging those associated with harm. It is also a source of income in rural communities and has health benefits (prevention of thrombosis of the heart [3]) if taken at desired levels. Alcoholism may be defined as the state of addiction to the consumption of alcoholic drinks which eventually accelerate to alcohol dependency (i.e., A disease in which a person has a physical or psychological dependence on drinks that contain alcohol).

M. M. Mayengo (✉)
Nelson Mandela-African Institute of Science and Technology, Arusha, Tanzania
e-mail: maranya.mayengo@nm-aist.ac.tz

M. Kgosimore
Botswana University of Agriculture and Natural Resources, Francistown, Botswana

S. Chakraverty
Department of Mathematics, National Institute of Technology Rourkela, Rourkela, Odisha, India

S. Chakraverty (ed.), *Soft Computing in Interdisciplinary Sciences*,
Studies in Computational Intelligence 988,
https://doi.org/10.1007/978-981-16-4713-0_5

Alcoholism is a precursor to injury and violence, its negative impacts can spread throughout a community or a country, and beyond, by influencing levels and patterns of alcohol consumption across borders [4]. Common symptoms of alcoholism include (i) craving (a strong compulsion to drink); (ii) impaired control (inability to limit one's drinking at any given time); (iii) physical dependence (nausea, sweating, shakiness, and anxiety); (iv) tolerance (Increased uptake of alcohol for optimum effects); and (v) health, family, and legal problems (injuries, receives multiple drunken driving citations, frequent arguments, and poor relationships in families).

The drinking limits or threshold to be referred to as an alcoholic is estimated to a maximum of 21 standard bottles per week for men and 14 drinks per week for women [2, 3, 5]. Each year, the harmful use of alcohol kills three million people through its associated health risks and 132.6 million disability-adjusted life years (DALYs) [6]. Comparatively, over-consumption of alcohol causes more harm to human health than tuberculosis, HIV/AIDS, and diabetes [6].

Substance abuse is one of the most sensitive and controversial issues in any society mainly due to its economic and detrimental health effects. For instance, alcohol drinkers or drug users are likely to discuss their drinking behaviors with fellow drinkers than acquaintances and are generally secretive in divulging information associated with their involvement in substance abuse [7]. Alcohol drinking can be characterized into the following categories: heavy, moderate, and low, with risk assessment levels high, moderate, and low of succumbing to health challenges such as liver and pancreatitis diseases [8–11].

Mathematical alcohol epidemic models have been formulated and analyzed in an attempt to provide insights on various effects of alcohol consumption on health and socio-economic aspects of society. For instance, [12] considered a link between drug addiction and infectious disease; [13] studied peer selection and influence effects on adolescent alcohol use; [14] considered effects of the spread of alcoholism on health and social phenomenon; [15] studied impact of public health educational campaigns in reducing alcoholism; [16] studied the dynamics of drinking epidemics; [10] studied alcoholic models on weighted network; and [17] studied the spread of alcoholism.

Fuzzy set theory is a tool for modeling and evaluating the influence of uncertainty and processing of vague or subjective information in mathematical models [18]. The notion of vague or subjective information and their associated variations may arise from various contexts among them insufficient knowledge of boundary conditions, simplifying assumptions, lack of precision in the characterization of objects, and randomness or stochasticity in the repeated occurrence of experimental outcomes. There are two kinds of uncertainties of variation, namely (i) variability generally addressed using probability theory approaches and (ii) ignorance (subjective) which requires the Bayesian analysis and/or multi-valued fuzzy logic approaches. These uncertainties play an important role in human thinking, particularly in the domains of pattern recognition, communication of information, and abstraction [19].

The concept of fuzzy sets first introduced by [20] led to the development of the theory and applications of fuzzy systems to the theory of differential equations with uncertainty [18]. The application involves modeling and evaluation of the influence of imprecise known parameters in mathematical, technical, and physical models. For instance, susceptibility and infectiousness are intrinsically vague modeling information or concepts ideal for characterization using fuzzy logic approaches. The approach characterizes vague data in intervals of variation.

Classical deterministic mathematical models generally do not incorporate a high degree of uncertainties or subjectivity associated with disease dynamics [21], and yet they play an important role in understanding biological phenomena. In this chapter, we assume initiation into alcoholism depends on the interaction between the proportion of alcohol and non-alcohol drinkers, where chances of initiation increase with high levels of alcoholism. To account for subjectivity in the model, we employ fuzzy theory techniques developed by [20]. This theory has been applied in mathematical model of disease epidemics [21–23] and worms in computer networks [24]. In this chapter, we apply the fuzzy set theory approach to develop a mathematical model to analyze health risks associated with alcoholism. The multi-valued logic approach is employed to characterize the population's drinking behaviors according to risks levels; mild, moderate, and high rate of consumption in the interval of variation.

1.1 Mathematical Preliminaries

Let X be a space of points (objects), with a generic element of X denoted by x. Thus, $X = z$. A fuzzy set (class) A in X is characterized by a membership (characteristic) function $f_A(x)$ which associates with each point in X a real number in the interval [0, 1], with the value of $f_A(x)$ at x representing the "grade of membership" of x in A. Thus, the nearer the value of $f_A(x)$ to unity, the higher the grade of membership of x in A. When A is a set in the ordinary sense of the term, its membership function can take only two values 0 and 1, with $f_A(x) = 1 \; or \; 0$ accordingly as x does or does not belong to A. Thus, in this case, $f_A(x)$ reduces to the familiar characteristic function of a set A [20].

2 The Mathematical Model

We develop and analyze a Fuzzy model for population health risks associated with alcoholism. We partition the population into six (6) distinct compartments based on the individual's level of risk corresponding to their drinking habits. The compartments are as follows: Susceptible, $S(t)$, comprising individuals at risk of engaging in alcohol drinking; $P(t)$ Protected, comprising individuals with strong cultural and belief systems; Low risk class, $L(t)$, consisting of individuals who drink responsibly; Moderate drinkers $M(t)$, individuals who are physically alcohol dependent or drink regularly; Heavy drinkers or Alcohol addicts $A(t)$, individuals who heavily depend on alcohol and have serious symptoms of alcoholism; and Recovered $R(t)$, individuals who have been treated or voluntarily quit drinking. The model considers multi-risk levels in a population with active religious beliefs using constant and variable parameters.

Understanding of the complex driving forces of substance abuse behaviors is critical in the design of intervention strategies to curb substance abuse. Factors which constitute peer influence include cultural norms, beliefs and social influence [25] and biological underpinnings or genetic factors [26, 27]. To account for uncertainties and vagueness susceptibility and initiation, we define the variable x as the degree of peer influence of a susceptible individual to initiate drinking behavior. In this case, we characterize membership functions $\beta = \beta(x)$ to capture the spread of health risks associated with alcoholism in the community and $\alpha = \alpha(x)$ to translate the consequences of health risks by means of additional death rate. The fuzzy numbers $\beta = \beta(x)$ and $\alpha = \alpha(x)$ represent the likelihood that a susceptible individual will drink alcohol after prolonged contact with drinking individuals and the additional death rate induced by alcoholism, respectively. This description leads to the development of a fuzzy system (1) of differential equations given by

$$\begin{aligned} \dot{S} &= (1-\phi)\pi + \omega R + \gamma_2 P - (\mu + \gamma_1 + \lambda(x))S, \\ \dot{P} &= \phi\pi + \gamma_1 S + \nu L + \tau M + \psi A - (\mu + \gamma_2)P, \\ \dot{L} &= \lambda(x)\rho S - (\mu + \sigma + \nu)L, \\ \dot{M} &= \lambda(x)(1-\rho)S + \sigma L - (\mu + \delta + \xi + \tau)M, \\ \dot{A} &= \delta M - (\mu + \eta + \psi + \alpha(x))A, \\ \dot{R} &= \xi M + \eta A - (\mu + \omega)R, \end{aligned} \tag{1}$$

with non-negative initial conditions of the state variables: $S(0) = S_0 > 0$, $P(0) = P_0 \geq 0$, $L(0) = L_0 \geq 0$, $M(0) = M_0 \geq 0$, $A(0) = A_0 \geq 0$, and $R(0) = R_0 \geq 0$, where the total population and changes in the total population are, respectively, governed by

$$N = S + P + L + M + A + R \tag{2}$$

and

$$\dot{N} = \pi - \mu N - \alpha(x)A. \tag{3}$$

Considering the fact that, at the initial stage, a non-alcoholic drinker requires alcohol drinking habits after sufficient regular social contacts with drinking individuals [17, 28, 29], we define the force of infection as

$$\lambda(x) = c\beta(x)\left(\frac{L + \theta_1 M + \theta_2 A}{N}\right). \tag{4}$$

We assume that the population is homogeneous and at initial level, drinking behavior is acquired by choice. Individuals in the recovered population do not develop permanent immunity against alcohol drinking. Similarly, individuals in the protected compartment acquire a non-permanent virtual protection from alcohol drinking for their entire life in the compartment. The table below presents the description of model parameters used in the model (Table 1).

Table 1 Model parameters and their description

Symbol	Descriptions
π	Per capita recruitment rate
ϕ	The proportion of recruitment joining the protected population
μ	Natural mortality rate
λ	Force of peer influence to induce drinking
β	The measure of influence of the risky individuals
c	The contact rate between a susceptible member and a drinker necessary to convince the susceptible member to drink
θ_1	The chances of becoming an alcoholic after successful influence of a moderate risk drinker
θ_2	The chances of becoming an alcoholic after successful influence of a high risk drinker
ρ	The proportion of susceptible individual recruited to the low risk drinking population
ω	The rate at which recovered individuals join susceptible compartment
γ_1	The rate at which susceptible population joins protected compartment
γ_2	Virtual protection wane rate
ν	The rate at which low risk population joins protected compartment
τ	The rate at which moderate risk population joins protected compartment
ψ	The rate at which high risk population joins protected compartment
σ	Progressive rate from low to moderate risk compartments
δ	Progressive rate from moderate to high risk compartments
ξ	Recovery rate for moderate risk population
α	Alcohol induced fatality rate
η	Recovery rate for high risk population

3 Model Analysis

3.1 *Steady State Solutions*

To obtain steady state solutions, we set the right hand side (RHS) of system (1) to zero and solve the following system of non-linear equations:

$$\begin{aligned}
&(1-\phi)\pi + \omega R^* + \gamma_2 P^* - (\mu + \lambda^*(x) + \gamma_1)S^* = 0,\\
&\phi\pi + \gamma_1 S^* + \nu L^* + \tau M^* + \psi A^* - (\mu + \gamma_2)P^* = 0,\\
&\rho\lambda^*(x)S^* - (\mu + \sigma + \nu)L^* = 0,\\
&(1-\rho)\lambda^*(x)S^* + \sigma L^* - (\mu + \delta + \xi + \tau)M^* = 0,\\
&\delta M^* - (\mu + \alpha(x) + \eta + \psi)A^* = 0,\\
&\xi M^* + \eta A^* - (\mu + \omega)R^* = 0.
\end{aligned} \tag{5}$$

From equations for L^*, M^*, and A^*, we have

$$L^* = Q_0\lambda^*(x)S^*,\ M^* = Q_1\lambda^*(x)S^*,\ \text{ and }\ A^* = Q_2\lambda^*(x)S^*, \tag{6}$$

where

$$Q_0 = \frac{\rho}{\mu+\nu+\sigma},\quad Q_1 = \frac{1}{(\mu+\tau+\delta+\xi)}\left(\sigma Q_0 + (1-\rho)\right),\ Q_2 = \frac{\delta}{(\mu+\eta+\psi+\alpha(x))}Q_1. \tag{7}$$

Substituting L^*, M^*, and A^*, we have

$$\lambda^*(x) = c\beta\left(\frac{L^*+\theta_1 M^* + \theta_2 A^*}{N^*}\right) = c\beta(Q_0+\theta_1 Q_1+\theta_2 Q_2)\left(\frac{S^*}{N^*}\right)\lambda^*(x)$$

or

$$\lambda^*(x)\left(\frac{S^*}{N^*} - \frac{1}{\mathcal{R}_0(x)}\right) = 0,\quad \text{where } \mathcal{R}_0(x) = c\beta(Q_0+\theta_1 Q_1+\theta_2 Q_2), \tag{8}$$

which reduces to

$$\lambda^*(x) = 0 \ \text{ or }\ N^* = \mathcal{R}_0(x)S^*. \tag{9}$$

Note that the model reproduction number $\mathcal{R}_0$ can be re-written as

$$\mathcal{R}_0(x) = \mathcal{R}_{01}(x) + \mathcal{R}_{02}(x) + \mathcal{R}_{03}(x) = \sum_{i=1}^{i=3} \mathcal{R}_{0i}(x) \tag{10}$$

where $\mathcal{R}_{0i}(x)s$ are contributions of the individual alcohol drinking groups to alcoholism which are defined as follows:

$$\mathcal{R}_{01}(x) = \frac{c\beta(x)\rho}{\mu+\nu+\sigma}, \tag{11}$$

$$\mathcal{R}_{02}(x) = \frac{c\beta(x)\theta_1\left((1-\rho)(\mu+\nu+\sigma)+\rho\sigma\right)}{(\mu+\nu+\sigma)(\mu+\tau+\delta+\xi)}, \tag{12}$$

$$\mathcal{R}_{03}(x) = \frac{c\beta(x)\left((1-\rho)(\mu+\nu+\sigma)+\rho\sigma\right)\delta\theta_2}{(\mu+\nu+\sigma)(\mu+\tau+\delta+\xi)\left(\mu+\eta+\psi+\alpha(x)\right)}. \tag{13}$$

The solution

$$\lambda^*(x) = 0 \tag{14}$$

yields the disease-free equilibrium $\mathcal{E}_0$.

$$\mathcal{E}_0 = (S^*, P^*, L^*, M^*, A^*, R^*) = (S_0, P_0, 0, 0, 0, 0), \tag{15}$$

where

$$S_0 = \left(\frac{\pi}{\mu}\right)\left(\frac{\gamma_2 + (1-\phi)\mu}{\mu + \gamma_1 + \gamma_2}\right) \quad \text{and} \quad P_0 = \left(\frac{\pi}{\mu}\right)\left(\frac{\gamma_1 + \phi\mu}{\mu + \gamma_1 + \gamma_2}\right).$$

At equilibrium, the total population is governed by

$$\begin{aligned} N^* &= S^* + P^* + L^* + M^* + A^* + R^* = S^* + P^* + Q\lambda^* S^* \\ &= (1 + Q\lambda^*(x))S^* + P^*, \end{aligned} \tag{16}$$

where

$$Q_3 = \frac{\xi Q_1 + \eta Q_2}{\mu + \omega} \quad \text{and} \quad Q = \sum_{k=0}^{k=3} Q_k.$$

Combining the results (9) and (16), we obtain

$$P^* = (\mathcal{R}_0(x) - 1 - Q\lambda^*(x))S^*. \tag{17}$$

The solution $N^* = \mathcal{R}_0(x)S^*$ leads to the endemic equilibrium given by

$$\mathcal{E}_1 = (S^*, P^*, L^*, M^*, A^*, R^*) \tag{18}$$

with coordinates

$$L^* = Q_0\lambda^*(x)S^*, \; M^* = Q_1\lambda^*(x)S^*, \; A^* = Q_2\lambda^*(x)S^*, \; \text{and} \; R^* = Q_3\lambda^*(x)S^* \tag{19}$$

in terms of $\lambda^*(x)$.

From the first and second equations of system (5) and result (17), we have

$$\begin{cases} (1-\phi)\pi + \omega Q_3\lambda^*(x)S^* + \gamma_2(\mathcal{R}_0(x) - 1 - Q\lambda^*(x))S^* - (\mu + \lambda^*(x) + \gamma_1)S^* = 0, \\ \phi\pi + \gamma_1 S^* + (\nu Q_0 + \tau Q_1 + \psi Q_2)\lambda^*(x)S^* - (\mu + \gamma_2)(\mathcal{R}_0(x) - 1 - Q\lambda^*(x))S^* = 0, \end{cases}$$

which results in

$$\lambda^*(x) = \frac{n_1(\mathcal{R}_0(x) - 1) + n_2}{n_3} = \frac{n_1}{n_3}\left(\mathcal{R}_0(x) - \mathcal{R}_0^c\right), \tag{20}$$

where

$$n_1 = \gamma_2 + (1-\phi)\mu, \;\; n_2 = -\gamma_1(1-\phi), \; n_3 = \phi(\gamma_2 Q - \omega Q_3 + 1) + (1-\phi)[(\mu + \gamma_2)Q + \nu Q_0 + \tau Q_1 + \psi Q_2(x)]$$

and

$$\mathcal{R}_0^c = \frac{\gamma_2 + (1-\phi)(\mu+\gamma_1)}{\gamma_2 + (1-\phi)\mu} = 1 + \frac{\gamma_1(1-\phi)}{\gamma_2 + (1-\phi)\mu}. \quad (21)$$

Observe that when the entire population is protected (i.e., if $\phi = 1$), the endemic equilibrium coalesce with the risk-free equilibrium. However, in the absence of any protection, a unique endemic equilibrium exists for

$$\mathcal{R}_0^c > 1 + \frac{\mu}{\mu + \gamma_2}. \quad (22)$$

The above results can be summarized with the following theorem:

Theorem 1 *The system (1) has a risk-free equilibrium* $\mathcal{E}_0 = (S_0, P_0, 0, 0, 0, 0)$ *and a unique risk endemic equilibrium* $\mathcal{E}_1 = (S^*, P^*, L^*, M^*, A^*, R^*)$.

4 Fuzzy Dynamical Systems

Following [19] and [22], we constructed and analyzed a fuzzy dynamical model of alcoholism. We define two fuzzy membership functions (or numbers) $\beta(x)$ and $\alpha(x)$ to account for individuals susceptibility and risks induced death due to alcoholism as functions of the degree of peer influence or factors leading to alcoholism. Thus, the value x represents the realization of driving elements usually called "grade of membership" or "degree of peer influence" caused by social and/or biological factors. In the context of this study, we assume that a drinking individual interacts with a susceptible member, and it requires $x = x_{min}$ as a threshold of the degree of peer influence to have an impact on a susceptible member. The impact of behavioral influence is considered negligible whenever when $x < x_{min}$. The quantity x_{min} is taken as a parameter whose exact value would depend upon both the attitude, public opinions toward the drinking behavior or the drinking individual, and willingness of a susceptible individual to conform with the peer pressures. The increase of x values increases the behavior inducement rate to a maximum value which is equal to one at $x \geq x_0$. Furthermore, it is assumed that the degree of peer influence is bounded above by $x = x_{max}$. Therefore, the values of x with an effect to the system lie in the interval of $x_{min} \leq x \leq x_{max}$. The fuzzy membership function for the fuzzy number $\beta(x)$ is given by

$$\beta(x) = \begin{cases} 0, & \text{if } x < x_{min}; \\ \dfrac{x - x_{min}}{x_0 - x_{min}}, & \text{if } x_{min} \leq x \leq x_0; \\ 1, & \text{if } x_0 < x < x_{max}. \end{cases} \quad (23)$$

Clearly, at low levels of peer influence, the manifestation of drinking habits of behaviors will be minimal, and there exist a number x_0 at which the peer influence is maximum and is unity.

Similarly, the parameter $\alpha(x)$, defined as alcohol induced death rate, characterizing the amount of health risks "transmission" associated with alcoholism making $\alpha(x)$ a fuzzy number. We assume that negligible amount of health risk transmission occurs when $x < x_{min}$ leading to the introduction of a minimum additional death rate, $\alpha(x < x_{min}) = \alpha_0$. The increase of peer influence x increases the additional death rate, which gets the highest value when $x = x_0$. The additional death rate may not reach $\alpha(x) = 1$ as its highest score; due to several limitations, we let the maximum value $\alpha(x) = 1 - u$ for some real number u such that $0 < u < 1 - \alpha(x)_0$. The fuzzy membership function of $\alpha(x)$ may be established as follows:

$$\alpha(x) = \begin{cases} \alpha_0, & \text{if } 0 \leq x < x_{min}; \\ \alpha_0 + \left(\dfrac{1 - u - \alpha_0}{x_0 - x_{min}}\right) x, & \text{if } x_{min} \leq x < x_0; \\ (1 - u), & \text{if } x_0 \leq x \leq x_{max}. \end{cases} \tag{24}$$

The graphs of two membership functions, $\beta(x)$ and $\alpha(x)$, are presented in Fig. 1a, b, respectively. To mimic the transmission of health risks associated with alcoholism, we assume the degree of peer influence in the community being studied, Π as critical in categorizing the population into different drinking levels depending on their social influence in the community. Thus, we consider the population Π being studied as a linguistic variable with varying classification. Using x_c and d, respectively, as the central value and dispersion of each of the fuzzy sets assumed by Π, we model each classification using a triangular fuzzy number whose membership function is given in (25) and the graph of membership function Π as presented in Fig. 2.

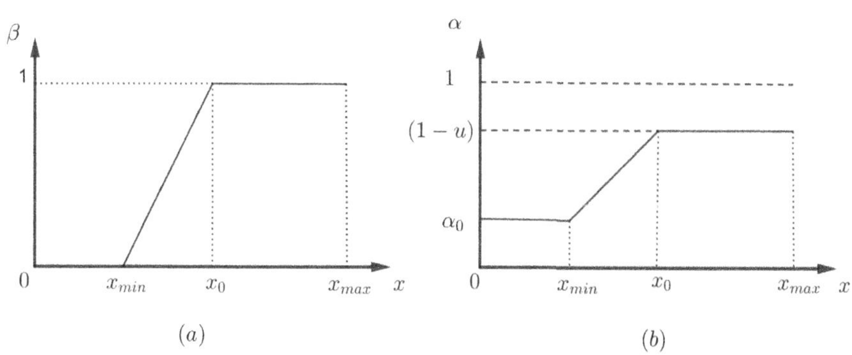

Fig. 1 The graph of membership functions β and α. *Source* [29]

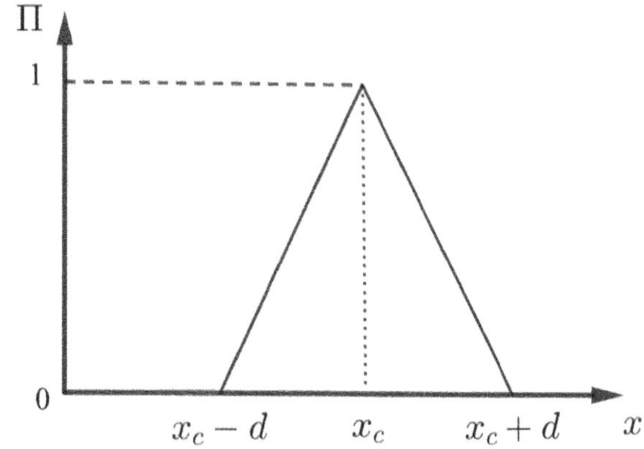

Fig. 2 The graph of membership function Π. *Source* [29]

$$\Pi(x) = \begin{cases} 0, & \text{if } x < x_c - d; \\ \dfrac{x - x_c + d}{d}, & \text{if } x_c - d \le x \le x_c; \\ -\dfrac{x - x_c - d}{d}, & \text{if } x_c < x \le x_c + d; \\ 0, & \text{if } x_c + d < x. \end{cases} \tag{25}$$

Suppose that $L(x,t)$, $M(x,t)$, and $A(x,t)$ are the family solutions of the given fuzzy model system. These are the numbers of risky population proportions created as the result of social interactions between the susceptible members and risky individuals with social influence x at time t. Now, $L(x,t)$, $M(x,t)$, and $A(x,t)$ are fuzzy numbers which lie in the interval [0, 1].

4.1 Fuzzy Model Risk Reproduction Number

Analogous to mathematical models for disease epidemics, we derive an equivalent threshold parameter $\mathcal{R}_0(x)$, which measures the severity of initiation into alcoholism and design of control strategies for alcoholism. This parameter gives a number of secondary cases caused by one alcohol drinker introduced in a population of non-alcohol drinkers. In this section, we first compute the health risk reproduction number, denoted as $\mathcal{R}_0(x)$, by using the Next Generation Matrix method [29–31]. Based on [31], we decompose the risk transmission model into the following system of equations:

$$\dot{Z} = \mathcal{F}(Z) - \mathcal{V}(Z) = \begin{bmatrix} \lambda(x)\rho S \\ \lambda(x)(1-\rho)S \\ 0 \end{bmatrix} - \begin{bmatrix} \kappa_3 L \\ -\sigma L + \kappa_4 M \\ -\delta M + (\kappa_5 + \alpha(x))\,A, \end{bmatrix} \tag{26}$$

where $\mathcal{V}(Z) = \mathcal{V}^-(Z) - \mathcal{V}^+(Z)$, $X = \{S, P, R\}^T \in \mathbb{R}^3$, $Z = \{L, M, A\}^T \in \mathbb{R}^3$, and $(\cdot)^T$ denotes transpose.

We then formulate the risk "*transmissions*" matrix F and risk "*transitions*" matrix V such that

$$F = \begin{bmatrix} c\beta(x)\epsilon\rho & c\beta(x)\epsilon\theta_1\rho & c\beta(x)\epsilon\theta_2\rho \\ c\beta(x)\epsilon(1-\rho) & c\beta(x)\epsilon\theta_1(1-\rho) & c\beta(x)\epsilon\theta_2(1-\rho) \\ 0 & 0 & 0 \end{bmatrix} \text{ and } V = \begin{bmatrix} \kappa_3 & 0 & 0 \\ -\sigma & \kappa_4 & 0 \\ 0 & -\delta & \kappa_5 + \alpha(x) \end{bmatrix} \tag{27}$$

where $\kappa_3 = \mu + \nu + \sigma$, $\kappa_4 = \mu + \tau + \delta + \xi$, $\kappa_5 = \mu + \eta + \psi$, and $\epsilon = \dfrac{S_0}{N_0} = \dfrac{\gamma_2 + (1-\phi)\mu}{\mu + \gamma_1 + \gamma_2}$ are the simplifying parameters. Thus, by direct computation, we have

$$V^{-1} = \begin{bmatrix} \dfrac{1}{\kappa_3} & 0 & 0 \\ \dfrac{\sigma}{\kappa_3\kappa_4} & \dfrac{1}{\kappa_4} & 0 \\ \dfrac{\sigma\delta}{\kappa_3\kappa_4(\kappa_5 + \alpha(x))} & \dfrac{\delta}{\kappa_4(\kappa_5 + \alpha(x))} & \dfrac{1}{(\kappa_5 + \alpha(x))} \end{bmatrix} \tag{28}$$

Now, we have the following next generation matrix (NGM)

$$FV^{-1} = \begin{bmatrix} \dfrac{c\beta(x)\epsilon\rho}{\kappa_3}\left(1 + \dfrac{\theta_1\sigma}{\kappa_4} + \dfrac{\theta_2\delta\sigma}{\kappa_4(\kappa_5 + \alpha(x))}\right) & \dfrac{c\beta(x)\epsilon\rho}{\kappa_4}\left(\theta_1 + \dfrac{\theta_2\delta}{(\kappa_5 + \alpha(x))}\right) & \dfrac{c\beta(x)\epsilon\theta_2\rho}{(\kappa_5 + \alpha(x))} \\ \dfrac{c\beta(x)\epsilon(1-\rho)}{\kappa_3}\left(1 + \dfrac{\theta_1\sigma}{\kappa_4} + \dfrac{\theta_2\delta\sigma}{\kappa_4(\kappa_5 + \alpha(x))}\right) & \dfrac{c\beta(x)\epsilon(1-\rho)}{\kappa_4}\left(\theta_1 + \dfrac{\theta_2\delta}{(\kappa_5 + \alpha(x))}\right) & \dfrac{c\beta(x)\epsilon\theta_2(1-\rho)}{(\kappa_5 + \alpha(x))} \\ 0 & 0 & 0 \end{bmatrix} \tag{29}$$

The risk reproduction number, $\mathcal{R}_0(x)$, is given by the dominant eigenvalue of matrix FV^{-1}. Therefore,

$$\mathcal{R}_0(x) = c\beta(x)\epsilon\left(\frac{\rho\kappa_4(\kappa_5 + \alpha(x)) + (\theta_1(\kappa_5 + \alpha(x)) + \theta_2\delta)((1-\rho)\kappa_3 + \rho\sigma)}{\kappa_3\kappa_4(\kappa_5 + \alpha(x))}\right) \tag{30}$$

which can be re-written as

$$\mathcal{R}_0(x) = \mathcal{R}_{01}(x) + \mathcal{R}_{02}(x) + \mathcal{R}_{03}(x), \tag{31}$$

where $\mathcal{R}_{0i}(x)s$ are as outlined in (10).

4.2 Stability Analysis of Risk-Free Equilibrium

Consider the Jacobian matrix evaluated at the risk-free equilibrium point below. Observe that

$$\left[\frac{\partial(\lambda S)}{\partial L}\right]_{\mathcal{E}_0} = \left[\frac{\partial}{\partial L}\left(c\beta(x)\left(\frac{L+\theta_1 M+\theta_2 A}{N}\right)S\right)\right]_{\mathcal{E}_0} = \frac{c\beta(x)}{(\pi/\mu)}, \tag{32}$$

$$[2ex]\left[\frac{\partial(\lambda S)}{\partial M}\right]_{\mathcal{E}_0} = \left[\frac{\partial}{\partial M}\left(c\beta(x)\left(\frac{L+\theta_1 M+\theta_2 A}{N}\right)S\right)\right]_{\mathcal{E}_0} = \frac{\theta_1 c\beta(x)}{(\pi/\mu)}, \tag{33}$$

$$[2ex]\left[\frac{\partial(\lambda S)}{\partial A}\right]_{\mathcal{E}_0} = \left[\frac{\partial}{\partial A}\left(c\beta(x)\left(\frac{L+\theta_1 M+\theta_2 A}{N}\right)S\right)\right]_{\mathcal{E}_0} = \frac{\theta_2 c\beta(x)}{(\pi/\mu)}. \tag{34}$$

Now, defining

$$\epsilon = \frac{S_0}{\pi/\mu} = \frac{\gamma_2 + (1-\phi)\mu}{\mu+\gamma_1+\gamma_2}, \tag{35}$$

we have the Jacobian matrix evaluated at the risk-free equilibrium point, $\mathcal{E}_0$ given by

$$J_0 = \begin{bmatrix} -\kappa_1 & \gamma_2 & -\epsilon c\beta(x) & -\theta_1\epsilon c\beta(x) & -\theta_2\epsilon c\beta(x) & \omega \\ \gamma_1 & -\kappa_2 & \nu & \tau & \psi & 0 \\ 0 & 0 & \rho\epsilon c\beta(x) - \kappa_3 & \theta_1\rho\epsilon c\beta(x) & \theta_2\rho\epsilon c\beta(x) & 0 \\ 0 & 0 & \sigma + (1-\rho)c\beta(x) & -\kappa_4 & 0 & 0 \\ 0 & 0 & 0 & \delta & -(\kappa_5+\alpha(x)) & 0 \\ 0 & 0 & 0 & \xi & \eta & -\kappa_6 \end{bmatrix}$$

where $\kappa_1 = \mu + \gamma_1, \quad \kappa_2 = \mu + \gamma_2$.

This matrix can be decomposed into block matrices as

$$J_0 = \begin{bmatrix} B_{11} & B_{12} & B_{13} \\ \mathbf{0} & B_{22} & \mathbf{0} \\ \mathbf{0} & B_{32} & B_{33} \end{bmatrix}$$

where $\mathbf{0}$ are zero matrices,

$$B_{11} = \begin{bmatrix} -(\mu+\gamma_1) & \gamma_2 \\ \gamma_1 & -(\mu+\gamma_2) \end{bmatrix}, \; B_{12} = \begin{bmatrix} -\epsilon c\beta(x) & -\theta_1\epsilon c\beta(x) & -\theta_2\epsilon c\beta(x) \\ \nu & \tau & \psi \end{bmatrix},$$

$$B_{13} = \begin{bmatrix} \omega \\ 0 \end{bmatrix}, \; B_{22} = \begin{bmatrix} \rho\epsilon c\beta(x) - (\mu+\sigma+\nu) & \theta_1\rho\epsilon c\beta(x) & \theta_2\rho\epsilon c\beta(x) \\ \sigma + (1-\rho)c\beta(x) & -(\mu+\delta+\xi+\tau) & 0 \\ 0 & \delta & -(\mu+\eta+\psi+\alpha(x)) \end{bmatrix},$$

$$B_{32} = \begin{bmatrix} 0 & \xi & \eta \end{bmatrix}, \text{ and } B_{33} = \begin{bmatrix} -(\mu+\omega) \end{bmatrix}$$

To study the stability of the risk-free equilibrium, it suffices to investigate the signs of real parts of the eigenvalues of B_{11}, B_{22}, and B_{33}. Clearly, the eigenvalue of B_{33} has negative real parts. The trace of B_{11} is

$$Tr(B_{11}) = -(\mu + \gamma_1) - (\mu + \gamma_2) < 0 \tag{36}$$

and the determinant of B_{11} is

$$Det(B_{11}) = \mu(\mu + \gamma_1 + \gamma_2) > 0. \tag{37}$$

Now, we can see clearly that the stability of the model $\mathcal{E}_0$ is solely determined by the signs of the real parts of the eigenvalues of B_{22}. The characteristic equation arising from B_{22} is a cubic equation

$$\lambda^3 + b_2\lambda^2 + b_1\lambda + b_0 = 0, \tag{38}$$

$$\begin{aligned}
b_0 &= \kappa_4(\kappa_5 + \alpha(x))(\kappa_3 - \rho\epsilon c\beta(x)) - (\sigma + (1-\rho)c\beta(x))((\kappa_5 + \alpha(x))\theta_1\rho\epsilon c\beta(x) - \sigma\theta_2\rho\epsilon c\beta(x)) \\
&= \kappa_3\kappa_4(\kappa_5 + \alpha(x))(1 - \mathcal{R}_0(x)), \\
b_1 &= -\theta_1\rho\epsilon c\beta(x)(\kappa_4 + (\kappa_5 + \alpha(x))) + \kappa_4(\kappa_5 + \alpha(x)) + \theta_1\rho\epsilon c\beta(x)(\sigma + (1-\rho)c\beta(x)), \\
b_2 &= \kappa_4 + (\kappa_5 + \alpha(x)) - \theta_1\rho\epsilon c\beta(x), \\
b_3 &= 1.
\end{aligned}$$

Using the Routh-Hurwitz stability criteria

$$b_0 > 0,\ b_1 > 0,\ b_2 > 0,\ \text{and } b_2b_1 > b_0,$$

it is easy to show that the local stability of the model risk-free equilibrium is established by Theorem 2 below.

Theorem 2 *The risk-free equilibrium $\mathcal{E}_0$ is locally asymptotic stable when $\mathcal{R}_0(x) < 1$ and unstable otherwise.*

The value of $\mathcal{R}_0(x)$ may vary significantly depending on different risk dynamics studied in the population or different populations involved in similar studies [23]. The mathematical model behavior involves the phenomenon of transcritical bifurcation which brings about the exchange of stability between the risk-free equilibrium which exists for all values of $\mathcal{R}_0(x)$ and the endemic risk equilibrium which exists for $\mathcal{R}_0(x)$ above unity. Thus, the system exhibits a transcritical bifurcation at the risk-free equilibrium when $\mathcal{R}_0(x) = 1$. Suppose that the bifurcation value occurs at x^*, (see Fig. 4) where x^* is given as

$$x^* = \frac{a_4 - \sqrt{a_2 - 2a_3 + (-1 + u + \alpha_0)^2 (x_0 - x_{min})^2}}{2a_1 (-1 + u + \alpha_0)(a_5\theta_1 + \rho\kappa_4)} \tag{39}$$

provided that $x_{min} \leq x^* \leq x_0$ and

$$
\begin{aligned}
a_1 &= \frac{c\epsilon}{\kappa_3\kappa_4},\\
a_2 &= (((-1+u+\alpha_0)\,x_{min} + (\alpha_0+\kappa_5)\,(x_{min}-x_0))\,(\theta_1 a_5 + \rho\kappa_4) + \theta_2\delta a_5\,(x_{min}+x_0))^2\,{a_1}^2,\\
a_3 &= (x_0 - x_{min})\,(-1+u+\alpha_0)\,(((-1+u+\alpha_0) + (\alpha_0+\kappa_5)\,(x_0+x_{min}))\,(\theta_1 a_5+\rho\kappa_4)\\
&\quad - (x_0+x_{min})\,\theta_2\,\delta a_5)\,a_1,\\
a_4 &= (((u-1)\,x_{min} + x_0\alpha_0)\,(\theta_1 a_5+\rho\kappa_4) + ((\theta_1 a_5+\rho\kappa_4)\,\kappa_5 + \theta_2\delta a_5)\,(x_0 - x_{min}))\,a_1\\
&\quad + (-1+u+\alpha_0)\,(\alpha_0 - x_{min})\,, a_5 = ((1-\rho)\,\kappa_3 + \rho\sigma)\,.
\end{aligned}
$$

The risk reproduction number presented in (30) is the function of the degree of social influence in the spread of the behavior. However, both $\beta(x)$ and $\alpha(x)$ incline to their maximum level whenever $x \geq x_0$ and as such we have

$$
\mathcal{R}_0(u) = c\epsilon\left(\frac{\rho\kappa_4\,(\kappa_5+(1-u)) + (\theta_1\,(\kappa_5+(1-u)) + \theta_2\delta)\,((1-\rho)\,\kappa_3+\rho\sigma)}{\kappa_3\kappa_4\,(\kappa_5+(1-u))}\right). \tag{40}
$$

We introduce a positive number ϵ_0 such that $\epsilon_0\mathcal{R}_0(x) \leq 1$, with an appropriate choices of ϵ_0, we have a fuzzy set $\epsilon_0\mathcal{R}_0(x)$ whose fuzzy expected value $FEV\,[\epsilon_0\mathcal{R}_0(x)]$ can be well defined. Therefore, the fuzzy risk reproduction number $\mathcal{R}_0^f(x)$, which can be defined as the average number of secondary risk cases caused by one infected node introduced into entirely susceptible nodes [24], is given by

$$
\mathcal{R}_0^f(x) = \frac{1}{\epsilon_0} FEV[\epsilon_0\mathcal{R}_0(x)]. \tag{41}
$$

According to Verma et al. and Nandi et al., the fuzzy expected value is defined by

$$
FEV\,[\epsilon_0\mathcal{R}_0(x)] = \sup_{0\leq y\leq 1}\, \inf\,[y, k(y)]\,, \tag{42}
$$

where $k(y) = \Phi\{z \in \Omega : \epsilon_0\mathcal{R}_0(x) \geq y\} = \Phi(\Omega)$ is a fuzzy measure. For the purpose of this study, the possibility measure is given by

$$
\Phi(\Omega) = \sup_{x\in X} \Pi(x), \quad \Omega \subset \mathbb{R}. \tag{43}
$$

From $FEV\,[\epsilon_0\mathcal{R}_0(x)]$, for a monotonic function $\mathcal{R}_0(x)$, the set X given as an interval $[x^*, x_{max}]$, we let $x^* \in X$ be the solution of the following equation:

$$
\epsilon_0 c\beta(x)\left(\frac{(\rho\kappa_4\,(\kappa_5+\alpha(x)) + (\theta_1\,(\kappa_5+\alpha(x)) + \theta_2\delta)\,((1-\rho)\,\kappa_3+\rho\sigma))\,\epsilon}{\kappa_3\kappa_4\,(\kappa_5+\alpha(x))}\right) = y. \tag{44}
$$

Thus,

$$
k(y) = \Phi\left[x^*, x_{max}\right] = \sup_{x^*\leq x\leq x_{max}} \Pi(x), \tag{45}
$$

where $k(0) = 1$ and $k(1) = \Pi(x_{max})$.

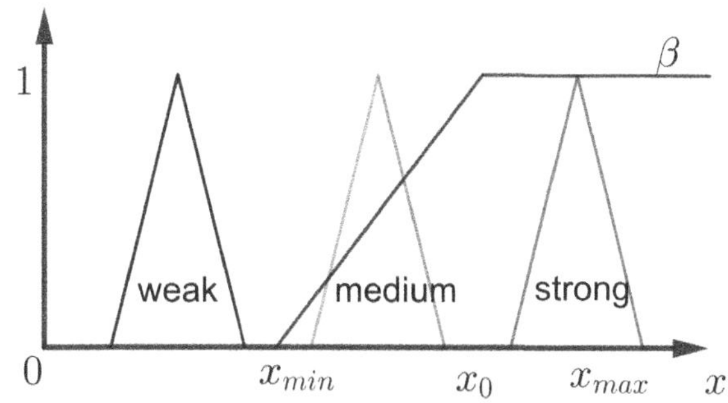

Fig. 3 Classification of linguistic variable Π. *Source* [29]

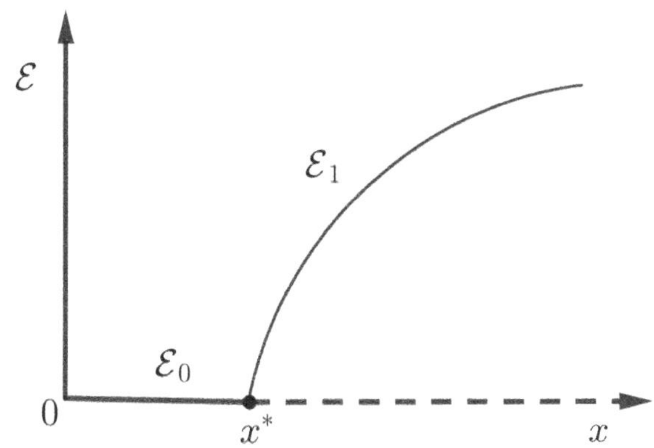

Fig. 4 Bifurcation diagram. *Source* [29]

Now, $FEV(\epsilon_0\mathcal{R}_0)$ can be determined by considering the linguistic variable Π in three classes: "weak Π_-", "medium Π_-^+", and "strong Π^+". Each of these classification is a fuzzy number based on x_{min}, x_0, and x_{max} as appeared in Fig. 3. The classification of the degree of social influences in the community can be explained in three different cases as follows (Fig. 4).

4.2.1 Case One: Weak Social Influence

In this case, we consider the weak degree of social influence (Π_-), where $x_c + d < x_{min}$. Suppose that there exists $\bar{x}$ such that $\mathcal{R}_0(\bar{x}) = \mathcal{R}_0^f(x)$, if $x_c + d < \bar{x}$, we have

$$k(y) = \sup_{\bar{x} \le x \le x_{max}} \Pi(x) = 0, \quad \forall y \in [0, 1]. \tag{46}$$

This implies that $FEV[\epsilon_0\mathcal{R}_0(x)] = 0 < \epsilon_0$. This can be translated that the fuzzy risk reproduction number $\mathcal{R}_0^f(x) < 1$ and hence the extinction of the health risk associated with alcoholism.

4.2.2 Case Two: Medium Social Influence

Here, we consider the medium degree of social influence (Π_{-}^{+}), where $x_c - d > x_{min}$ and $x_c + d < x_0$ giving

$$k(y) = \begin{cases} 1, & \text{if } 0 \leq y < \epsilon_0 \mathcal{R}_0(x_c); \\ \Pi(\bar{x}), & \text{if } \epsilon_0 \mathcal{R}_0(x_c) \leq y \leq \epsilon_0 \mathcal{R}_0(x_c + d); \\ 0, & \text{if } \epsilon_0 \mathcal{R}_0(x_c + d) < y \leq 1. \end{cases} \tag{47}$$

For $d > 0$, $k(y)$ is the continuous function with $k(0) = 1$ and $k(1) = 0$. This translates $\mathcal{R}_0^f(x)$ as the fixed point k and $\mathcal{R}_0(x_c) < \mathcal{R}_0^f(x) < \mathcal{R}_0(x_c + d)$. Since $\mathcal{R}_0(x)$ is a continuous monotonic function, by the Intermediate Value Theorem it follows that there exists $\bar{x}$ with $x_c < \bar{x} < x_c + d$ such that the values of $\mathcal{R}_0(\bar{x})$ and $\mathcal{R}_0^f(x)$ coincide to yield the result $\mathcal{R}_0^f(x) = \mathcal{R}_0(\bar{x}) > \mathcal{R}_0(x_c)$. Furthermore, the average number of fuzzy risk reproduction number $\mathcal{R}_0^f(x)$ is higher than the number of secondary risk cases $\mathcal{R}_0(x_c)$ due to the medium level of social influence implying that the health risk associated with alcoholism is endemic.

4.2.3 Case Three: Strong Social Influence

Finally, the strong degree of social influence (Π^{+}) is considered on $x_c - d > x_0$ and $x_c + d < x_{max}$, and analysis of the problem results in (48) stated below

$$k(y) = \begin{cases} 1, & \text{if } 0 \leq y < \epsilon_0 \mathcal{R}_0(x_c); \\ \Pi(\bar{x}), & \text{if } \epsilon_0 \mathcal{R}_0(x_c) \leq y < \epsilon_0 \mathcal{R}_0(x_c + d); \\ 0, & \text{if } \epsilon_0 \mathcal{R}_0(x_c + d) \leq y \leq 1. \end{cases} \tag{48}$$

For any given $d > 0$, $k(y)$ is a monotonically decreasing and continuous function with $k(0) = 1$ and $k(1) = 0$. Therefore, $FEV[\epsilon_0 \mathcal{R}_0(x)]$ is established as a fixed point such that

$$\epsilon_0 \mathcal{R}_0(x_c) < FEV[\epsilon_0 \mathcal{R}_0(x)] < \epsilon_0 \mathcal{R}_0(x_c + d). \tag{49}$$

Dividing by ϵ_0 throughout, we establish the result

$$\mathcal{R}_0(x_c) < \mathcal{R}_0^f(x) < \mathcal{R}_0(x_c + d). \tag{50}$$

Since $\mathcal{R}_0^f(x) > 1$, this result establishes persistence of health risks associated with alcoholism.

4.3 Risk Control in Fuzzy Epidemic System

In this section, we perform the control analysis of the risk estimation in the population using the fuzzy risk threshold $\mathcal{R}_0^f(x) = \mathcal{R}_0(\bar{x})$. The spread of health risk in the proposed fuzzy model (1) depends on the degree of the social influence x as an input value of the transmission factor $\beta(x)$. The description of the existence and stability of the risk in the system is case-wise presented hereunder. Since the proposed fuzzy system represents a family of systems depending on the parameter x, this family of systems can be simplified by a unique system of equations with the same results. It is shown that the bifurcation occurs at $x = x^*$, corresponding to $\mathcal{R}_0(x^*) = 1$.

1. **Weak influence:** In this case, we have $x < x_{min}$ giving $\mathcal{R}_0(x) = 0$ suggesting the extinction of the health risks associated with alcoholism in the community.
2. **Medium influence:** In this case, three possibilities may arise as follows:

   (i) If $x < x^*$, then $\mathcal{R}_0(x) < 1$, suggesting the risk-free community.
   (ii) If $x = x^*$, then $\mathcal{R}_0(x) = 1$, an indication of risk bifurcation.
   (iii) If $x > x^*$, then $\mathcal{R}_0(x) > 1$, implying the risk endemic state in the community.

3. **Strong influence:** In this case, we have $x \in [x_0, x_{max}]$ giving

$$\mathcal{R}_0(u) = c\epsilon \left(\frac{\rho\kappa_4 (\kappa_5 + (1-u)) + (\theta_1 (\kappa_5 + (1-u)) + \theta_2\delta) ((1-\rho)\kappa_3 + \rho\sigma)}{\kappa_3\kappa_4 (\kappa_5 + (1-u))} \right).$$

   The spread of the health risks depends upon the parameter u. Let u^* be an improved value of u, we can establish three possibilities in which the spread of health risks takes as follows:

   (i) If $0 \le u < u^*$, then $\mathcal{R}_0(u) < 1$, suggesting the health risks would be cleared in the community.
   (ii) If $0 \le u = u^*$, then $\mathcal{R}_0(u) = 1$, implying that the system passes through a bifurcation state.
   (iii) If $0 \le u^* < u$, then $\mathcal{R}_0(u) > 1$, suggesting that the health risk problem would spread out in the system.

5 Discussion and Conclusion

In this chapter, the fuzzy model is proposed and analyzed. Although all the parameters associated with the model system are important, in an uncertain environment, we have considered only the two most important parameters β and α representing the probability that a susceptible individual will drink alcohol after prolonged contact with drinking individuals and the additional death rate induced by alcoholism, respectively. In particular, the two parameters are considered fuzzy numbers and are functions of x (the degree of force of influence, whose membership functions are well defined). The fuzzy risk reproduction number $\mathcal{R}_0^f(x)$ was computed and analyzed to determine conditions for risk clearance and persistence as well as point of exchange of stability.

The risk reproduction number $\mathcal{R}_0(x)$ is directly related to varied factors of drinking settings, defining $\beta(x)$ as a fuzzy set enable analysis to capture risk behavior as characterized by the degree of influence. Analysis of the fuzzy risk reproduction number $\mathcal{R}_0^f(x)$ provided additional information regarding the health risk dynamics. The dynamics of health risks associated with alcoholism can be effectively controlled by controlling the value $\mathcal{R}_0^f(x)$. The analysis of fuzzy model suggests that $\mathcal{R}_0^f(x)$ can be reduced by increasing the value of x_{min}. This can be enhanced through the provision of public health education and reinforcement of social or cultural beliefs which increase the resistance of susceptible individuals and prevent them from initiation into drinking behavior which has the consequences of accelerating their health conditions into more risky states. It is generally observed that if the amount of degree of the peer influence of an individual is low, then the alcohol-related health risks in the community may be reduced. Social norms and beliefs provide virtual immunity from engaging into alcoholic behaviors. On the other hand, if the perceived most influential people in the community engage in alcoholism, it leads to an increased degree of force of influence and consequently persistent endemic health risk.

Future work will extend this work to incorporate some intrinsic parameters associated with different functionalities under an uncertain environment. The present model can be applied to those types of diseases or conditions which spread through direct contact between susceptible and infected individuals.

References

1. Ellison RC, Martinic M (2007) The harms and benefits of moderate drinking: findings of an international symposium. Elsevier
2. WHO (2014) The Global Status Report on Alcohol and Health 2014. World Health Organization
3. Grönbaek M (2009) The positive and negative health effects of alcohol-and the public health implications. J. Int. Med. 265:407–420
4. WHO (2011) World Health Statistics 2011. World Health Organization
5. Health risks and benefits of alcohol consumption (2000) National Institute on Alcohol Abuse and Alcoholism. Alcohol Res Health 24:5–11
6. WHO (2018) The Global Status Report on Alcohol and Health 2018. World Health Organization
7. O'Dwyer C, Mongan D, Millar SR, Rackard M, Galvin B, Long J, Barry J (2019) Drinking patterns and the distribution of alcohol-related harms in Ireland: evidence for the prevention paradox. BMC Publ Health 19:1–9
8. Murray Christopher JL, Lopez, (1996) Alan D and WHO and others.: the global burden of disease: a comprehensive assessment of mortality and disability from diseases, injuries
9. Jernigan D (2004) Alcohol use. Comparative quantification of health risks. Ann 959
10. Huo HF, Liu Y-P (2016) The analysis of the SIRS alcoholism models with relapse on weighted networks. SpringerPlus, p 5
11. Griswold MG, Fullman N, Hawley C, Nicholas A, Zimsen SRM, Tymeson HD, Venkateswaran V, Tapp AD, Forouzanfar MH, Salama JS, others (2018) Alcohol use and burden for 195 countries and territories, 1990–2016: a systematic analysis for the Global Burden of Disease Study 2016. The Lancet *392*, 1015–1035
12. Rossi C (2003) The role of dynamic modelling in drug abuse epidemiology. Bull Narcot 33–44

13. Mundt Marlon P, Mercken L, Zakletskaia L (2012) Peer selection and influence effects on adolescent alcohol use: a stochastic actor-based model. BMC Pediatr 12
14. Wang XY, Huo HF, Kong QK, Shi W-X (2014) Optimal control strategies in an alcoholism model. Abstract and applied analysis. Hindawi Publishing Corporation
15. Xiang H, Song N-N, Huo H-F (2016) Modelling effects of public health educational campaigns on drinking dynamics. J Biol Dyn 10:164–178
16. Adu IK, Mojeeb AL, Yang C (2017) Mathematical model of drinking epidemic. Int J Hum Soc Sci 22
17. Bhunu CP (2012) A mathematical analysis of alcoholism. WJMS 8:124–134
18. Oberguggenberger M, Pittschmann S (1999) Differential equations with fuzzy parameters. Math Comput Model Dyn Syst 5(3):181–202
19. Verma R, Tiwari SP, Upadhyay RK (2017) Dynamical behaviors of fuzzy SIR epidemic model, pp 482–492
20. Sets F (1965) Lotfi Zadeh. Inf Control 8:338–353
21. Barros LD, Leite MF, Bassanezi RC (2003) The SI epidemiological models with a fuzzy transmission parameter. Comput Math Appl 45:1619–1628
22. Nandi SK, Jana S, Manadal M, Kar TK (2018) Analysis of a fuzzy epidemic model with saturated treatment and disease transmission. Int J Biomath 11
23. Verma R, Tiwari SP, Upadhyay RK (2018) Fuzzy modeling for the spread of influenza virus and its possible control. Comput Ecol Soft 8
24. Mishra BK, Pandey SK (2010) Fuzzy epidemic model for the transmission of worms in computer network. Nonlinear Anal: Real World Appl 11(5):4335–4341
25. Morris H, Larsen J, Catterall E, Moss AC, Dombrowski SU (2020) Peer pressure and alcohol consumption in adults living in the UK: a systematic qualitative review. BMC Publ Health 20(1):1–13
26. Anacker AMJ, Ryabinin AE (2010) Biological contribution to social influences on alcohol drinking: evidence from animal models. Int J Environ Res Publ Health **7**(2):473–493
27. Gordis E (1997) The etiology, consequences, and treatment of alcoholism. Liver Transpl Surg 3(3):199–205
28. Sánchez F, Wang X, Castillo-Chávez C, Gorman Dennis M, Gruenewald Paul J (2007) Drinking as an epidemic–a simple mathematical model with recovery and relapse. Therapist's guide to evidence-based relapse prevention. Elsevier, pp 353–368
29. Mayengo Maranya M, Kgosimore M, Chakraverty S (2020) Fuzzy modeling for the dynamics of alcohol-related health risks with changing behaviors via cultural beliefs. J Appl Math 1–9
30. van den Driessche P, Watmough J (2002) Reproduction numbers and sub-threshold endemic equilibria for compartmental models of disease transmission. Math Biosci 180:29–48
31. Diekmann O, Heesterbeek JAP, Roberts MG (2009) The construction of next-generation matrices for compartmental epidemic models. J R Soc Interface

Curriculum Learning-Based Artificial Neural Network Model for Solving Differential Equations

Arup Kumar Sahoo and S. Chakraverty

Abstract This chapter is dedicated to studying the impact of the curriculum learning process and the Swish activation function for finding the solution of Differential Equations (DEs) with initial conditions. Then we have compared the result of the proposed training algorithm and the usual training algorithm. Also, we have compared the neural result using Swish, Tanh, and Sigmoid activation functions. The artificial neural network (ANN) trial solution of the differential equation is written as a sum of two terms. Here the first term satisfies boundary conditions and the second term involves containing adjustable parameters so as the trial solution can solve the differential equations. In this investigation, first we have trained our neural network in a small domain and gradually expanded the domain. Feedforward neural network (FFNN) and error backpropagation algorithm have been used to minimize the error function and modification of weights and biases. Finally, several problems have been solved to illustrate the proposed training method, and analytical results have been compared with neural results.

1 Introduction

A wide-scale application of various phenomena in physics, chemical, medicine, finance, trading, economics, and engineering is well described by ANN. To find the solution of differential equations (DEs) is another field where ANN has been convincingly used in the past few years. DEs play a vital role in solving real-world problems. In past decades, scientists and engineers have used traditional numerical methods, such as finite difference, predictor–corrector, Euler, Runge–Kutta, Monte Carlo, finite element, etc., to solve DEs. These methods provide a satisfactory approximation to the solution for DEs [1].

But in comparison to traditional numerical methods, ANN-based solutions have advantages. In the conventional numerical method, the solution is discrete in nature,

A. K. Sahoo (✉) · S. Chakraverty
Department of Mathematics, National Institute of Technology Rourkela, Rourkela 769008, India

S. Chakraverty (ed.), *Soft Computing in Interdisciplinary Sciences*,
Studies in Computational Intelligence 988,
https://doi.org/10.1007/978-981-16-4713-0_6

where the ANN model-based solution is continuous over the given domain of integration. We can use it as a black box to get numerical solutions at any arbitrary point in the given domain after training the ANN model. Like human brain neural network provides a better result when we started training from small and gradually increase the difficulty level [2]. In the field of machine learning, this is known as "Curriculum Learning" [3]. In this investigation, we discuss the effect of curriculum learning in solving DEs.

First, we go through a few investigations made by researchers and scientists using the ANN model to solve DEs in the last few years.

In 1943, a neuroscientist Warren s. McCulloch and logician Walter Pits developed the first elementary model of ANN [4]. In their historical paper, they described a logical calculus of nervous activities that merged the studies of neurophysiology and mathematical logic. In their model, neurons escort by the "all-or-none" process. Lee and Kang [5] first introduced a neural model using Hopfield neural network to solve first-order DEs. Neural network methods for solving boundary value problems with irregular boundaries were first introduced by Lagaris et al. [6]. Many essential works on neural network model to solve DEs have been done in past years, and numerous research papers had been developed by various authors [6–14]. In recent work, Chakraverty and Mall [15] solved nonlinear ordinary differential equations using single-layer ChNN with regression-based weights.

2 Artificial Neural Network

Artificial neural network (ANN) is an exciting form of artificial intelligence (AI) that mimics the human brain's training process to predict patterns from the given factual data. Neural networks are processing devices of mathematical algorithms that can be implemented by computer languages.

ANN depends upon various learning processes and different parameters [16–18]. The neural network is made up of layers, and layers are made up of several neurons/nodes. The input layer receives signals and is multiplied by the interneuron connections called weights, then the signals are summed and passed to one or more hidden layers. The summed value passes through an activation function and gives an output of the hidden layer. Hidden layers send links to the output layer.

ANN similarly processes data as the human brain. It reads input, output values of the training data set, and update numerical values of weights to increase the accuracy of neural predicted value. One complete cycle of passing training algorithm and update weights is called epoch or iteration. The error is minimized using many epochs. However, if a network trains for an extended period, it losses its ability to generalize.

3 Curriculum Learning

Learning becomes easy for a human when a teacher has started from easy examples and gradually makes the task difficult. Similarly, the network fails to train correctly when the entire dataset is presented simultaneously and succeeds when the data are presented incrementally [2]. Using this training strategy in the context of machine learning is known as "Curriculum Learning."

Especially when the training dataset is large, curriculum learning shows better convergence than normal training. In our experiment, we trained the network in a small domain and gradually expanded the domain. Comparison results between curriculum training and unsupervised normal training to solve different DEs are given in tables.

4 General Formulation for Differential Equations

Let us consider the general form of ODEs as

$$G\big(x, \phi(x), \phi'(x), \phi''(x).......\big) = 0, \quad x \in \overline{D} \subset R^n \tag{1}$$

where G denotes a function defining the structure of a differential equation,

$\phi(x)$ denotes the solution,

and $\overline{D}$ is the discretized domain over a finite set of points.

Let $\phi_t(x, p)$ indicate the ANN trail solution of ODE with adjustable parameters p (weights and biases), then Eq. (1) can be rewritten as

$$G\left(x_i, \phi_t(x_{i.}p), \phi_t'(x_{i.}p), \phi_t''\left(x_{i.}p\right).......\right) = 0 \tag{2}$$

The corresponding error function of ANN can be written as

$$\text{Min} \sum_{x_i \in \overline{D}}^{n} \left(G\left(x_i, \phi_t(x_{i.}p), \phi_t'(x_{i.}p), \phi_t''\left(x_{i.}p\right).......\right)\right)^2 \tag{3}$$

Here ANN trail solution $\phi_t(x, p)$ can be written as the sum of two terms

$$\phi_t(x, p) = \alpha + F\left(x, N(x, p)\right) \tag{4}$$

where α satisfies initial or boundary conditions without adjustable parameters. $N(x, p)$ is an output of feedforward neural network with parameters, and the second term $F(x, N(x, p))$ does not contribute to initial and boundary conditions. It is a neural network model whose weights and biases are adjusted to deal with the minimization of the error function.

Now consider a multilayer neural network with a single input node, one hidden layer (m nodes), and a linear output unit. For a given input $x \in R^n$, the output of ANN is denoted by

$$N(x, p) = \sum_{j=1}^{m} v_j \sigma(z_j) \tag{5}$$

where $z_j = \sum_{i=1}^{n} w_{ji} x_i + b_j$

Here v_j represents the weight from the hidden unit j to the output unit,
w_{ji} represents the weight from the input unit i to the hidden unit j,
b_j represents the biases, and
$\sigma(z_j)$ represents an activation function.

4.1 Construction for First-Order IVP

Let us consider the first-order ordinary differential equation as given below:

$$\frac{d\phi}{dx} = \mathrm{f}\,(\mathrm{x},\phi\,)\,,\ \mathrm{x} \in [\mathrm{a},\mathrm{b}] \tag{6}$$

with initial condition $\phi(a) = \alpha$

Here the ANN trail solution may be written as

$$\phi_t(x, p) = \alpha + (x - a)N(x, p) \tag{7}$$

Also, Mattheakis et al. [19] proposed ANN trail solution which can be written as

$$\phi_t(x, p) = \alpha + (1 - e^{-(x - x_0)})N(x, p) \tag{8}$$

where $N(x, p)$ is the neural output of the FFNN with one input data x with parameters p and the ANN trial solution $\phi_t(x, p)$ satisfies the initial condition of given DE.

The error function for the first-order ODE may be calculated as

$$E(p) = \sum_{i=1}^{n} \left(\frac{d\phi_t(\mathrm{x_i}, \mathrm{p})}{\mathrm{dx}} - f(x_i, \phi_t(x_i, p)) \right)^2 \tag{9}$$

4.2 Construction for Second-Order IVP

Let us consider second-order ordinary differential equation as given below:

$$\frac{d^2\phi}{dx^2} = \mathrm{f}\,(\mathrm{x},\phi,\frac{d\phi}{dx}),\ \mathrm{x}\ \in\ [\mathrm{a},\mathrm{b}], \tag{10}$$

with the initial condition $\phi(a) = \alpha, \phi'(a) = \alpha'$.

Here the ANN trail solution may be written as

$$\phi_t(x,p) = \alpha + \alpha'(x-a) + (x-a)^2 N(x,p), \tag{11}$$

where $N(x, p)$ is the neural output of the FFNN with one input data x with parameters p. The ANN trial solution $\phi_t(x, p)$ satisfies the initial condition of the given DE.

The error function for second-order ODE may be calculated as

$$E(p) = \sum_{i=1}^{n} \frac{1}{2}\left(\frac{d^2\phi_t(\mathrm{x_i},\mathrm{p})}{\mathrm{dx}^2} - f\left(x_i, \phi_t(x_i,p), \frac{d\phi}{dx}\right)\right)^2 \tag{12}$$

4.3 Construction for Second-Order BVP

Let us consider second-order ordinary differential equation as given below:

$$\frac{d^2\phi}{dx^2} = \mathrm{f}\,(\mathrm{x},\phi,\frac{d\varphi}{dx}),\ \mathrm{x}\ \in\ [\mathrm{a},\mathrm{b}], \tag{13}$$

with the initial condition $\varphi(a) = \alpha, \varphi(b) = \beta$.

Here the ANN trail solution may be written as

$$\varphi_t(x,p) = \frac{b\,\alpha - a\beta}{b-a} + \frac{\beta-\alpha}{b-a} + (x-a)(x-b)N(x,p) \tag{14}$$

where $N(x, p)$ is the neural output of the FFNN with one input data x with parameters p. The ANN trial solution $\varphi_t(x, p)$ satisfies the boundary conditions of the given DE.

As such, the error function for second-order ODE may be calculated as

$$E(p) = \sum_{i=1}^{n} \frac{1}{2}\left(\frac{d^2\varphi_t(\mathrm{x_i},\mathrm{p})}{\mathrm{dx}^2} - f\left(x_i, \varphi_t(x_i,p), \frac{d\varphi}{dx}\right)\right)^2 \tag{15}$$

In all the above cases, weights are updated as the unsupervised backpropagation learning algorithm.

5 First-Order ODEs

Here we have taken two first-order ODEs to show the reliability of curriculum training and Swish activation function [20]. The accuracy of the result has been shown in the tables and graphs. The mean squared error (MSE) for every case has been calculated and written in the table.

Example 1 Let us consider a differential equation

$$\frac{d\gamma}{dt} = \cos t$$

with initial conditions $\gamma(0) = 0$.

We have trained the network for ten equispaced points in [0,5] using curriculum learning and a normal training process. A comparison between analytical results, ANN results of normal training, and using curriculum training is given in Table 1. Also, we compared between ANN results using Swish activation function, Sigmoid activation function, and tanh activation function as given in Table 2.

Analytical and neural results with normal training and curriculum training are delineated in Fig. 1. Analytical results and ANN results using Swish activation func-

Table 1 Analytical and neural results of normal training and using curriculum training (Example 1)

Neural results			
Input data	Analytical result	Normal training	Curriculum training
0	0.0	0.0	0.0
0.5	0.479426	0.475761	0.479996
1.0	0.841471	0.843328	0.841693
1.5	0.997495	1.003361	0.997806
2.0	0.909297	0.914455	0.910308
2.5	0.598472	0.603841	0.598301
3.0	0.141120	0.146295	0.141045
3.5	−0.350783	−0.349553	−0.349735
4.0	−0.756803	−0.757479	−0.757773
4.5	−0.977530	−0.970576	−0.978441
5.0	−0.958924	−0.944094	−0.959348

Normal training mean squared error = 0.000036705974496466 32
Curriculum training mean squared error = 0.0000004160270524825938

Table 2 Analytical and neural results using Tanh, Sigmoid, and Swish activation functions (Example 1)

Neural results				
Input data	Analytical result	Tanh	Sigmoid	Swish
0	0.0	0.0	0.0	0.0
0.5	0.479426	0.481343	0.481357	0.481357
1.0	0.841471	0.843613	0.843636	0.843636
1.5	0.997495	0.994682	0.997478	0.997478
2.0	0.909297	0.906310	0.909830	0.909830
2.5	0.598472	0.602723	0.601319	0.601319
3.0	0.141120	0.147248	0.143174	0.143174
3.5	−0.350783	−0.353585	−0.351767	−0.351767
4.0	−0.756803	−0.761499	−0.757179	−0.757179
4.5	−0.977530	−0.965392	−0.971852	−0.971852
5.0	−0.958924	−0.941236	−0.950415	−0.950415

Activation function Tanh mean squared error = 0.000051895508050002679
Activation function Sigmoid mean squared error = 0.0010148524221728931
Activation function Swish mean squared error = 0.0000115255968653419 82

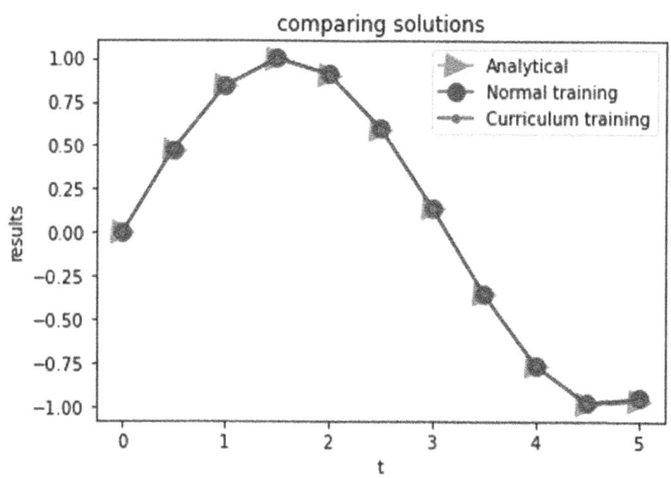

Fig. 1 Plot of analytical and neural results using normal training and curriculum training (Example 1)

tion, Sigmoid activation function, and tanh activation functions are compared in Fig. 2.

Example 2 We consider following first-order ODE:

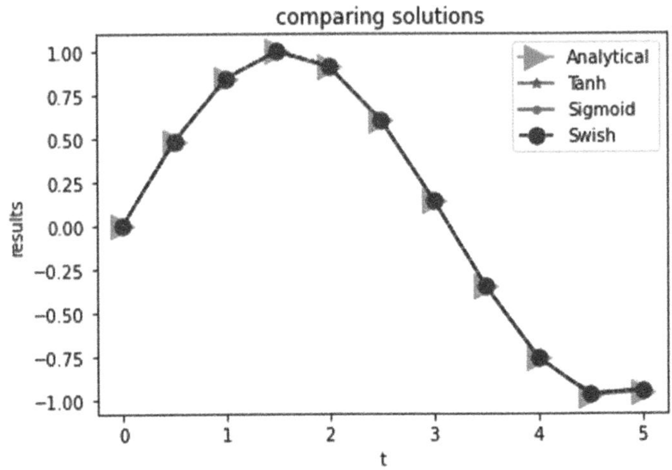

Fig. 2 Plot of analytical and neural results using Tanh, Sigmoid, and Swish activation functions (Example 1)

$$\frac{d\gamma}{dt} + t\gamma + \left(\frac{1+3t^2}{1+t+t^3}\right)\gamma - t^3 - 2t - t^2\left(\frac{1+3t^2}{1+t+t^3}\right) = 0, \quad t \in [0, 3]$$

with the initial condition $\gamma(0) = 1$.

The network has been trained for ten equispaced points in [0,3] using curriculum learning and a normal training process. Table 3 shows a comparison between analytical results, ANN results of normal training, and using curriculum training. Table 4 shows a comparison between ANN results using Swish activation function, Sigmoid activation function, and tanh activation function.

One very well observes that curriculum training (fourth column, Table 3) has a better result than normal training from Table 3. The Swish activation function (fifth column, Table 4) has a better impact than the other two activation functions.

Analytical solutions are compared with neural solutions with normal training and curriculum training in Figs. 3 and 4, respectively. Analytical results and ANN results using different activation functions are compared in Fig. 5.

6 Higher Order ODEs

In this section, two higher order ODEs have been taken to show the reliability of curriculum training and Swish activation function [20]. The mean squared error (MSE) for every case has been calculated and is written in the table.

Example 3 We consider the following second-order damped free vibration equation:

Table 3 Analytical and neural results of normal training and using curriculum training (Example 2)

Neural results			
Input data	Analytical result	Normal training	Curriculum training
0	1.0	1.0	1.0
0.3	0.81042	0.756999	0.809361
0.6	0.819951	0.80905	0.820128
0.9	1.0637	1.100237	1.064433
1.2	1.563919	1.603659	1.572241
1.5	2.30526	2.313963	2.311399
1.8	3.262926	3.239454	3.257197
2.1	4.418919	4.39109	4.406916
2.4	5.763259	5.766145	5.764725
2.7	7.291117	7.32798	7.315835
3.0	9.000358	8.991758	8.996914

Normal training mean squared error = 0.0007935675674768702
Curriculum training mean squared error = 0.00008277253457239567

Table 4 Analytical and neural results using Tanh, Sigmoid, and Swish activation functions (Example 2)

Neural results				
Input data	Analytical result	Tanh	Sigmoid	Swish
0	1.0	1.0	1.0	1.0
0.3	0.810420	0.756999	0.600148	0.824880
0.6	0.819951	0.809050	0.590097	0.837393
0.9	1.063700	1.100237	0.897763	1.073747
1.2	1.563919	1.603659	1.475683	1.551851
1.5	2.305260	2.313963	2.291868	2.275423
1.8	3.262926	3.239454	3.321133	3.236872
2.1	4.418919	4.391090	4.536651	4.418765
2.4	5.763259	5.766145	5.902577	5.794302
2.7	7.291117	7.327980	7.370000	7.328375
3.0	9.000358	8.991758	8.878694	8.980761

Activation function Tanh mean squared error = 0.0007935675674768702
Activation function Sigmoid mean squared error = 0.017293658420688387
Activation function Swish mean squared error = 0.00046044716962774794

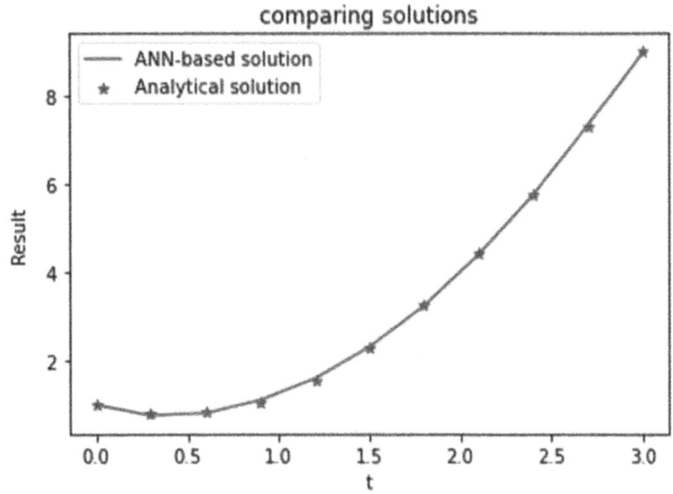

Fig. 3 Plot of analytical and neural results using normal training (Example 2)

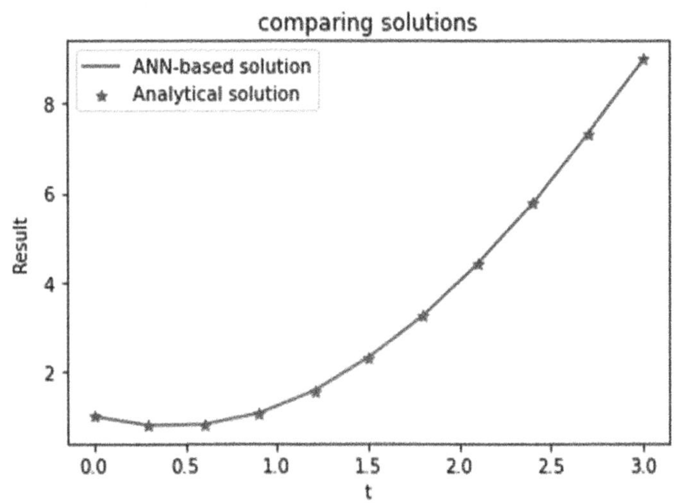

Fig. 4 Plot of analytical and neural results using curriculum training (Example 2)

$$\frac{d^2\lambda}{dt^2} + 4\frac{d\gamma}{dt} + 4\gamma = 0, \quad t \in [0, 4].$$

With initial conditions $\gamma(0) = 1, \ \gamma'(0) = 1$.

The network is trained for ten equispaced points in [0,4] using curriculum learning and a normal training process. ANN results of normal training, and using curriculum training is given in Table 5 and result is depicted in Figs. 6 and 7. Also, ANN results

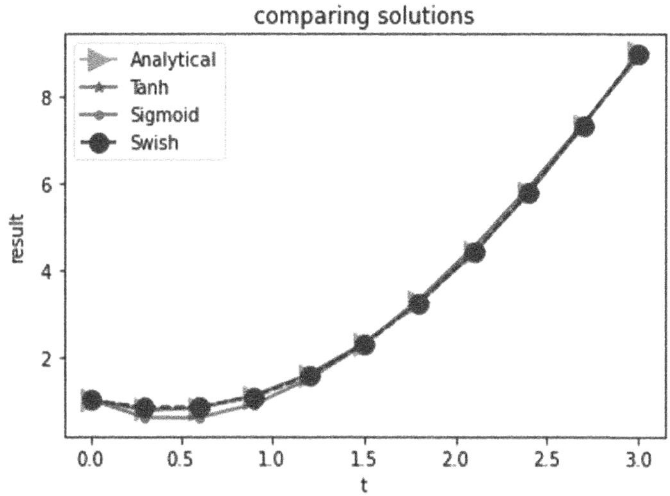

Fig. 5 Plot of analytical and neural results using Tanh, Sigmoid, and Swish activation functions (Example 2)

Table 5 Analytical and neural results of normal training and using curriculum training (Example 3)

Neural results			
Input data	Analytical result	Normal training	Curriculum training
0	1.0	1.0	1.0
0.4	0.988524	0.991530	0.988296
0.8	0.686448	0.683186	0.685969
1.2	0.417303	0.415273	0.417167
1.6	0.236421	0.239771	0.236632
2.0	0.128209	0.133688	0.127396
2.4	0.067484	0.069790	0.066542
2.8	0.034760	0.030796	0.035096
3.2	0.017613	0.007611	0.018230
3.6	0.008810	-0.003037	0.007657
4.0	0.004361	0.000062	0.002402

Normal training mean squared error = 0.000031354284410 55342.
Curriculum training mean squared error = 0.0000006869530293823444.

using Swish activation function, Sigmoid activation function, and tanh activation function are given in Table 6, and the result is plotted in Fig. 8.

For the experiment of Table 6, the authors have taken the same number of nodes in the hidden layer (here 10 nodes) and the same number of epoch (here 5000) for all three activation functions.

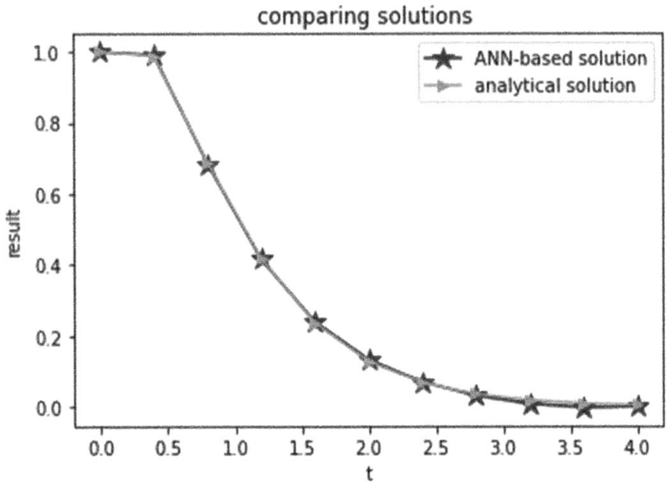

Fig. 6 Plot of analytical and neural results using normal training (Example 3)

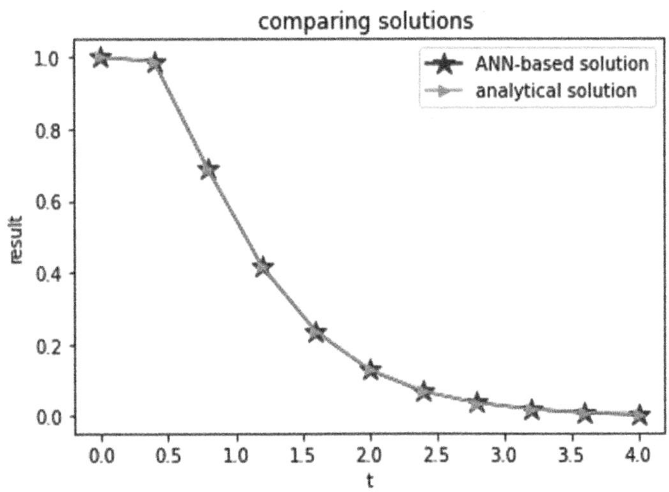

Fig. 7 Plot of analytical and neural results using curriculum training (Example 3)

Example 4 Consider a second-order ODE.

$$2\frac{d^2\gamma}{dt^2} + \frac{d\gamma}{dt} = 2t^2 + 3t + 1$$

with initial conditions $\gamma(0) = 0,\ \gamma'(0) = 0.$

Here, the network is trained for ten equispaced points in [0,3] using curriculum learning and a normal training process. A comparison of analytical results, ANN

Table 6 Analytical and neural results using Tanh, Sigmoid, and Swish activation functions (Example 3)

Neural results				
Input data	Analytical result	Tanh	Sigmoid	Swish
0	1.0	1.0	1.0	1.0
0.4	0.988524	1.007866	1.048775	0.985585
0.8	0.686448	0.693143	0.770235	0.673432
1.2	0.417303	0.390335	0.455287	0.401844
1.6	0.236421	0.190283	0.202331	0.228593
2.0	0.128209	0.089403	0.030744	0.131001
2.4	0.067484	0.052837	−0.068424	0.076688
2.8	0.034760	0.042274	−0.110371	0.043208
3.2	0.017613	0.030168	−0.107583	0.019729
3.6	0.008810	0.009075	−0.067205	0.003563
4.0	0.004361	−0.002477	0.008279	−0.003788

Activation function Tanh mean squared error = 0.00047783892427601425
Activation function Sigmoid mean squared error = 0.007614245109648199
Activation function Swish mean squared error = 0.00006732651464612434

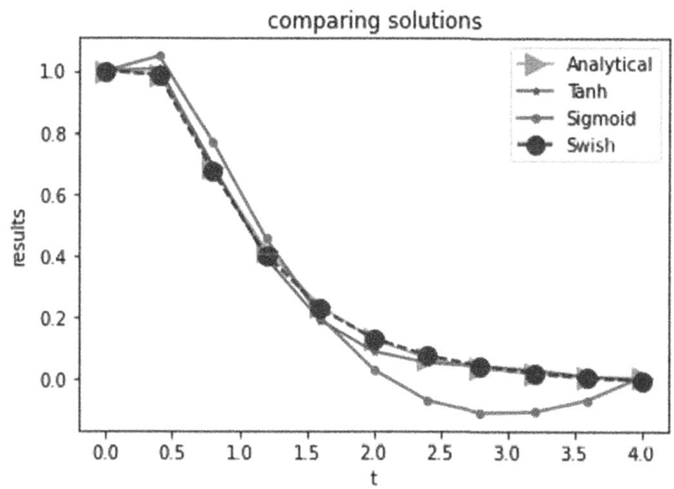

Fig. 8 Plot of analytical and neural results using Tanh, Sigmoid, and Swish activation functions (Example 3)

results of normal training, and using curriculum training is incorporated as given in Table 7. And also, a comparison of ANN results using Swish activation function, Sigmoid activation function, and tanh activation function is incorporated as given in Table 8, and the result is plotted in Fig. 11.

Table 7 Analytical and neural results of normal training and using curriculum training (Example 4)

Neural results			
Input data	Analytical result	Normal training	Curriculum training
0	0.0	0.0	0.0
0.3	0.028575	0.030396	0.028062
0.6	0.142001	0.145363	0.143478
0.9	0.388819	0.391383	0.391883
1.2	0.825856	0.827222	0.827463
1.5	1.517064	1.512888	1.516775
1.8	2.532533	2.521408	2.533330
2.1	3.947630	3.935866	3.949491
2.4	5.842273	5.830725	5.839088
2.7	8.300286	8.278417	8.292136
3.0	11.408864	11.383437	11.405857

Normal training mean squared error = 0.00014188686845 62258
Curriculum training mean squared error = 0.0000094714910 31338204

Table 8 Analytical and neural results using Tanh, Sigmoid, and Swish activation functions (Example 4)

Neural results				
Input data	Analytical result	Tanh	Sigmoid	Swish
0	0.0	0.0	0.0	0.0
0.3	0.028575	0.029796	0.030971	0.019575
0.6	0.142001	0.142990	0.158257	0.127004
0.9	0.388819	0.385243	0.421536	0.380271
1.2	0.825856	0.813904	0.862798	0.825551
1.5	1.517064	1.492158	1.538205	1.516860
1.8	2.532533	2.493676	2.522899	2.520322
2.1	3.947630	3.897338	3.910893	3.916728
2.4	5.842273	5.776253	5.807618	5.802074
2.7	8.300286	8.209490	8.309469	8.276902
3.0	11.408864	11.309785	11.465826	11.419028

Activation function Tanh mean squared error = 0.0024760455507982708
Activation function Sigmoid mean squared error = 0.0008294929669745246
Activation function Swish mean squared error = 0.0003408381206203553 3

Analytical and neural results with normal training and curriculum training are delineated in Figs. 9 and 10, respectively. We had also examined neural results with normal training and curriculum training using different activation functions with the same number of nodes in the hidden layer and the same epoch. We find neural results using curriculum training have better convergence.

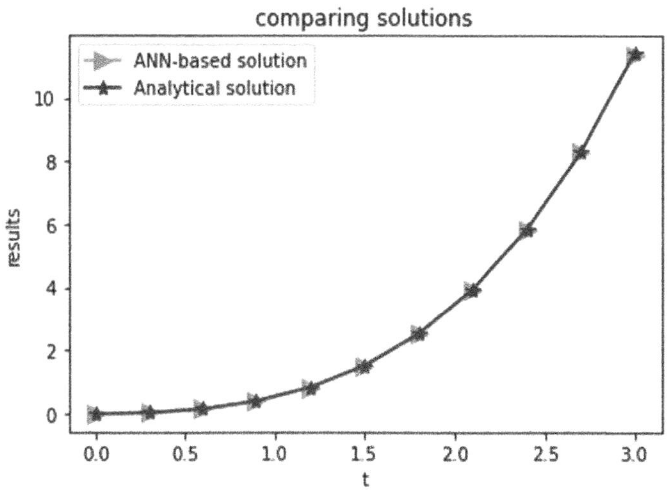

Fig 9 Plot of analytical and neural results using normal training (Example 4)

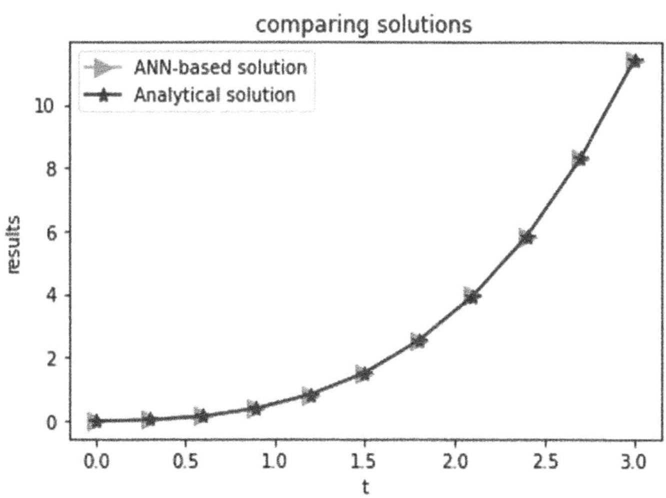

Fig 10 Plot of analytical and neural results using curriculum training (Example 4)

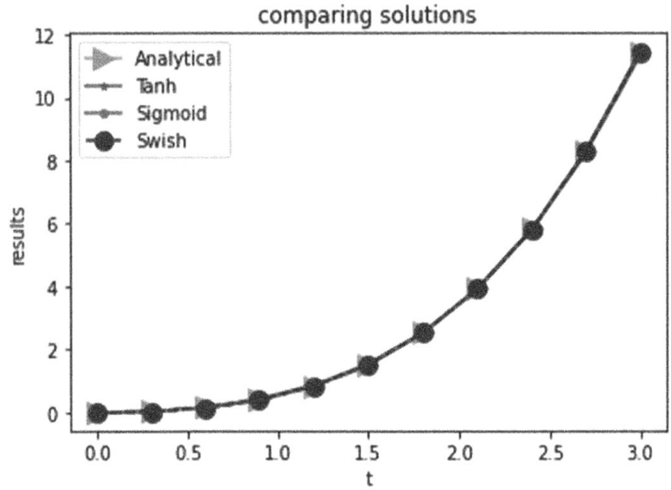

Fig 11 Plot of analytical and neural results using Tanh, Sigmoid, and Swish activation functions (Example 4)

7 Conclusion

This chapter shows the effect of curriculum learning to train neural networks for solving DEs. It has also investigated better use of activation function for the training of a neural network. The advantages of this learning method were examined by first-order DEs and second-order damped free vibration equations. It may be seen from the tables and calculated MSE that curriculum learning and Swish activation function make the ANN results more accurate. Lastly, it may be mentioned that though curriculum learning takes more time for execution, it is easy to implement and is computationally efficient.

Acknowledgements The first author is thankful to the Council of Scientific and Industrial Research (CSIR), New Delhi, India, for the support and funding to carry out the present research work.

References

1. Chakraverty S, Mall S (2017) Artificial neural networks for engineers and scientists: solving ordinary differential equations. Taylor and Francis, Boca Raton
2. Elman JL (1993) Learning and development in neural network: The importance of starting small. Cognition 48:781–799
3. Bengio Y, Louradour J, Collobert R, Weston J (2009) Curriculum learning. In: Proceeding of the 26th International Conference on Machine Learning. ACM, New York, , 41–48.
4. McCulloch WS, Pitts W (1943) A logical calculus of the ideas immanent in nervous activity. Bull Math Biophys 5(4):115–133

5. Lee H, Kang IS (1990) Neural algorithms for solving differential equations. J Comput Phys 91(1):110–131
6. Lagaris IE, Likas A, Fotiadis DI (1998) Artificial neural networks for solving ordinary and partial differential equations. IEEE Trans Neural Networks 9(5):987–1000
7. Lagaris IE, Likas AC, Papageorgiou DG (2000) Neural network methods for boundary value problems with irregular boundaries. IEEE Trans Neural Networks 11(5):1041–1049
8. Parisi DR, Mariani MC, Laborde MA (2003) Solving differential equations with unsupervised neural networks. Chem Eng Process 42(8–9):715–721
9. Tsoulos IG, Lagaris IE (2006) Solving differential equations with genetic programming. Genet Program Evolvable Mach 7(1):33–54
10. Choi B, Lee J-H (2009) Comparison of generalization ability on solving differential equations using back-propagation and reformulated radial basis function networks. Neurocomputing 73(1–3):115–118
11. Mall S, Chakraverty S (2013) Comparison of artificial neural network architecture in solving ordinary differential equations. Adv Artif Neural Syst 1–24:2013
12. Mall S, Chakraverty S (2013) Regression-based neural network training for the solution of ordinary differential equations. Int J Math Model Numer Optim 4(2):136–149
13. Chakraverty S, Mall S (2014) Regression based weight generation algorithm in neural network for solution of initial and boundary value problems. Neural Comput Appl 25(3):585–594
14. Panghal S, Kumar M (2020) Optimization free neural network approach for solving ordinary and partial differential equations. Eng Comput. https://doi.org/10.1007/s00366-020-00985-1
15. Chakraverty S, Mall S (2020) Single layer Chebyshev neural network model with regression-based weights for solving nonlinear ordinary differential equations. Evol Intell 1–8.
16. Zurada JM (1992) Introduction to artificial neural systems. West Publishing, St. Paul
17. Haykin SS (1999) Neural networks: a comprehensive foundation. Prentice Hall Inc., Upper Saddle River
18. Chakraverty S, Sahoo DM, Mahato NM (2019) Concepts of soft computing: fuzzy and ANN with programming. Springer, Singapore
19. Mattheakis M, Protopapas P, Sondak D, Di Giovanni M, Kaxiras E (2020) Physical symmetries embedded in neural networks. Preprint at https://arxiv.org/abs/1904.08991
20. Ramachandran P, Zoph B, Le QV (2017) Searching for activation functions. Preprint at https://arxiv.org/abs/1710.05941

Analysis of EEG Signal for Drowsy Detection: A Machine Learning Approach

B Venkata Phanikrishna and Suchismita Chinara

Abstract Human body functions require brain activities. Neurons present in the brain control these activities. There are over 86 billion neurons in the brain, and these neuron's interactions are responsible for the brain activities. Neurocomputing is a computational process to investigate the pattern of brain signals for various purposes. It is an exciting and interesting research topic for neuroscience and computer science field researchers to visualize human behavioral actions through neural activity. Neurocomputing has established its identity by analyzing brain signals for clinical and non-clinical applications using machine learning and deep learning. Here, brain signals are extracted, and relevant features are collected to classify. There are some brain imaging signals, such as Electroencephalogram (EEG), Positron Emission Tomography (PET)—obtrusive atomic imaging strategy based on gamma radiation, metabolic screen action, Magnetoencephalography (MEG)—gather the Magnetic fields created by neural movement, and Functional Magnetic Resonance Imaging (fMRI)—measures changes in the bloodstream related with neural activity to perceive brain activities. In non-clinical applications such as motor imaging, drowsiness detection and emotion detection, electroencephalography (EEG) signals are used more frequently than other brain signals. EEG signals have a good temporal resolution, versatile, lightweight, and easy to acquire and use. Moreover, these signals refer to nerve activity without surgery and can mobilize the mental activity. The main theme of this chapter is to understand how the EEG-based methods for detecting drowsiness were developed for sleep analysis and prevent drowsiness-related risks by combining neuroscience and computer science knowledge.

B. Venkata Phanikrishna (✉) · S. Chinara
Computer Science & Engineering, NIT Rourkela, Rourkela 769008, Odisha, India
e-mail: b.phanikrishna@gmail.com

S. Chinara
e-mail: suchismita@nitrkl.ac.in

S. Chakraverty (ed.), *Soft Computing in Interdisciplinary Sciences*,
Studies in Computational Intelligence 988,
https://doi.org/10.1007/978-981-16-4713-0_7

1 Introduction

Drowsiness is an inattentiveness state, and it is a very common as well as short period state during the transition from being awake to a sleepy state. Drowsiness decreases an individual's strength and alertness, which raises the chances of accidents while they are involved in personal or professional activities such as driving a vehicle, operating a crane, working with heavy machinery in large industries such as steel plants, mine blasts, and so on [1]. The Drowsiness Detection (DD) method is helpful in preventing accidents since a person with drowsiness can regain alertness with regular stimulation. However, human drowsiness is a vague event that does not occur at specific timings. Therefore, it is necessary to monitor the human mental state continuously [2]. The brain is one of the most critical organs in the human body. The signals produced in the brain are more correlated to every function of human activities and are reliable to determine the onset of drowsiness [3]. Furthermore, brain signal acquisition is easy and can be in the form of

- Fashion cap used to protect against heat and sunlight.
- Wearing safety mine hats and helmets while driving.
- Headphones while listening to songs.

To acquire brain signals, Kaplan et al. [4] have suggested the following clinical brain analysis techniques:

- **EEG**: Electroencephalogram is the process of gathering and monitoring electrical motions of the cerebrum.
- **PET**: Positron Emission Tomography is the obtrusive atomic imaging strategy based on gamma radiation, screen metabolic action.
- **MEG**: Magnetoencephalography is the collection of magnetic fields created by nerve motion.
- **fMRI**: Functional Magnetic Resonance Imaging measures changes in blood stream related with neural activity.

Among these brain diagnostic techniques, EEG is proven to be a flexible procedure for both therapeutic and non-clinical work [5]. Hans Berger invented electroencephalography in 1924, a method of obtaining brain waves [6]. He also discovered how EEG signals are associated with various brain diseases. This discovery paves the way for researchers to understand brain wave activity in relation to multiple conditions and explore multiple possibilities of using them in different applications. The EEG is used in many research areas in cognitive, human-robot interaction, epilepsy, and sleep-related problems [7]. EEG benefits are given as follows [8]:

- EEG specifically measures neural activity.
- EEG is economically inexpensive, lightweight, and additionally versatile (adaptable for gathering signals in real time), and easy to obtain.
- EEG refers to the bio-electrical potential of the mind, which changes to various parameters of conditions inside and outside the human being. It helps to monitor what is happening before, during, and after a specific action in the brain.

- Due to the high sampling rate, more information can be obtained with EEG in a short time. It helps to disclose much about the sequence of mental processes.
- EEG can capture basic and physiological changes with high time resolution. Intellectual (logical) approaches are better compared to other cerebrum imaging techniques such as MRI or PET scanners.

After reading this chapter, the reader will know an overview of the single-channel EEG-based drowsiness detection method through EEG signal analysis and its processing methods. It also understands the role of machine learning and deep learning techniques using hand-engineered and automated features in detecting drowsiness.

2 Background: Drowsiness Detection

DD methods developed for defense purposes are categorized into four groups based on the parameters used in that method. Those categories are as follows:

- Subjective based: In subjective-based methods, a person's drowsiness is measured using different sleep scalings, with answers that come from questioning with direct interactions, as indicated in [9–11]. These methods are low-cost and easy to implement. However, frequent questionnaires and interactions would always make a drowsy person active. So the detection of drowsiness through this method is unrealistic from an implementation point of view.
- Vehicle based: In vehicle-based DD methods, driver drowsiness is detected based on vehicle movement patterns while driving the vehicle. This monitoring is done by some sensors which are attached to certain parts of the vehicle [12]. These include Steering Wheel Movement (SWM), Speed Variability (SV), and Standard Deviation of Lateral Position (SDLP). The parameters of these methods are related to vehicles only and are harmless and flexible to the users. However, these methods depend on the vehicle type and road geometry. Therefore, it is expensive and impossible to use them on ships and aircraft. Moreover, the current driver's driving style needs to be trained before using the model. It is not possible to find micro-sleep conditions that are more frequent on straight roads.
- Behavioral or Visual based: In Behavioral-based (also known as visual-based) methods, individual behavioral movements are captured by the camera to detect the drowsiness [13]. Here, cameras are used to collect subject behavior while driving [14]. Drowsiness-related patterns are collected from these images. These methods are non-intrusive; there is no need to interrupt the person physically or mentally. Behavioral techniques can be generalized compared to subjective- and vehicle-based methods. However, these methods are not found to be reliable in some cases. Especially in mining areas, in certain lighting, and in varying foggy conditions, it is difficult to capture images and videos using cameras. It is also expensive to process and classify the captured images and videos for DD. On the potholed roads, it is very difficult to detect drowsiness based on head movement, as well.

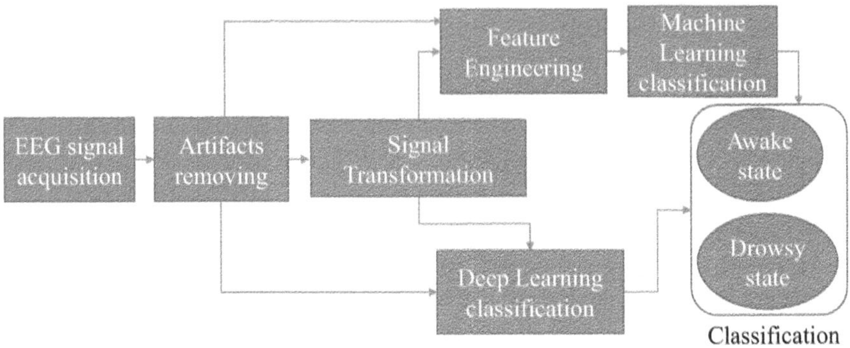

Fig. 1 Typical process flow of an EEG-based drowsiness detection system

- Physiological or Non-visual based: Physiological-based (also known as non-visual-based) methods identify drowsiness based on the analysis of the physiological signals produced by the human body. Here, bio-sensors are used to obtain non-visual signals related to the heart, skin, blood, muscles, head, and other organs that are familiar to the physical and mental activity of the human being. Drowsiness is detected by continually monitoring and analyzing these non-visual signals during work, like driving, operating a crane, etc.

Electroencephalogram (EEG)-based physiological methods are considered more effective, instantaneous, and promising than other drowsiness detection methods due to EEG signal benefits (specified in the previous session). EEG-based models are classified as multi-channel or single-channel, depending on the number of EEG channels used. More than one channel signal is analyzed in multi-channel EEG-based DD. Due to the connection of multiple electrodes in the human body, these are inconvenient to use in real-time applications. Furthermore, manufacturing sensors to collect multi-channel data is costly. It's a difficult job to keep track of multiple sensor's connectivities. Drowsiness detection methods based on single-channel EEG have been given high priority because they are easier to use in real-time applications [1, 2, 15]. Therefore, advanced single-channel EEG-based drowsiness detection methods have been proposed. Figure 1 illustrates the functional flow steps in a system that detects drowsiness with a single-channel EEG, from signal acquisition to classification.

3 EEG Signal Acquisition and Artifact Removing

Electrodes used for acquiring the EEG signal from the brain must be disposable, recyclable, and doesn't harm the human scalp [8]. The placement of the electrodes follows some topographic pattern for getting tomography of the brain area for covering all parts of the brain. The initial topographic system is a 10–20 system which is designed by Jasper in 1958 [16] as shown in Fig. 2 subplot A. Dr. Jasper built up

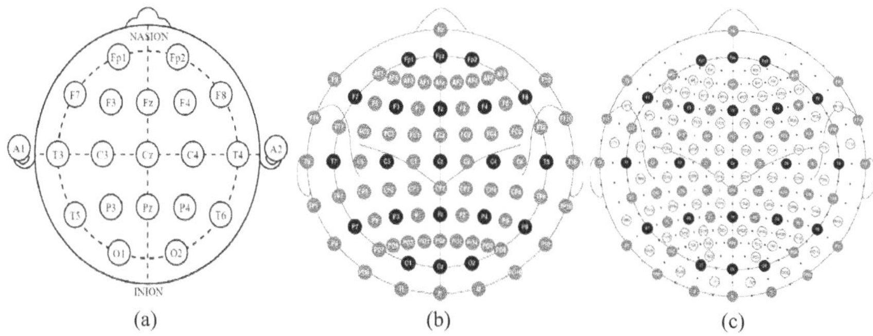

Fig. 2 Topographic models to specifying the channel/sensor location to obtain the EEG signal are **a** 10–20 system, **b** 10-10 system, and **c** 10-5 system

a few rules for setting these electrodes. Those are as follows: Each electrode used in an EEG device has a specified name. These names not only identify the electrodes but also specify which part of the brain it covered [17]. The number of electrodes required to gather the EEG signal is dependent on its requirement for experimental or clinical use. As an extension to 10–20 topographic systems, 10-10 and 10-5 systems were invented with more sensors to completely cover the brain area. The 10-10 system is shown in Fig. 2 subplot B, also known as the "10% system". There are many 10-10 systems. Chatrian 10-10 system has an additional 60 electrodes along with 21 electrodes in the Jasper 10–20 system [18], as well as the IFCN 10-10 system with 64 electrodes, the ACNS 10-10 system with 75 electrodes, and the Ostenveld 10-10 system with 85 electrodes are also available. In some laboratories, it is compulsory to get high-resolution ERP-based EEG signals for analyzing brain activity, in such cases they may require more than 100s of channel EEG signals. To accomplish these requirements, there is a new system called a 10-5 system [19] which is an extension to the 10-10 system. This system also called "5%-system" is shown in Fig. 2 subplot C. For academic research purposes, we have some wearable headbands and electrode caps which were designed based on a 10–20 system [20, 21]. In existing drowsiness detection methods, EEG acquisition was done in Multi-channel and single-channel EEG devices. The authors of [22–24] used a 32-channel cap EEG device, while [25–27] used 15, 12, and four channel devices, respectively. Single-channel EEG devices, such as the TGAM headset Neurosky Mindset, Mindwave Mobile, and MUSE, were used in [28–32]. Some other researchers [2, 15, 33–35] have used free EEG datasets available online.

It is important to search for any artifacts in the EEG signal after obtaining EEG data because artifacts reduce the EEG signal's intended information. Based on the occurrence of artifacts, those categories are of two types. Those are

- **Hardware artifacts**: These are caused by hardware equipment issues such as improper contact with the scalp and power fluctuations in the EEG unit. By manually checking the signal amplitude values for each epoch (1s time-window) signal, hardware errors can be avoided [1, 2, 15].

- **Physiological artifacts**: These artifacts include eye movement and blinking, muscle-related artifacts, including individual movements caused by the forehead muscles. Physiological artifacts are removed using 50 Hz notch filter and a 0.15Hz 45 Hz bandpass filter [23, 31, 33].

4 Machine Learning-Based Drowsiness Detection

Without being precisely programmed, machine learning is used to automate EEG signals' classification into awake and drowsy states using existing datasets. The machine learning classifier does not classify the raw EEG data because EEG signals are large. Furthermore, there is no discernible difference between the awake and drowsy states in the raw EEG signal. Before performing the classification of EEG signals, the EEG signals are converted into finite variables called features. Features decrease the burden on machine learning and improve the discriminative property of the awakening and drowsiness of the EEG signal. These features may be extracted from a raw EEG signal or after it has been transformed to a different domain signal, such as a frequency domain or time-frequency domain signal. The authors of [22, 23, 32, 36] used direct EEG signals to extract features. Most of these features are statistical features such as standard deviation, and minimum and maximum values. Entropy-based features such as entropy, fuzzy entropy, and spectrum entropy have been used by some authors [22–24]. In the case of converted domain features, some signal conversion methods are used, of which Fourier Transform (FFT) [37], Short-Term Fourier Transform (STFT) [38], Discrete Wavelet Transform (DWT) [39], and Wavelet Packet Transform (WPT) [39] are the most widely used.

- FFT is mainly used to obtain frequency-related information from the EEG signal. In [31, 35, 40–42], the authors used the FFT method to extract frequency-related features such as the power spectrum density from the main EEG signal or its subbands. Subbands include Delta, Theta, Alpha, Beta, and Gamma.
- The STFT method is similar to the FFT method. It provides time-related information by providing frequency information over a fixed period of time, rather than providing frequency information for the entire signal. In [34, 43–45], the authors used STFT to extract time-frequency-related features from EEG signals.
- Similar to the STFT, DWT provides time-frequency information. Here, DWT uses variable-length time instead of a fixed time window. The authors of [46–49] have used DWT for extracting time-frequency domain features. DWT has two parameters such as decomposition levels and wavelet function. The decomposition levels were considered based on the required frequency band signals and sampling frequency of EEG signal. There are many wavelet function libraries. Daubechies wavelet functions were used heavily [47–49].
- WPT and DWT are similar in that they use the same parameters. For the appropriate frequency band signal, WPT considers multiple coefficient values, while DWT only considers single coefficient values. As compared to DWT, the frequency

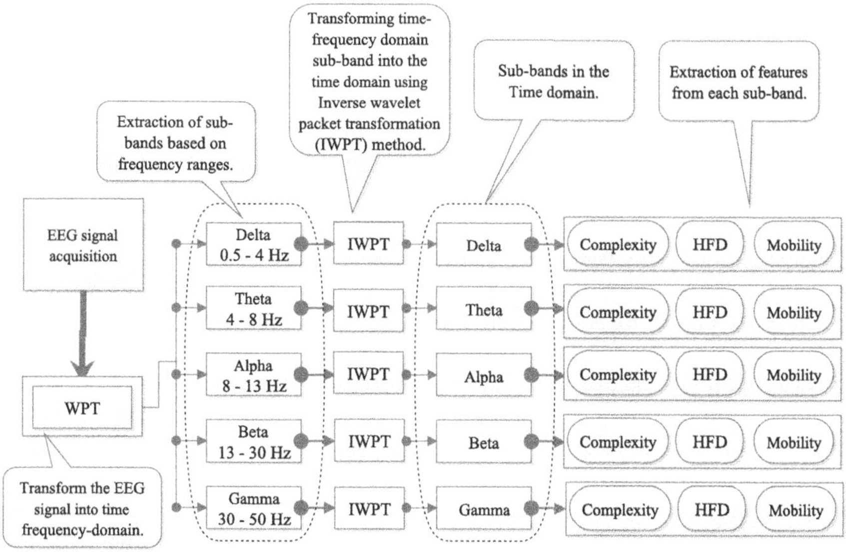

Fig. 3 Extraction of features from time-domain EEG subbands

extraction process can be achieved with the appropriate frequency ranges [1]. WPT was used by the authors of [1, 33] in their hand-engineered-based drowsiness detection methods.

The majority of current drowsiness identification feature extraction processes have been conducted on EEG subbands. The majority of these features were derived from EEG subbands in the transformed domain. The authors used the frequency domain in [31, 41, 49, 50], and time-frequency subbands in [33, 48, 49]. In contrast to these methods, the authors of [1] used wavelet pack transformation to extract subbands in the time domain. They extracted Higuchi fractal dimension, complexity, and mobility features from these time-domain subbands, as shown in Fig. 3.

The collected features are given to the classifier to perform the final classification. In machine learning, there are various classification algorithms. There is no such thing as a good or bad classifier. Each classifier is tried and evaluated to determine which classifier is appropriate for the dataset features. As a result, each author's features were classified using different classifiers. For example, the gradient boosting decision tree classification used to classify the dataset features in [22, 23] reported 94% accuracy. In [24], 96% accuracy was achieved using the random forest classification process. Current state-of-the-art DD methods mostly use Support Vector Machine (SVM) and Artificial Neural Networks (ANN) machine learning classifications for FFT-based features. The authors of [28, 29, 31, 42, 51] used FFT-based EEG subbands with SVM classification to classify drowsiness, while the authors of [26, 52] used the same SVM classifier with entropy-based features and Tunable Q wavelet transform (TQWT)-based features, respectively. Some authors have used

FFT-based features with ANN [32, 40, 49, 53] to detect drowsiness. The authors of [47, 48] classified awake and drowsy state signals using DWT-based features and ANN classifiers.

- **SVM classifier**: The support vector machine (SVM) is a common supervised learning algorithm for binary classification that was developed in the 1990s. It creates an ideal hyper-plane to separate the awake and drowsy state EEG signals using input features. To optimize the margin between awake and drowsy state features, SVM has some hyper-parameters (C or gamma values) and a kernel function for balancing bias and variance. C parameter is used to control decision boundary error by considering misclassified data (feature) points. Gamma decides how much curvature is needed in a decision boundary. SVM is a linear separable classifier. If the features aren't linearly separable, the kernel function is applied to render them so. In comparison to the other kernels, the radial base function (RBF) kernel is more common. Both the C and gamma parameters are optimized at the same time in RBF.
- **ANN classifier**: Artificial Neural Networks, or ANN, is a data processing model inspired by how the brain's biological nervous system processes the data. It is made up of a large number of processing elements (neurons) that work together to solve a problem. If any ANN has a single neuron, then it is called the perceptron. The following is how a neuron processes data: The neuron is given the input features "$x_1, x_2, \ldots, x_n$", and each feature value is multiplied by some scalar weight "$w_1, w_2, \ldots, w_n$", resulting in the product as scalar "$w_1x_1 + w_2x_2, \ldots + w_nx_n$" The net input N is formed by adding the scalars "$w_1x_1 + w_2x_2, \ldots + w_nx_n$" to the scalar bias b. The scalar output is generated by passing the net input through the transfer function "f." In drowsiness detection, a multilayer perceptron (MLP) with one input layer, one or more hidden layers, and one output layer is used. Via the hidden layers, information is propagated from the input layer to the output layer. The hidden layer's neurons are bound to the previous layer's output and the next layer's input, while the output layer's neurons decide the input feature vector class. The relation between two neurons in two different layers is called a synapse, and each link is given a weight w [55].

Drowsiness detection methods based on hand-engineered features with SVM and ANN machine learning classifiers are summarized in Table 1.

5 Deep Learning-Based Drowsiness Detection Methods

Deep learning is a subfield of artificial intelligence that uses more data processing than machine learning to perform tasks (classification or regression) automatically. Deep neural networks, convolutional networks, and recurrent neural networks are three classifier techniques in deep learning. These three strategies have been used on hand-engineered features [56, 57] and automatic features [2, 44] in single-channel EEG-based drowsiness detection methods. In [56], the authors used a deep neural

Table 1 Drowsiness detection with SVM and ANN machine learning classifications based on hand-engineered features

SVM				ANN			
Author	Signal Transform method	Features	Accuracy (%)	Author	Signal Transform method	Features	Accuracy (%)
[52]	TQWT	Statistical features	96	[40]	FFT	EEG subbands	89
[42]	FFT	Frequency and amplitudes	87.5	[53]	FFT	EEG subbands	86.5
[28]	FFT	EEG subbands	91	[32]	FFT, Time	Power spectral density, statistical	93
[29]	FFT	Power Spectral density	72.7	[47]	Time, DWT	statistical, zero cross value , etc.	90.27
[54]	FFT	EEG subbands	83.3	[48]	Time, DWT, FFT	Statistical, zero cross value, spectrogram, etc.	87
[31]	FFT	EEG subbands	74	[49]	FFT	Power spectral density	81.5
[26]	Time	Entropy	93.6				
[51]	FFT	EEG subbands	88.6				

network on ten hand-engineered features, which were extracted using filters and the entropy principle. The author of [44] used a recurrent neural network's long short-term memory classifier to extract automatic features from EEG spectrogram images collected using the STFT technique and achieved 94% accuracy. Venkat et al. [2] used a convolutional network to extract automatic features from direct EEG signals and recorded 95% accuracy.

6 Experimental Results

Drowsiness detection is performed using a single-channel EEG signal and classification processes such as machine learning and deep learning. This research made use of free data from the National Institutes of Health (NIH), also known as Physionet [58]. This database contains EEG of two nights' recordings of awake, drowsiness, and other sleep states. The EEG data were obtained using two sensors, Pz-Oz and Fpz-Cz, at a sampling frequency of 100 Hz. For single-channel EEG-based drowsiness detection approaches, many researchers [2, 15, 33, 34, 40, 48, 49] have used the same dataset. As stated in [2], first-night recordings of Pz-Oz sensor data from 23 subjects were used in this analysis.

The SVM machine learning classification, which has been widely used in the past, has been used to identify drowsiness using hand-crafted features, and another approach to CNN deep learning classification with automated features. This chapter uses EEG data from 23 different topics. Ten minutes of EEG signals were assessed from each subject. To measure classifier output, one sec time window (epochs) signals representing 13,800 (i.e., $23 \times 10 \times 60$) epoch drowsy state EEG and another 13,800 epoch awake state EEG signals were taken from the physionet dataset Pz-Oz channel. These two 13800 epoch data are analyzed using a cross-fold process. In the cross-fold operation, the considered data is supplied to the classifier model. The cross-fold value in this experiment is set to ten. That is, overall data is split into ten folds, with one fold being used for test and the other nine folds for training. The procedure is replicated ten times until each fold of data is considered test data. The final accuracy was determined by taking the sum of all of these accuracies. The output of these two classification processes is evaluated using four criteria, as shown in [1, 2]: total accuracy, sensitivity, precision, and F1 score. The following are the components of the metrics:

- True Positive (TP): The classifier correctly predicts the state of the EEG as drowsy when it is drowsy.
- True Negative (TN): The classifier correctly predicts the state of the EEG as not drowsy when it is not drowsy.
- False Positive (FP): The classifier incorrectly predicts the state of the EEG as drowsy when it is not drowsy.
- False Negative (FN): The classifier incorrectly predicts the state of the EEG as not drowsy when it is drowsy.

Thus, the metrics are calculated as follows:

Total Accuracy (Acc): It's the ratio of the number of correct predictions to the total number of predictions.

$$Acc = \frac{TP + FN}{TP + TN + FP + FN}$$

Sensitivity (Sen): It tests the model's capacity to detect drowsiness when a subject is actually drowsy.

$$Sen = \frac{TP}{TP + FN}$$

Precision (Pre): It examines the model's ability to detect activity when a subject is actually active.

$$Pre = \frac{TP}{TP + FP}$$

$\mathbf{F1_{Score}}$ (F1S): It balances the right rates of drowsy and awake states by integrating precision and sensitivity information.

$$F1S = 2 \times \frac{Precision \times Sensitivity}{Precision + Sensitivity}$$

6.1 Machine Learning Classification

Eight hand-engineered features used in several drowsiness detection techniques [1, 8, 15, 23, 35] were considered for machine learning classification, including Higuchi Fractal Dimension (HFD), Hjorth parameters such as Mobility and Complexity, Detrended Fluctuation Analysis (DFA), Energy, Approximate Entropy, Sample Entropy, and Fuzzy Entropy.

The computation procedure of these eight features are as follows:

1. **Higuchi fractal dimension (HFD)**: It is a quick computational technique that can trace the modifications in a biosignal from a measure of its complication, working directly in time-space. There may consist some common observations at regular intervals. To extract such observations, these datasets make it into "d" new sub-datasets as "X_d^m" with $m = 1, 2, \ldots, d$. where m is the starting time of each subset d. For this work, the d value is half of the length of the signal. X_d^m are constructed as

$$X_d^m = x_m, x_{m+d}, x_{m+2d}, \ldots, x_{m+[\frac{N-m}{d}].d} \quad (1)$$

For each X_d^m, the length $L_m(d)$ is computed as

$$L_m(d) = \frac{\left\{\left[\sum_{i=1}^{[\frac{N-m}{d}]} \left|x_{m+id} - x_{m+(i-1).d}\right|\right] \frac{N-1}{[\frac{N-m}{d}].d}\right\}}{d} \quad (2)$$

The mean of $L_m(d)$ is computed to find the HFD as shown below

$$L(d) = \frac{1}{d} \sum_{M=1}^{d} L_m(d) \quad (3)$$

2. **Complexity**: It is measured by the square root of differences of two ratios as shown in Eq. 4.

$$Complexity = \sqrt{\left[\frac{rms(\frac{d(\frac{dX}{dt})}{dt})}{rms(\frac{d(X)}{dt})}\right] - \left[\frac{rms(\frac{d(X)}{dt})}{rms(X)}\right]} \quad (4)$$

where $\frac{dX}{dt}$ is the rate of change of EEG signal (X) with respect to time (t) and rms is the root mean square.

3. **Mobility**: Like complexity, mobility is another parameter to analyze the time series. Mobility and complexity also called as Hjorth parameters. Mobility is measured as shown in Eq. 5.

$$Mobility = \frac{rms(\frac{d(x)}{dt})}{rms(X)} \tag{5}$$

4. **Detrended Fluctuation Analysis (DFA)**: It is a method for determining the statistical self-correlation of a time-domain signal. Its computation process is as follows:

$$C(k) = \sum_{i=1}^{k}(x_i - mean(X)) \tag{6}$$

where k=1, 2, ..., N. The 1s signal C_k is divided into equal sub-parts with a length of 'n' (duration of each subpart is 0.25s). It is represented as $y_n(k)$ for n=5, then fluctuations is calculated as

$$F(n) = \sqrt{\frac{1}{N}\sum_{k=1}^{N}[C(k) - y_n(k)]} \tag{7}$$

Polynomial curve fit is computed on a logarithmic values of "$log(F(n))$" and "n" with a degree one.

$$P(x) = p_1(x) + c \tag{8}$$

DFA is the coefficient (p_1) of a polynomial curve $(P(x))$.

5. **Energy** (Eng):

$$Eng = \sum_{i=1}^{N}(x_i)^2 \tag{9}$$

6. **Approximate Entropy:** (AE)

$$AE = \Phi^m(r) - \Phi^{m+1}(r) \tag{10}$$

where similarity tolerance (r) and probability $(\Phi^m(r))$ are

$$r = 0.2 \times Standard\ Deviation(X) \tag{11}$$

$$\Phi^m(r) = \frac{1}{N-m+1} \cdot \sum_{i=1}^{N-m+1} ln(C_i^m(r)) \tag{12}$$

$$C_i^m(r) = \frac{(number\ of\ y(j), d[y(i), y(j)] \leqslant r)}{N-m+1} \tag{13}$$

7. **Sample Entropy:** (SE)

$$\Phi^m(r) = \frac{1}{N-m} \cdot \sum_{i=1}^{N-m} C_i^m(r) \tag{14}$$

$$SE = ln\left(\frac{\Phi^m(r)}{\Phi^{m+1}(r)}\right) \tag{15}$$

8. **Fuzzy Entropy:** (FE)

$$y(i) = [x_i - \overline{x}_i, \ldots, x_{i+m-1} - \overline{x}_i] \tag{16}$$

$$y(j) = [x_j - \overline{x}_j, \ldots, x_{j+m-1} - \overline{x}_j] \tag{17}$$

are two subsets of given EEG signal, where

$$\overline{x}_i = \frac{1}{m} \sum_{k=0}^{m-1} x_{i+k} \tag{18}$$

Fuzzy entropy (FE) represented as

$$FE = ln\left(\frac{\Phi^m(r)}{\Phi^{m+1}(r)}\right) \tag{19}$$

where probability $(\Phi^m(r))$ is

$$\Phi^m(r) = \frac{1}{N-m} \sum_{i=1}^{N-m} \sum_{j=1, j\neq i}^{N-m} \frac{D_{i,j}^m}{N-m-1} \tag{20}$$

$D_{i,j}^m$ is the Fuzzy membership matrix, which is defined by the following equation:

$$D_{i,j}^{m} = \mu(d(y_i^{-m}, y_j^{-m})) \tag{21}$$

μ fuzzy is the membership function, which is defined by the following equation:

$$\mu(x) = e^{-(\frac{y}{r})^n} \tag{22}$$

6.2 Deep Learning Classification

The Convolution neural network from deep learning was used to predict drowsiness using automated features. The three key layers in CNN are the Convolution layer with filters and kernel weights, the pooling layer to reduce the Convolution layer's filter maps, and the final fully connected layer with an activation function.

The 1D CNN model was used here, as stated in [2]. It's a six-layer, one-dimensional CNN model. The following is a list of each layer:

- Layer 1: 1D Convolution Layer with 2 filters and each filter kernel size is 2.
- Layer 2: Batch Normalization layer followed by an ELU activation.
- Layer 3: Max pooling layer.
- Layer 4: Drop out layer with a probability of 0.2 to fight overfitting.
- Layer 5: Dense layer with ReLU activation function.
- Layer 6: Fully connected layer with Softmax activation function.

7 Discussion

The chapter's main focus is on how machine learning and deep learning were used to develop single-channel EEG-based DD methods. The following is a description of the work:

- Single-channel EEG-based approaches are more easy to use in real time than other drowsiness detection methods. The combination of neuroscience and computer science knowledge makes this a reality.
- Hand-engineered feature-based machine learning methods take less time to train, and a normal CPU sufficient in execution. Automated feature-based deep learning methods require a high computational Graphical processing unit because it requires large amounts of data for training and large-scale matrix multiplication operations to extract automatic features.
- Hand-engineered features are limiting human knowledge. To extract features, it's also important to understand the EEG signal analysis mechanism and signal transform methods. Automated feature extraction methods solve these limitations by considering multiple large-scale matrix multiplications.

Table 2 SVM and CNN classifiers results (all values in (%))

Classifier	Features extraction mode	Total	Sen	Pre	F1s
SVM	Hand-engineered	89.84	93.19	87.84	90.44
CNN	Automated	84.03	85.55	81.01	83.22

- For both machine learning and deep learning, the majority of current drowsiness prediction approaches use transformed domain signals. This chapter examines direct domain features for hand-engineered features as well as direct domain signals for automatic feature extraction. As compared to CNN's automatic features, the hand-engineered-based SVM classifier performs better. Table 2 illustrates this.

8 Conclusion

Drowsiness is one of the main causes of decreasing strength and alertness, which may lead to increases in the accidents in personal or professional activities such as driving a vehicle, operating a crane, working with heavy machinery in large industries such as steel plants, mine blasts, and so on. Drowsiness detection using neuroscience and computer science knowledge is a safeguarded technique to prevent accidents and improve safety. This chapter discusses the drowsiness detection methods based on single-channel EEG using hand-engineered feature-based machine learning and automated feature-based deep learning. More research will be required in developing new deep learning methods to extract automated features from single-channel EEG signals without any signal transform methods.

References

1. Venkata P, Chinara S (2021) Automatic classification methods for detecting drowsiness using wavelet packet transform extracted time-domain features from single-channel EEG signal. J Neurosci Methods 347(4):1–15
2. Balam VP, Sameer VU, Chinara S (2021) Automated classification system for drowsiness detection using convolutional neural network and electroencephalogram. IET Intell Transp Syst 15(4):514–524
3. Wei C, Chen L-l, Song Z-Z, Lou X-G, Li D-d (2020) Eeg-based emotion recognition using simple recurrent units network and ensemble learning. Biomed Signal Process Control 58(2):101756
4. Kaplan HI, Sadock BJ (1989) Comprehensive textbook of psychiatry. Williams & Wilkins Co Vols, pp 1–2

5. Bashashati A, Fatourechi M, Ward RK, Birch GE (2007) A survey of signal processing algorithms in brain–computer interfaces based on electrical brain signals. J Neural Eng 4(2):R32
6. Jung R, Berger W (1979) Hans bergers entdeckung des elektrenkephalogramms und seine ersten befunde 1924–1931. Archiv für Psychiatrie und Nervenkrankheiten 227(4):279–300
7. Kanda PAM, Aguiar ADAX, Miranda JL, Falcao AL, Andrade CS, Reis LNdS, Almeida EWRB, Bello YB, Monfredinho A, Kanda RG (2018) Sleep EEG of microcephaly in zika outbreak. Neurodiagnostic J 58(1):11–29
8. Venkata PCS (2020) Time domain parameters as a feature for single-channel EEG-based drowsiness detection method. In: 2020 IEEE international students' conference on electrical, electronics and computer science (SCEECS), pp 1–5
9. Hoddes E, Zarcone V, Dement W (1972) Stanford sleepiness scale. Enzyklopädie der Schlafmedizin 1184(1)
10. Epworthsleepinessscale (2013). http://epworthsleepinessscale.com/about-the-ess/
11. Miley AÅ, Kecklund G, Åkerstedt T (2016) Comparing two versions of the karolinska sleepiness scale (KSS). Sleep Biol Rhythm 14(3):257–260
12. Thiffault P, Bergeron J (2003) Monotony of road environment and driver fatigue: a simulator study. Accid Anal Prev 35(3) 381–391
13. Flores MJ, Armingol JM, de la Escalera A (2010) Real-time warning system for driver drowsiness detection using visual information. J Intell Robot Syst 59(2):103–125
14. Sahayadhas A, Sundaraj K, Murugappan M (2012) Detecting driver drowsiness based on sensors: a review. Sensors 12(12):16937–16953
15. Venkata PCS (2018) Drowsiness detection by analysis of EEG signal with the help of machine learning. In: 24th annual international conference on advanced computing and communications (ADCOM 2018), pp 22–27
16. Jurcak V, Tsuzuki D, Dan I (2007) 10/20, 10/10, and 10/5 systems revisited: their validity as relative head-surface-based positioning systems. Neuroimage 34(4):1600–1611
17. Klem GH, Lüders HO, Jasper H, Elger C et al (1999) The ten-twenty electrode system of the international federation. Electroencephalogr Clin Neurophysiol 52(3):3–6
18. Chatrian G, Lettich E, Nelson P (1985) Ten percent electrode system for topographic studies of spontaneous and evoked EEG activities. Am J EEG Technol 25(2):83–92
19. Oostenveld R, Praamstra P (2001) The five percent electrode system for high-resolution EEG and ERP measurements. Clin Neurophysiol 112(4):713–719
20. Liao L-D, Lin C-T, McDowell K, Wickenden AE, Gramann K, Jung T-P, Ko L-W, Chang J-Y (2012) Biosensor technologies for augmented brain–computer interfaces in the next decades. In: Proceedings of the IEEE 100 (Special Centennial Issue) 1553–1566
21. Lee S, Shin Y, Woo S, Kim K, Lee H-N (2013) Review of wireless brain-computer interface systems. In: Brain-computer interface systems-recent progress and future prospects, pp 215–238
22. Hu J, Min J (2018) Automated detection of driver fatigue based on EEG signals using gradient boosting decision tree model. Cogn Neurodynamics 12(4):431–440
23. Wang P, Min J, Hu J (2018, 2018,) Ensemble classifier for driver's fatigue detection based on a single EEG channel. IET Intell Transp Syst 12(10):1322–1328
24. Hu J (2017) Comparison of different features and classifiers for driver fatigue detection based on a single EEG channel. Comput Math Methods Med 2017(1):1–9
25. Shabani H, Mikaili M, Noori SMR (2016) Assessment of recurrence quantification analysis (RGA) of EEG for development of a novel drowsiness detection system. Biomed Eng Lett 6(3):196–204
26. Xiong Y, Gao J, Yang Y, Yu X, Huang W (2016) Classifying driving fatigue based on combined entropy measure using EEG signals. Int J Control Autom 9(3):329–338
27. Picot A, Charbonnier A (2009) Sylvie. Monitoring drowsiness on-line using a single encephalographic channel. In: Recent Advances in Biomedical Engineering. HAL, pp 145–164
28. Nissimagoudar PC, Nandi AV (2020) Precision enhancement of driver assistant system using EEG based driver consciousness analysis & classification. Comput Netw Appl Tools Perform Manag 2020:247–257 Springer

29. Ogino M, Mitsukura Y (2018) Portable drowsiness detection through use of a prefrontal single-channel electroencephalogram. Sensors 18(12):4477–4496
30. Stanley PK, Prahash TJ, Lal SS, Daniel PV (2017) Embedded based drowsiness detection using EEG signals. In: 2017 IEEE international conference on power, control, signals and instrumentation engineering (ICPCSI). IEEE, pp 2596–2600
31. Rohit F, Kulathumani V, Kavi R, Elwarfalli I, Kecojevic V, Nimbarte A (2017) Real-time drowsiness detection using wearable, lightweight brain sensing headbands. IET Intell Transp Syst 11(5):255–263
32. Abdel-Rahman A, Seddik AF, Shawky DM (2015) An affordable approach for detecting drivers' drowsiness using EEG signal analysis. In: 2015 international conference on advances in computing, communications and informatics (ICACCI). IEEE, pp 1326–1332
33. da Silveira TL, Kozakevicius AJ, Rodrigues CR (2016) Automated drowsiness detection through wavelet packet analysis of a single EEG channel. Expert Syst Appl 55(1):559–565
34. Jalilifard A, Pizzolato EB (2016, 2016,) An efficient k-nn approach for automatic drowsiness detection using single-channel EEG recording. In: Engineering in Medicine and Biology Society (EMBC), pp 820–824
35. Venkata Phanikrishna B, Jaya Prakash A, Suchismitha C (2021) Deep review of machine learning techniques on detection of drowsiness using EEG signal. IETE J Res 1–16. https://doi.org/10.1080/03772063.2021.1913070
36. Dey I, Jagga S, Prasad A, Sharmila A, Borah SK, Rao G (2017) Automatic detection of drowsiness in EEG records based on time analysis. In: 2017 innovations in power and advanced computing technologies (i-PACT). IEEE, pp 1–5
37. Cooley JW, Tukey JW (1965) An algorithm for the machine calculation of complex fourier series. Math Comput 19(90):297–301
38. Allen J (1977) Short term spectral analysis, synthesis, and modification by discrete fourier transform. IEEE Trans Acoust, Speech, Signal Process 25(3):235–238
39. Percival DB, Walden AT (2000) Wavelet methods for time series analysis, vol 4. Cambridge University Press
40. Belakhdar I, Kaaniche W, Djemal R, Ouni B (2018) Single-channel-based automatic drowsiness detection architecture with a reduced number of EEG features. Microprocess Microsyst 58(2):13–23
41. Lin C-T, Chang C-J, Lin B-S, Hung S-H, Chao C-F, Wang I-J (2010,) A real-time wireless brain-computer interface system for drowsiness detection. IEEE Trans Biomed 4(4):214–222
42. Anitha C (2019) Detection and analysis of drowsiness in human beings using multimodal signals. In: Digital business. Springer, pp 157–174
43. Picot A, Charbonnier S, Caplier A (2008) On-line automatic detection of driver drowsiness using a single electroencephalographic channel. In: 2008 30th Annual International Conference of the IEEE Engineering in Medicine and Biology Society. IEEE, pp 3864–3867
44. Budak U, Bajaj V, Akbulut Y, Atila O, Sengur A (2019) An effective hybrid model for eeg-based drowsiness detection. IEEE Sens J 19(17):7624–7631
45. Khalifa KB, Bedoui MH, Dogui M, Alexandre F (2004) Alertness states classification by som and lvq neural networks. Int J Inf Technol 1(4):228–231
46. Silveira TD, Kozakevicius AdJ, Rodrigues CR (2015) Drowsiness detection for single channel eeg by dwt best m-term approximation. Res Biomed Eng 31(2):107–115
47. Boonnak N, Kamonsantiroj S, Pipanmaekaporn L (2015, 2015,) Wavelet transform enhancement for drowsiness classification in eeg records using energy coefficient distribution and neural network. Int J Mach Learn Comput 5(4):288–293
48. Correa AG, Orosco L, Laciar E (2014) Automatic detection of drowsiness in EEG records based on multimodal analysis. Med Eng Phys 36(2):244–249
49. Correa AG, Leber EL (2010) An automatic detector of drowsiness based on spectral analysis and wavelet decomposition of EEG records. In: 2010 Annual international conference of the IEEE engineering in medicine and biology. IEEE, pp 1405–1408
50. L.-W. Ko, W.-K. Lai, W.-G. Liang, C.-H. Chuang, S.-W. Lu, Y.-C. Lu, T.-Y. Hsiung, H.-H. Wu, C.-T. Lin, Single channel wireless eeg device for real-time fatigue level detection. In: 2015 international joint conference on neural networks (IJCNN). IEEE, pp 1–5

51. Li G, Lee B-L, Chung W-Y (2015) Smartwatch-based wearable EEG system for driver drowsiness detection. IEEE Sens J 15(12):7169–7180
52. Khare S, Bajaj V (2020) Optimized tunable q wavelet transform based drowsiness detection from electroencephalogram signals. IRBM 41(4):1–7
53. Belakhdar I, Djmel WKR, Ouni B (2016) Detecting driver drowsiness based on single electroencephalography channel. In: 2016 13th international multi-conference on systems, signals & devices (SSD). IEEE, pp 16–21
54. Albalawi H, Li X (2018) Single-channel real-time drowsiness detection based on electroencephalography. In: 2018 40th annual international conference of the IEEE engineering in medicine and biology society (EMBC). IEEE, pp 98–101
55. Belakhdar I, Kaaniche W, Djmel R, Ouni B (2016) A comparison between ANN and SVM classifier for drowsiness detection based on single EEG channel. In: 2016 2nd international conference on advanced technologies for signal and image processing (ATSIP). IEEE, pp 443–446
56. Tripathy AK, Chinara S, Sarkar M (2016) An application of wireless brain-computer interface for drowsiness detection. Biocybern Biomed Eng 36(1):276–284
57. Khessiba S, Blaiech AG, Khalifa KB, Abdallah AB, Bedoui MH (2020) Innovative deep learning models for EEG-based vigilance detection. Neural Comput Appl 20(1):1–17
58. Kemp B, Zwinderman AH, Tuk B, Kamphuisen HA, Oberye JJ (2000) Analysis of a sleep-dependent neuronal feedback loop: the slow-wave microcontinuity of the EEG. IEEE Trans Biomed Eng 47(9):1185–1194

Author Biographies

Venkata Phanikrishna Balam received his doctorate in computer science and engineering from the National Institute of Technology Rourkela, India, in 2021. His master's degree (M.Tech.) was completed in Computer Science and Technology from Andhra University, Visakhapatnam, India, in 2013. He completed his graduation (B.Tech.) degree in Information Technology from Jawaharlal Nehru Technological University-Kakinada (JNTUK) at Kakinada, India, in 2010. Further, his research interests including bio-signal processing, pattern recognition, computer vision, the Internet of Things, and machine learning

Suchismitha Chinara received her Ph.D. degree in computer science and engineering from the National Institute of Technology Rourkela, India, in 2011. She is currently working as an Assistant Professor with the National Institute of Technology at Rourkela. She has published more than 50 research articles in national and international journals and conferences. Her research interests include Computer Networks, Internet of Things, Data Communication, and Machine Learning.

Uncertain Structural Parameter Identification by Intelligent Neural Training

Deepti Moyi Sahoo and S. Chakraverty

Abstract System identification problems have been investigated by different authors using ANN by considering the exact or crisp form of data. But the data obtained from experiments are not always in crisp form; they may have some errors. The errors may be due to humans or equipment errors which give the uncertain form of data. Here, uncertainty has been taken as intervals. Hence, the present model is developed to handle these uncertainties. Our aim is to identify the uncertain structural parameters such as mass, stiffness and damping matrices from dynamic responses of the structure. This chapter proposes an interval neural network-based strategy for the simultaneous identification of mass, stiffness and damping of multi-storey shear buildings. Classical methods are used in governing equations of motion to find the response of consecutive stories. Then these governing equations of motion are modified based on relative responses of consecutive stories in such a way that the new set of equations can be implemented in a cluster of Interval Neural Networks (INNs). Cluster of INNs to estimate uncertain structural parameters where initial weights are taken as the design parameters in interval form. Converged cluster of INNs directly identify structural parameters in interval form. Interval estimates are certainly useful for civil engineers by knowing the lower and upper bounds of the parameters.

Keywords System identification · Mass · Stiffness · Damping · Interval neural network · Shear buildings

1 Introduction

The process of modelling an unknown system or developing a mathematical description from a known system based on a set of inputs and resulting outputs is known as System Identification (SI). Structural dynamic problems can be categorized into two ways, direct and indirect. In the direct approach, the governing equations and parameters of the system are known. Using these parameters, we can find the response

D. M. Sahoo · S. Chakraverty (✉)
National Institute of Technology Rourkela, Rourkela 769008, Odisha, India

S. Chakraverty (ed.), *Soft Computing in Interdisciplinary Sciences*,
Studies in Computational Intelligence 988,
https://doi.org/10.1007/978-981-16-4713-0_8

of the system for a specific input. But in the case of inverse problems, the output response for a given input is known, but either the governing equation or some of the parameters of the physical process are unknown. In structural system identification problems, we identify the structural parameters such as stiffness, frequencies, mode shapes, damping ratios and structural response. For health monitoring, damage assessment and safety evaluation of existing engineering structures, system identification techniques needs to be studied.

There are different methods to study the dynamic behaviour of structures. Structural behaviour can be studied analytically, numerically and/or experimentally. Though there exist different methods to study the dynamic behaviour of the structures, each method has its own advantages and disadvantages. In order to cope with these limitations and to determine the dynamic properties of a structure, a model correlation and model updating procedure should be performed. In this regard, a few authors have worked on model updating of structural systems which may be mentioned as [1–7].

Studies related to structural damage detection using ANN have been done by various researchers. A nonparametric structural damage detection methodology based on nonlinear system identification approaches has been given by Masri et al. [8] for health monitoring of structure-unknown systems. Kao and Hung [9] used two step neural system identification networks (NSINs) to identify the undamaged and damaged states of a structural system. A multistage identification scheme for structural damage detection with the use of modal data using a hybrid neural network strategy is presented by Pillai and Krishnapillai [10]. Bakhary et al. [11] detected structural damage using the ANN method with progressive substructure zooming. Zhang et al. [12] studied the application of neural networks to damage detection in structures. To avoid the false positives of damages in the deterministic identification method induced by uncertainties in measurement noise, Zhang et al. [13] proposed a probabilistic method to identify damages of the structures with uncertainties under unknown input. The application of artificial neural networks and wavelet analysis to develop an intelligent and adaptive structural damage detection system has been investigated by Shi and Yu [14]. A probabilistic approach for damage identification considering measurement noise uncertainties has been given by Lei et al. [15]. The probability of identified structural damage is further derived based on the reliability theory.

Some workers gave various methodologies for different types of problems in system identification [16–20]. Chakraverty [16] gave a procedure which systematically modifies and identifies the structural parameters from known dynamic data. Nicoud et al. [17] presented a system identification methodology that involves the identification of model characteristics including boundary conditions rather than simply determining parametric values. A structural parameter identification and damage detection approach using displacement measurement time series is proposed by Xu et al. [18]. A novel inverse scheme based on consistent mass Transfer Matrix (TM) to identify the stiffness parameters of structural members has been given by Nandakumar and Shankar [19]. Billmaier and Bucher [20] used the selective sensitivity analysis method to solve system identification problems. Using different methods, system identification problems have also been solved by [21–25].

Structural identification problems have also been solved using ANN which may be mentioned as [26–30] have used ANN for solving. A neural network-based substructural identification for the estimation of the stiffness parameters of a complex structural system has been proposed by Yun and Bahng [26]. A concept of decentralized parametric evaluation neural networks for the parametric identification of substructures is developed by Wu et al. [27]. The dynamic characteristics of a steel frame using the backpropagation neural network have been identified by Huang et al. [28]. Direct identification of structural parameters from the time domain dynamic responses of a structure were developed by Xu et al. [31]. Chen [32] gave a neural network-based method to determine the modal parameters of structures from field measurement data. Chakraverty [29] proposed an iterative training procedure for the identification of structural parameters of two-storey shear buildings using neural networks. The above literature review shows that Artificial Neural Networks (ANNs) may very well be used to solve system identification problems. Due to its excellent learning capacity and high tolerance to partially inaccurate data, they are successfully applied for the identification and control of dynamic systems in various fields of engineering.

We have seen from the above literature that ANN developed to deal system identification problems have data in exact or crisp form. The experimental data obtained are not always in exact or crisp form. This happens due to human or equipment errors. A mathematical model has to be developed accordingly if the data available are uncertain or non-probabilistic in nature. Hence, a neural network has to be developed that can handle uncertain or interval data, and we call the neural network Interval Neural Network. Interval Neural Networks have also been used by a few researchers in different fields. Beheshti et al. [33] categorized general three-layer interval neural network training problems into type1 and type2. Garczarczyk [34] presented an algorithm for Interval Neural Networks. Chetwynd et al. [35] presented an application of interval neural networks to a regression problem. The interval analysis technique for structural damage identification has been proposed by Wang et al. [36]. Zhang et al. [37] proposed a numerically efficient approach to treat modelling errors as intervals which results in bounded values for obtaining the identified parameters. Okada et al. [38] proposed an extension of a genetic algorithm for neuroevolution of interval-valued neural networks. Sahoo et al. [39] used Interval Artificial Neural Network (IANN) which can estimate the structural parameters. An Interval neural network technique for system identification of multi-storey buildings using interval response data has been done by Chakraverty and Sahoo [40].

In this chapter, uncertain structural parameters such as mass, stiffness and damping matrices of a multi-story shear building in interval form are identified using a single-layer interval neural network. The governing equations of motion are systematically arranged in a series of interval neural networks in order to get the physical parameters in interval form. To get the responses of the consecutive stories, governing equations of motion are initially solved by classical method. Then the governing equations of motion are modified based on relative responses of consecutive stories in such a way that the new set of equations can be implemented in a cluster of Interval

Neural Networks (INNs). As such, the model starts solving the nth floor by INN modelling to estimate the structural parameters. Subsequently, a series of INN models are used to estimate the parameters for (n−1)th storey to the first storey. Here, single-layer interval neural networks have been used for training for each cluster of the INN such that the converged weights give the uncertain structural parameters. The initial weights in the INN architecture are taken as the design parameters in uncertain (interval) form. The present model is validated by considering various example problems of different multi-storey shear structures. The results are shown in tables and graphs. Comparisons among theoretical and identified results are shown and the results are found to be in good agreement.

2 Interval Arithmetic

Let us assume A and B as numbers expressed as intervals. For all $\underline{a}, \overline{a}, \underline{b}, \overline{b} \in R$ where $A = [\underline{a}, \overline{a}]$, $B = [\underline{b}, \overline{b}]$, the main operations of the intervals may be written as Lee [41]

(1) Addition

$$[\underline{a}, \overline{a}] (+) [\underline{b}, \overline{b}] = [\underline{a} + \underline{b}, \overline{a} + \overline{b}]$$

(2) Subtraction

$$[\underline{a}, \overline{a}] (-) [\underline{b}, \overline{b}] = [\underline{a} - \overline{b}, \overline{a} - \underline{b}]$$

(3) Multiplication

$$\begin{aligned}&[\underline{a}, \overline{a}] (\times) [\underline{b}, \overline{b}] \\ &\quad = [\min(\underline{a} \times \underline{b}, \underline{a} \times \overline{b}, \overline{a} \times \underline{b}, \overline{a} \times \overline{b}), \max(\underline{a} \times \underline{b}, \underline{a} \times \overline{b}, \overline{a} \times \underline{b}, \overline{a} \times \overline{b})]\end{aligned}$$

(4) Division

$$\begin{aligned}&[\underline{a}, \overline{a}] (\div) [\underline{b}, \overline{b}] \\ &\quad = [\min(\underline{a} \div \underline{b}, \underline{a} \div \overline{b}, \overline{a} \div \underline{b}, \overline{a} \div \overline{b}), \max(\underline{a} \div \underline{b}, \underline{a} \div \overline{b}, \overline{a} \div \underline{b}, \overline{a} \div \overline{b})]\end{aligned}$$

excluding the case $\underline{b} = 0$ or $\overline{b} = 0$.

3 Learning Algorithm for Single-Layer Interval Neural Network

A Neural Network in which at least one of its input, output and weight have interval values is said to be an Interval Neural Network; Escarcina et al. [42]. Interval Neural Network (INN) is formed by processing units called interval neurons. In Interval Neural Networks, neurons are connected in a similar way as they are connected in traditional Neural Networks. The structure of a typical single-layer INN is shown in Fig. 1. The stepwise algorithm for the present INN model with interval computation (defined above) is shown below.

Step 1: The input weights $\left(\tilde{W}_{ij}\right)$ and bias weights $\tilde{\theta}_i$ in interval form are initialized.

Step 2: The training pairs are considered in the form of $\left\{\tilde{Z}_1, \tilde{d}_1; \tilde{Z}_2, \tilde{d}_2; \ldots.\tilde{Z}_I, \tilde{d}_I\right\}$ where $\tilde{Z}_I = \left(\left[\underline{z}_1, \overline{z}_1\right], \left[\underline{z}_2, \overline{z}_2\right], \ldots, \left[\underline{z}_n, \overline{z}_n\right]\right)$ are the inputs and $\tilde{d}_I = \left(\left[\underline{d}_1, \overline{d}_1\right], \left[\underline{d}_2, \overline{d}_2\right], \ldots, \left[\underline{d}_n, \overline{d}_n\right]\right)$ are the desired values for the given inputs in interval form.

Step 3: The output of the network is calculated for the input $\tilde{Z}_I$ as

$$\tilde{O}_J = \left[f\left(\underline{Y}_J\right), f\left(\overline{Y}_J\right)\right]$$

where $\left[\underline{Y}_J, \overline{Y}_J\right] = \left[\underline{W}_{ij}, \overline{W}_{ij}\right] \cdot \left[\underline{Z}_I, \overline{Z}_I\right] + \left[\underline{\theta}_i, \overline{\theta}_i\right]$,

and f is the unipolar activation function defined by

$$f(\tilde{Y}_J) = 1 \Big/ \left[1 + \exp\left(-\gamma\, \tilde{Y}_J\right)\right].$$

Step 4: The weight is modified as

$$\tilde{W}_{ij}^{(New)} = \left[\underline{W}_{ij}^{(New)}, \overline{W}_{ij}^{(New)}\right] = \left[\underline{W}_{ij}^{(Old)}, \overline{W}_{ij}^{(Old)}\right] + \left[\Delta\underline{W}_{ij}, \Delta\overline{W}_{ij}\right]$$

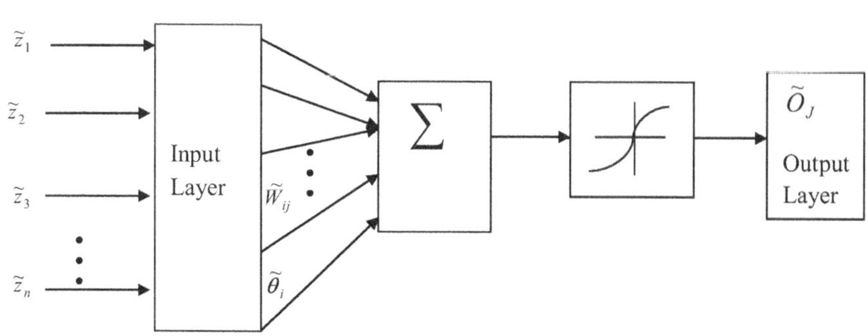

Fig. 1 Single-layer interval neural network

where a change in weights is calculated as

$$\Delta \tilde{W}_{ij} = \left[\Delta \underline{W}_{ij}, \Delta \overline{W}_{ij}\right] = \left[-\eta \frac{\partial \tilde{E}}{\partial \underline{W}_{ij}}, -\eta \frac{\partial \tilde{E}}{\partial \overline{W}_{ij}}\right].$$

In a similar fashion, the bias weights are also updated.

Step 5: The error value is computed as

$$\tilde{E} = \frac{1}{2}\left[(\underline{d}_J - \underline{O}_J)^2 + (\overline{d}_J - \overline{O}_J)^2 \right].$$

4 System Identification of Interval Structural Parameter

Let us consider a shear building with n storey structural system governed by the following set of linear differential equations in interval form as

$$[\tilde{M}]\{\ddot{\tilde{Y}}\}_t + [\tilde{C}]\{\dot{\tilde{Y}}\}_t + [\tilde{K}]\{\tilde{Y}\}_t = \{\tilde{F}\}_t \tag{1}$$

where $\{\ddot{\tilde{Y}}\}_t$ and $\{\dot{\tilde{Y}}\}_t$ are the known acceleration and velocity vectors in interval form, respectively. Moreover, $\left\{\tilde{M}\right\} = [\underline{M}, \overline{M}]$ is an $n \times n$ interval mass matrix of the structure and is given by

$$\left\{\tilde{M}\right\} = \begin{bmatrix} [\underline{m}_1, \overline{m}_1] & 0 & \cdots & \cdots & 0 \\ 0 & [\underline{m}_2, \overline{m}_2] & 0 & \cdots & 0 \\ \cdots & \cdots & \cdots & \cdots & \cdots \\ \cdots & \cdots & 0 & [\underline{m}_{n-1}, \overline{m}_{n-1}] & 0 \\ 0 & \cdots & \cdots & 0 & [\underline{m}_n, \overline{m}_n] \end{bmatrix},$$

$\left\{\tilde{C}\right\} = [\underline{C}, \overline{C}]$ represents *an* $n \times n$ damping matrix of the structure in interval form and is written as

$$\left\{\tilde{C}\right\} = \begin{bmatrix} [\underline{c}_1, \overline{c}_1] + [\underline{c}_2, \overline{c}_2] & -[\underline{c}_2, \overline{c}_2] & 0 & \cdots & 0 \\ -[\underline{c}_2, \overline{c}_2] & [\underline{c}_2, \overline{c}_2] + [\underline{c}_3, \overline{c}_3] & -[\underline{c}_3, \overline{c}_3] & \cdots & 0 \\ \cdots & \cdots & & & \cdots \\ 0 & \cdots & -[\underline{c}_{n-1}, \overline{c}_{n-1}] & [\underline{c}_{n-1}, \overline{c}_{n-1}] + [\underline{c}_n, \overline{c}_n] & -[\underline{c}_n, \overline{c}_n] \\ 0 & & \cdots & -[\underline{c}_n, \overline{c}_n] & [\underline{c}_n, \overline{c}_n] \end{bmatrix},$$

and $\left\{\tilde{K}\right\} = \left[\underline{K}\,,\ \overline{K}\right]$ is $n \times n$ stiffness matrix of the structure in interval form which may be obtained as

$$\left\{\tilde{K}\right\} = \begin{bmatrix} [\underline{k}_1, \overline{k}_1] + [\underline{k}_2, \overline{k}_2] & -[\underline{k}_2, \overline{k}_2] & 0 & \cdots & 0 \\ -[\underline{k}_2, \overline{k}_2] & [\underline{k}_2, \overline{k}_2] + [\underline{k}_3, \overline{k}_3] & -[\underline{k}_3, \overline{k}_3] & \cdots & 0 \\ \cdots & \cdots & & & \cdots \\ 0 & \cdots & -[\underline{k}_{n-1}, \overline{k}_{n-1}] & [\underline{k}_{n-1}, \overline{k}_{n-1}] + [\underline{k}_n, \overline{k}_n] & -[\underline{k}_n \overline{k}_n] \\ 0 & & \cdots & -[\underline{k}_n, \overline{k}_n] & [\underline{k}_n, \overline{k}_n] \end{bmatrix}$$

The solution of the free vibration equation, i.e. for Eq. (1) with uncertain (interval) mass and stiffness, gives the corresponding interval eigenvalues and eigenvectors; Chakraverty [43]. The eigenvalues and eigenvectors are denoted by $\tilde{\lambda}_i$ and $\left\{\tilde{A}\right\}_i = \left\{\underline{A}\,,\ \overline{A}\right\}_i$, $i = 1\,, \ldots,\ n$, respectively, where $\tilde{\omega}_i^2 \left(= \tilde{\lambda}_i\right)$ are the system's interval natural frequency. The above free vibration equation will be an interval eigenvalue problem. The interval eigenvalue and vector are obtained by considering different sets of lower and upper stiffness and mass values. Although there exist different methods to handle interval eigenvalue problems, here the above procedure has been followed to handle the inverse of the matrices in the crisp form separately as lower and upper values. And that is why now we will replace the ' ~ ' from all notations and will consider the case for lower form first and similarly for upper form. Hence, the modal matrix for lower form $\left\{\underline{A}\right\}$ is written as

$$\left[\underline{A}\right] = \left[\left\{\underline{A}\right\}_1 \ \left\{\underline{A}\right\}_2 \ \cdots \ \left\{\underline{A}\right\}_n\right]$$

The diagonal matrix consisting of eigenvalues in lower form is denoted as $\underline{\lambda}_i$, as $\left[\underline{\lambda}\right]_{n \times n}$, a new set of co-ordinates in lower form $\left\{\underline{x}\right\}$ related to the co-ordinates $\left\{\underline{Y}\right\}$, is introduced by the well-known transformation

$$\left\{\underline{Y}\right\} = \left[\underline{A}\right]\left\{\underline{x}\right\}.$$

Proceeding by transforming into normal coordinates $\left(\left\{\underline{Y}\right\} = \left[\underline{A}\right]\left\{\underline{x}\right\}\right)$ and premultiplying Eq. (1) with $\left[\underline{A}\right]^T$, we get

$$\left[\underline{A}\right]^T\left[\underline{M}\right]\left[\underline{A}\right]\left\{\ddot{\underline{x}}\right\} + \left[\underline{A}\right]^T\left[\underline{C}\right]\left[\underline{A}\right]\left\{\dot{\underline{x}}\right\} + \left[\underline{A}\right]^T\left[\underline{K}\right]\left[\underline{A}\right]\left\{\underline{x}\right\} = \left[\underline{A}\right]^T\left\{\underline{F}\right\}. \quad (2)$$

Rewriting Eq. (2) in terms of generalized mass and spectral matrices, we obtain

$$\left[\underline{P}\right]\left\{\ddot{\underline{x}}\right\} + \left[\underline{A}\right]^T\left[\underline{C}\right]\left[\underline{A}\right]\left\{\dot{\underline{x}}\right\} + \left[\underline{S}\right]\left\{\underline{x}\right\} = \left[\underline{A}\right]^T\left\{\underline{F}\right\} \quad (3)$$

where $\left[\underline{P}\right] = \left[\underline{A}\right]^T\left[\underline{M}\right]\left[\underline{A}\right]$ and $\left[\underline{S}\right] = \left[\underline{A}\right]^T\left[\underline{K}\right]\left[\underline{A}\right]$.
Thus, we get the uncoupled equation as

$$\underline{P}_i\ddot{\underline{x}}_i + \overline{\underline{C}}_i\dot{\underline{x}}_i + \underline{S}_i\underline{x}_i = \overline{\underline{F}}_i \text{ for } i = 1, 2, \ldots, n \quad (4)$$

where $[\underline{A}]^T[\underline{C}][\underline{A}] = [\overline{\underline{C}}]$ and $[\underline{A}]^T\{\underline{F}\} = \{\overline{\underline{F}}\}$.

The final solution may be obtained by solving the above differential equations and is written in the form

$$(\{\underline{Y}\} = [\underline{A}]\{\underline{x}\}).$$

Similarly, we can get the solution for the upper form too. After getting the solution using the above classical method, Eq. (1) is rewritten to get the following set of equations in interval form

$$\tilde{m}_1\ddot{\tilde{y}}_1 + (\tilde{c}_1 + \tilde{c}_2)\,\dot{\tilde{y}}_1 - \tilde{c}_2\dot{\tilde{y}}_2 + \left(\tilde{k}_1 + \tilde{k}_2\right)\tilde{y}_1 - \tilde{k}_2\,\tilde{y}_2 = \tilde{f}_1 \tag{5}$$

$$\tilde{m}_2\ddot{\tilde{y}}_2 - \tilde{c}_2\,\dot{\tilde{y}}_1 + (\tilde{c}_2 + \tilde{c}_3)\,\dot{\tilde{y}}_2 - \tilde{c}_3\,\dot{\tilde{y}}_3 - \tilde{k}_2\,\tilde{y}_1 + \left(\tilde{k}_2 + \tilde{k}_3\right)\tilde{y}_2 - \tilde{k}_3\,\tilde{y}_3 = \tilde{f}_2 \tag{6}$$

$$\vdots$$

$$\vdots$$

$$\begin{aligned}\tilde{m}_{n-1}\ddot{\tilde{y}}_{n-1} - \tilde{c}_{n-1}\,\dot{\tilde{y}}_{n-2} + (\tilde{c}_{n-1} + \tilde{c}_n)\,\dot{\tilde{y}}_{n-1} - \tilde{c}_n\,\dot{\tilde{y}}_n - \tilde{k}_{n-1}\,\tilde{y}_{n-2}\\ + \left(\tilde{k}_{n-1} + \tilde{k}_n\right)\tilde{y}_{n-1} - \tilde{k}_n\,\tilde{y}_n = \tilde{f}_{n-1}\end{aligned} \tag{7}$$

$$\tilde{m}_n\ddot{\tilde{y}}_n - \tilde{c}_n\,\dot{\tilde{y}}_{n-1} + \tilde{c}_n\,\dot{\tilde{y}}_n - \tilde{k}_n\,\tilde{y}_{n-1} + \tilde{k}_n\,\tilde{y}_n = \tilde{f}_n \tag{8}$$

Equation (8) is then written as

$$\tilde{m}_n\ddot{\tilde{y}}_n + \tilde{c}_n\left[\dot{\tilde{y}}_n - \dot{\tilde{y}}_{n-1}\right] + \tilde{k}_n\left[\tilde{y}_n - \tilde{y}_{n-1}\right] = \tilde{f}_n. \tag{9}$$

The above equation may now be presented as

$$\tilde{m}_n\ddot{\tilde{y}}_n + \tilde{c}_n\dot{\tilde{d}}_n + \tilde{k}_n\tilde{d}_n = \tilde{f}_n \tag{10}$$

where $\dot{\tilde{d}}_n = \left[\dot{\tilde{y}}_n - \dot{\tilde{y}}_{n-1}\right]$ and $\tilde{d}_n = \left[\tilde{y}_n - \tilde{y}_{n-1}\right]$.

Here, $\dot{\tilde{d}}$ and $\tilde{d}_n$ define the known relative velocity and displacement in interval form for n^{th} storey which is calculated by the above classical method. Single-layer interval neural network is used to solve Eq. (10). For this interval neural network, the inputs are taken as structural acceleration, relative velocity and relative displacement for the nth storey, and outputs are taken as the applied force at time t. Using this single-layer interval neural network, Eq. (10) is solved by a continuous training process with n training patterns, and the converged weight matrix is obtained. This converged weight matrix gives the corresponding physical parameters such as $\tilde{m}_n$, $\tilde{c}_n$ and $\tilde{k}_n$ in interval form. Identified parameters of nth storey are then used to identify the parameters of the $(n-1)^{\text{th}}$ storey using Eq. (7). Proceeding in the same manner, the parameters of the $(n-2)^{\text{th}}$ storey are determined and the process goes on till the unknown parameters for the first storey are obtained. The cluster of interval neural network diagram for n storey structure is shown in Fig. 2.

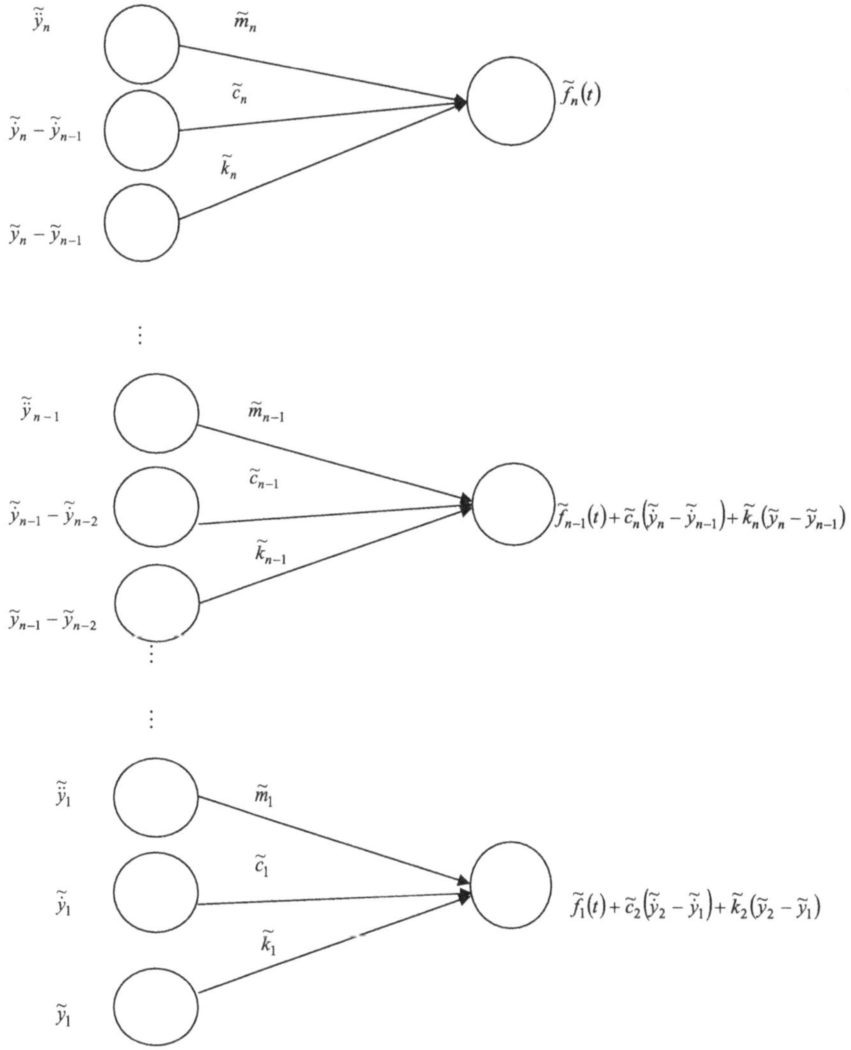

Fig. 2 Proposed cluster interval neural network model for n storey shear structure

5 Results and Discussion

The above method has been developed for different storey shear structures with damping and without damping cases. Interval neural network is trained till the desired accuracy is reached. The methodology has been discussed below by giving the results for the following two cases.

Case (i) Without Damping: In this case, three problems are considered which are

(a) Two storey;
(b) Four storey;
(c) Eight storey.

Case (ii) With Damping: In this case, only one problem of a three-storey shear structure has been considered

(a) Three storey.

Case (i) (a) Two-storey shear building:

Structural parameter of the shear building has been identified using the direct method where the data are considered to be in interval form. The data are initially generated by taking the theoretical structural parametric values. These generated data are used first to train the neural network for n training patterns thus establishing the converged weight matrix of the neural network. The corresponding component of the converged weight matrix gives the unknown or present structural parameters. Then the trained and theoretical data has been compared to show the efficiency of the proposed method. These ideas have been applied in all the cases. The initial structural parameter matrices in interval form are taken as storey masses $\tilde{m}_1 = [1.5,\ 2.5]$, $\tilde{m}_2 = [1,\ 2]$ kNs^{2}m^{-1}, and the storey stiffnesses $\tilde{k}_1 = [390,\ 410]$ and $\tilde{k}_2 = [290,\ 310]$ kNm^{-1}. The harmonic force exerted in the shear building is assumed in interval form, viz., $\tilde{f}_1(t) = [90\sin(1.6\pi t) + \pi,\ 110\sin(1.6\pi t) + \pi]$ and $\tilde{f}_2(t) = [90\sin(1.6\pi t),\ 110\sin(1.6\pi t)]$ kN. Table 1 includes the comparison of identified structural parameters with the theoretical parameters. The epoch versus mass and stiffness for two storeys are plotted in Figs. 3, 4, 5, 6 to show how the structural parameters converge.

(b) Four-storey shear building:

Table 1 Identified mass and stiffness parameters of two-storey shear building in interval form under the forced-vibration test (without damping)

Parameter	Storey	Theoretical	Identified	Centered theoretical	Centered identified
Mass (kN s^2 m^{-1})	M_1	[1.5, 2.5]	[1.6136, 2.4988]	2	1.9866
	M_2	[1, 2]	[1.0915, 2.2869]	1.5	1.6859
Stiffness (kN m^{-1})	K_1	[390, 410]	[389.9947,409.9999]	400	399.9994
	K_2	[290, 310]	[290.008, 309.9951]	300	299.1889

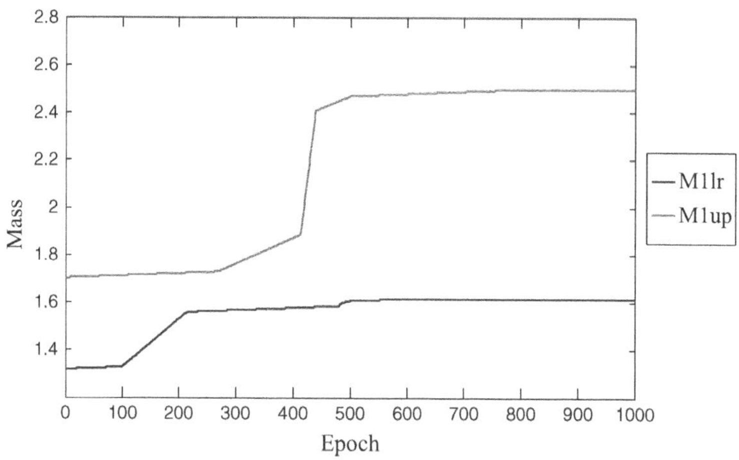

Fig. 3 Convergence of interval mass parameter (M_1) with respect to the number of epochs for two-storey shear structure (without damping)

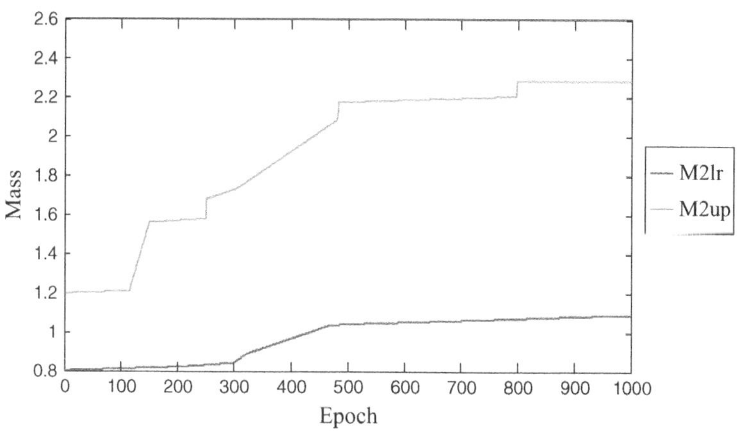

Fig. 4 Convergence of interval mass parameter (M_2) with respect to the number of epochs for two-storey shear structure (without damping)

In this problem, the structural parameters in interval form are the storey masses $\tilde{m}_1 = [3, 4]$, $\tilde{m}_2 = [3, 4]$, $\tilde{m}_3 = [2.5, 3.5]$, $\tilde{m}_4 = [2, 3]$ kNs^{2}m^{-1}, and the storey stiffnesses $\tilde{k}_1 = [1190, 1210]$, $\tilde{k}_2 = [1190, 1210]$, $\tilde{k}_3 = [790, 810]$ and $\tilde{k}_4 = [590, 610]$ kNm^{-1}. The harmonic forces exerted in the shear building are taken as $\tilde{f}_1(t) = [90\sin(1.6\pi t) + \pi,\ 110\sin(1.6\pi t) + \pi]$, $\tilde{f}_2(t) = [90\sin(1.6\pi t),\ 110\sin(1.6\pi t)]$, $\tilde{f}_3(t) = [0.5\sin(3.2\pi t),\ 1.5\sin(3.2\pi t)]$ and $\tilde{f}_4(t) = [1.0\sin(3.2\pi t),\ 3\sin(3.2\pi t)]$ kN. Dynamic displacements of four DOFs with a duration of 1 s are stored and used to train the interval neural network

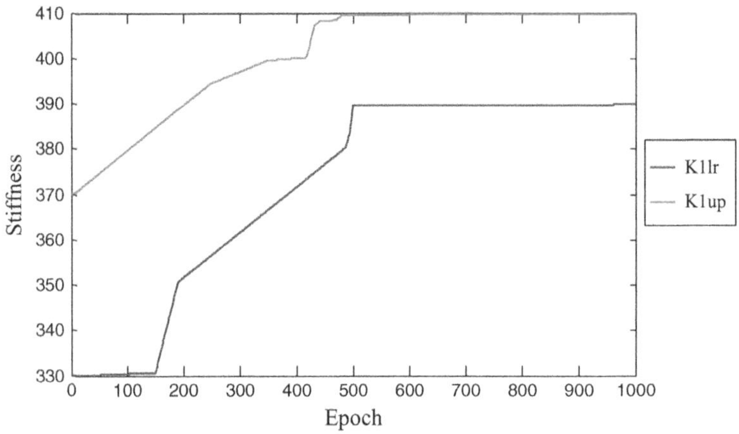

Fig. 5 Convergence of interval stiffness parameter (K_1) with respect to the number of epochs for two-storey shear structure (without damping)

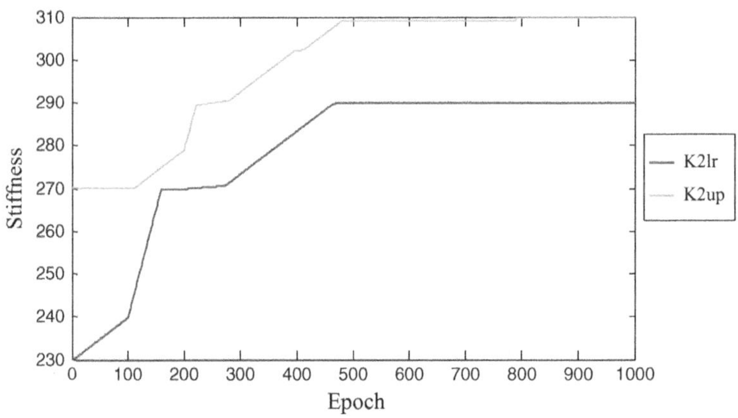

Fig. 6 Convergence of interval stiffness parameter (K_2) with respect to the number of epochs for two-storey shear structure (without damping)

for the identification procedure. Comparisons between the identified and theoretical parameters of the structures in interval form are incorporated in Table 2.

(c) Eight-storey shear building:

In this case, the structural parameters, viz., the storey masses and stiffnesses of the structure in interval form, are given in Table 3. The harmonic force in interval form exerted in the eight storeys of the building is $\tilde{f}_8(t) = [1.0\sin(3.2\pi t),\ 3.0\sin(3.2\pi t)]$ kN. The dynamic displacements of eight DOFs with a duration of 1 s are stored and used to train the interval neural network for the identification procedure. The identified structural parameters are compared with the

Table 2 Identified mass and stiffness parameters of four-storey shear building in interval form under the forced-vibration test (without damping)

Parameter	Storey	Theoretical	Identified	Centered theoretical	Centered identified
Mass (kN s^2 m^{-1})	M_1	[3, 4]	[2.8, 3.999]	3.5	3.5000
	M_2	[3, 4]	[2.9, 3.899]	3.5	3.5000
	M_3	[2.5, 3.5]	[2.3577, 3.3336]	3	3.0825
	M_4	[2, 3]	[1.7690, 2.9828]	2.5	2.4627
Stiffness (kN m^{-1})	K_1	[1190, 1210]	[1190.00, 1210.00]	1200	1199.0988
	K_2	[1190, 1210]	[1190.00, 1210.00]	1200	1199.0008
	K_3	[790, 810]	[789.998, 809.999]	800	799.9976
	K_4	[590, 610]	[589.9983, 610.000]	600	599.9999

theoretical structural parameters of this structural system. The corresponding results are shown in Table 3.

Case (ii) (a) Three-storey shear building:
Structural parameters in interval form are taken as the storey masses $\tilde{m}_1 = [2.5,\ 3.5]$, $\tilde{m}_2 = [2.5,\ 3.5]$, $\tilde{m}_3 = [1.5,\ 2.5]$ kNs^{2}m^{-1} and the storey stiffnesses $\tilde{k}_1 = [1190,\ 1210]$, $\tilde{k}_2 = [790,\ 810]$ and $\tilde{k}_3 = [590,\ 610]$ kNm^{-1}; the damping values $\tilde{c}_1 = [7,\ 9]$, $\tilde{c}_2 = [4,\ 6]$ and $\tilde{c}_3 = [3,\ 5]$ kN s m^{-1}. The harmonic forces exerted in the shear building in interval form are considered as $\tilde{f}_1(t) = [90\sin(1.6\pi t) + \pi,\ 110\sin(1.6\pi t) + \pi]$, $\tilde{f}_2(t) = [90\sin(1.6\pi t),\ 110\sin(1.6\pi t]$ and $\tilde{f}_3(t) = [0.5\sin(3.2\pi t),\ 1.5\sin(3.2\pi t)]$ kN. Comparisons between the identified and theoretical structural parameters are incorporated in Table 4.

6 Conclusions

The present chapter uses the powerful soft computing technique, viz., single-layer Interval Neural Network (INN) to identify the uncertain structural parameters. If the available information and/or data are uncertain or non-probabilistic in nature, then the mathematical model needs to be developed accordingly. As such, an interval neural network has been developed which can handle uncertain or interval data. Here, a direct method for system identification of uncertain multi-storey shear structures from their dynamic responses has been proposed in interval form. Governing equations of motion are modified based on relative responses of consecutive stories

Table 3 Identified mass and stiffness parameters of eight-storey shear building in interval form under the forced-vibration test (without damping)

Parameter	Storey	Theoretical	Identified	Centered theoretical	Centered identified
Mass (kN s^2 m^{-1})	M_1	[3, 4]	[2.899, 3.999]	3.5	3.5000
	M_2	[2.5, 3.5]	[2.500, 3.500]	3	3.0000
	M_3	[2.5, 3.5]	[2.1858, 3.500]	3	2.9435
	M_4	[2.5, 3.5]	[2.4453, 3.500]	3	2.8900
	M_5	[2, 3]	[1.6021, 3.00]	2.5	2.4378
	M_6	[2, 3]	[1.6644, 3.00]	2.5	2.4998
	M_7	[2, 3]	[1.8965, 3.00]	2.5	2.3492
	M_8	[1.5, 2.5]	[1.2234, 2.500]	2	1.9928
Stiffness (kN m^{-1})	K_1	[1190, 1210]	[1190.00, 1210.00]	1200	1199.9879
	K_2	[1190, 1210]	[1190.00, 1210.00]	1200	1200.0000
	K_3	[790, 810]	[789.998, 810.000]	800	799.9985
	K_4	[790, 810]	[789.999, 810.000]	800	799.9997
	K_5	[790, 810]	[789.9993, 810.000]	800	799.9992
	K_6	[590, 610]	[589.999, 610.000]	600	599.9955
	K_7	[590, 610]	[589.9997, 610.00]	600	599.8989
	K_8	[590, 610]	[589.9999, 610.00]	600	599.9998

and are implemented in a cluster of interval neural network models. Various example problems have been solved and related results are reported to show the reliability and efficacy of the present method. It is worth mentioning that the cluster of INN may directly estimate the structural parameters in interval form. Interval estimates are certainly useful for civil engineers by knowing the lower and upper bounds of the parameters.

Table 4 Identified mass and stiffness parameters of three-storey shear building in interval form under the forced-vibration test (with damping)

Parameter	Storey	Theoretical	Identified	Centered theoretical	Centered identified
Mass (kN s^2 m^{-1})	M_1	[2.5, 3.5]	[2.50, 3.50]	3	3.0000
	M_2	[2.5, 3.5]	[2.2451, 3.3727]	3	2.8881
	M_3	[1.5, 2.5]	[1.4321, 2.331]	2	1.6813
Stiffness (kN m^{-1})	K_1	[1190, 1210]	[1190.00, 1210.00]	1200	1200.000
	K_2	[790, 810]	[789.991, 809.992]	800	799.9997
	K_3	[590, 610]	[589.997, 609. 998]	600	599.9990
Damping (kN s m^{-1})	C_1	[7, 9]	[7.00, 9.00]	8	8.0000
	C_2	[4, 6]	[3.9858, 5.9903]	5	4.9948
	C_3	[3, 5]	[2.9991, 4.9938]	4	3.9835

References

1. Friswell MI, Mottershead JE (1995) Finite element model updating in structural dynamics. Kluwer Academic Publishers, Dordrecht, the Netherlands
2. Alvin KF, Robertson AN, Reich GW, Park KC (2003) Structural system identification: from reality to models. Comput Struct 81:1149–1176
3. Fassois SD, Sakellariou JS (2007) Time-series methods for fault detection and identification in vibrating structures. Philos Trans R Soc A 365:411–448
4. Khanmirza E, Khaji N, Majd VJ (2011) Model updating of multistory shear buildings for simultaneous identification of mass, stiffness and damping matrices using two different soft-computing methods. Expert Syst Appl 38(5):5320–5329
5. Mahmoudabadi M, Ghafory-Ashtiany M, Hosseini M (2007) Identification of modal parameters of non-classically damped linear structures under multicomponent earthquake loading. Earthquake Eng Struct Dynam 36:765–782
6. Hegde G, Sinha R (2008) Parameter identification of torsionally coupled shear buildings from earthquake response records. Earthquake Eng Struct Dynam 37:1313–1331
7. Hong AL, Betti R, Lin CC (2009) Identification of dynamic models of a building structure using multiple earthquake records. Struct Control Health Monit 16:178–199
8. Masri SF, Smyth AW, Chassiakos AG, Caughey TK, Hunter NF (2000) Application of neural networks for detection of changes in nonlinear systems. J Eng Mech 126(7):666–676
9. Kao CY, Hung SL (2003) Detection of structural damage via free vibration responses generated by approximating artificial neural networks. Comput Struct 81(28–29):2631–2644
10. Pillai P, Krishnapillai S (2007) A hybrid neural network strategy for identification of structural parameters. Struct Infrastruct Eng 6(3):379–391
11. Bakhary N, Hao H, Deeks AJ (2009) Structure damage detection using neural with multi-stage substructuring. Adv Struct Eng 13(1):1–16
12. Zhang S, Wang H and Wang W (2010) Damage detection in structures using artificial neural networks. In: International conference on artificial intelligence and computational intelligence, Sanya, 1:207–210
13. Zhang K, Li H, Duan Z, Law SS (2011) A probabilistic damage identification approach for structures with uncertainties under unknown input. Mech Syst Signal Process 25(4):1126–1145

14. Shi A and Yu XH (2012) Structural damage detection using artificial neural networks and wavelet transform. In: IEEE international conference on computational intelligence for measurement systems and applications (CIMSA) proceedings, pp 7–11
15. Lei Y, Su Y, Shen W (2013) A probabilistic damage identification approach for structure under unknown excitation and with measurement uncertainties. J Appl Math 2013:1–7
16. Chakraverty S (2004) Modelling for identification of stiffness parameters of multi-storey structure from dynamic data. J Sci Ind Res 63(2):142–148
17. Nicoud YR, Raphael B, Smith IF (2005) System identification through model composition and stochastic search. J Comput Civ Eng 19(3):239–247
18. Xu B, Song G, Masri SF (2012) Damage detection for a frame structure model using vibration displacement measurement. Struct Health Monit 11(3):281–292
19. Nandakumar P, Shankar K (2013) Identification of structural parameters using consistent mass transfer matrix. Inverse Probl Sci Eng 24:1–22
20. Billmaier M, Bucher C (2013) System identification based on selective sensitivity analysis: a case-study. J Sound Vib 332(11):2627–2642
21. Brownjohn JMW (2003) Ambient vibration studies for system identification of tall buildings. Earthquake Eng Struct Dynam 32(1):71–96
22. Yang JN, Lei Y, Pan SW, Huang N (2003) System identification of linear structures based on Hilbert-Huang spectral analysis; Part 1: normal modes. Earthquake Eng Struct Dynam 32:1443–1467
23. Chakraverty S (2005) Identification of structural parameters of multistorey shear buildings from modal data. Earthquake Eng Struct Dynam 34:543–554
24. Yoshitomi S, Takewaki I (2009) Noise-bias compensation in physical-parameter system identification under microtremor input. Eng Struct 31:580–590
25. Beskhyroun S, Wotherspoon L, Ma QT (2013) System identification of a 13- story reinforced concrete building through ambient and forced vibration. In: 4th ECCOMAS thematic conference on computational methods in structural dynamics and earthquake engineering. Greece, pp 1–11
26. Yun CB, Bahng EY (2000) Substructural identification using neural networks. Comput Struct 77(1):41–52
27. Wu ZS, Xu B, Yokoyama K (2002) Decentralized parametric damage based on neural networks. Comput Aided Civ Infrastruct Eng 17(3):175–184
28. Huang CS, Hung SL, Wen CM, Tu TT (2003) A neural network approach for structural identification and diagnosis of a building from seismic response data. Earthquake Eng Struct Dynam 32(2):187–206
29. Chakraverty S (2007) Identification of structural parameters of two-storey shear buildings by the iterative training of neural networks. Archit Sci Rev 50(4):380–384
30. Pizano DG (2011) Comparison of frequency response and neural network techniques for system identification of an actively controlled structure. Dyna 78(170):79–89
31. Xu B, Wu Z, Chen G, Yokoyama K (2004) Direct identification of structural parameters from dynamic responses with neural networks. Eng Appl Artif Intell 17(8):931–943
32. Chen CH (2005) Structural identification from field measurement data using a neural network. Smart Mater Struct 14:104–115
33. Beheshti M, Berrached A, Korvin AD, Hu C, Sirisaengtaksin O (1998) On interval weighted three-layer neural networks. In: Proceedings of the 31st annual simulation symposium. Boston, MA, pp 188–194
34. Garczarczyk ZA (2000) Interval neural networks. In: IEEE international symposium on circuits and systems. Geneva, Switzerland, 3:567–570
35. Chetwynd D, Worden K, Manson G (2006) An application of interval-valued neural networks to a regression problem. Proc R Soc A 462(2074):3097–3114
36. Wang X, Yang H, Qiu Z (2010) Interval analysis method for damage identification of structures. AIAA J 48(6):1108–1116
37. Zhang MQ, Beer M, Koh CG (2012) Interval analysis for system identification of linear MDOF structures in the presence of modeling errors. J Eng Mech 138(11):1326–1338

38. Okada H, Matsuse T, Wada T, Yamashita A (2012) Interval GA for evolving neural networks with interval weights and biases. In: SICE annual conference ankita university. Ankita, Japan, pp 1542–1545
39. Sahoo DM, Das A, Chakraverty S (2014) Interval data based system identification of multi-storey shear buildings by artificial neural network modelling. Archit Sci Rev 58:1–11
40. Chakraverty S, Sahoo DM (2014) Interval response data based system identification of multi storey shear buildings using interval neural network modelling. Comput Assist Methods Eng Sci 21:123–140
41. Lee KH (2009) First course on fuzzy theory and applications. Springer International Edition, pp 1–333
42. Escarcina REP, Bedregal BRC, Lyra A (2005) Interval computing in neural networks: one layer interval neural networks. In: Proceedings of intelligent information technology, 7th international conference on information technology. CIT, Hyderabad, India, vol 3356, pp 68–75
43. Chakraverty S (2009) Vibration of plates. CRC Press, Taylor and Francis Group, pp 1–411

Security Issues on IoT Communication and Evolving Solutions

Uddalak Chatterjee and Sangram Ray

Abstract Internet of Things (IoT) is an interconnected wireless network where smart nodes (IoT devices) interact with each other in order to exchange data through the communicating medium. IoT technology has become important for people to build smart systems upon the use of technology. Internet of things is realized by the idea of free flow of information among various low-power embedded devices that use the Internet to communicate with one another. In the recent past IoT have grown rapidly and have become an extension of existing universal Internet. It can be easily anticipated that large scale systems equipped with numerous sensors will prevail in our society. With the rise of the Internet of Things (IoT) technology, the number of IoT devices/sensors has also increased significantly Billions of smart devices connected to IoT environment can communicate among themselves using sensors and actuators. This rapid growth and inclusion of IoT technologies in our daily life are facing challenges since most of the devices specially sensors in IoT network are resource constraint in terms of energy, computation capability, etc. Data collected from these sensors sent through the middleware like gateway, routers, etc. to the cloud servers or toward various analytical engine for meaningful knowledge discovery. These processed data and knowledge have lately attracted huge attention, and organizations are excited about the business value of the data that will be generated by deploying such networks. With this advent of IoT it has also attracted various security and privacy concerns. Due to the structurally open IoT architecture and the tremendous usage of the paradigm itself, causes to generate many unconventional security issues for the existing networking technologies. Moreover, since sensor nodes are cooperative within the IoT network, this sharing of data can create new challenges that can disrupt the systems' regular functionalities and operations. In another aspect the growth of IoT technologies has enhanced by assimilating them with cloud computing and the era IoT-cloud has emerged. With these, some new class of security and privacy issues have also introduced. Furthermore, the commercialization of the IoT has led to several public security concerns including threats of cyber-attacks, privacy issues, and fraud crimes. This chapter gathers the needed

U. Chatterjee · S. Ray (✉)
Department of Computer Science and Engineering, National Institute of Technology Sikkim, Ravangla 737139, Sikkim, India

S. Chakraverty (ed.), *Soft Computing in Interdisciplinary Sciences*,
Studies in Computational Intelligence 988,
https://doi.org/10.1007/978-981-16-4713-0_10

information to give a complete picture of security issues and problems faced in IoT communication. In this chapter, we detail the major security as well as privacy issues more specifically. An extensive description of security threats and challenges across the different layers of the architecture of IoT systems is represented. The issues related to IoT-cloud is also highlighted. The light will be shed on the state-of-the-art solutions to the emerging and latest security issues in this field. We hereby present the evolving resolve policies as mined from the research work of various authors in this field that will expose the several research areas in IoT-cloud era.

Keywords Internet of Things · IoT security · Privacy · Cloud server

1 Introduction

Internet of Things is one of the current top tech buzzwords. IoT comprises things that have unique identities and are connected to internet. By 2022 there will be a total of 50 billion devices or things connected to internet. IoT is not limited to just connecting things to the Internet but also allow things to communicate and exchange data. So, IoT can be defined as a dynamic global network infrastructure with self-configuring capabilities based on standard and interoperable communication protocols where physical and virtual things have identities, physical attributes and virtual personalities and use intelligent interfaces, and are seamlessly integrated into information network, often communicate data associated with users and their environments. IoT network has distinct characteristics such as they are dynamic and self-adopting, i.e., IoT devices and systems may have the capability to dynamically adapt with the changing contexts and take actions based on their operating conditions, user's context or sensed environment. It is self-configuring and support a number of interoperable communication protocols and can communicate with other devices and also with outside infrastructure. With the advent of ipv6 and 6lowpan technology each IoT device has a unique identity. IoT will get absolute power to connect billions of devices all over the globe while operating on IPV6 as a backbone to it. More no of smart devices can communicate seamlessly among themselves using IoT services supported by IPV6. Figure 1 depicts an overview of architecture and various application domains of IoT such as smart home, smart cities, energy sector, retail and Agriculture sector, etc.

A. **Evolution of IoT communication**: The first step toward IoT evolution was started in mid 1990s when Internet connectivity began to gain popularity and when first sensors were introduced and started getting embedded with the various devices. In 1999 Kevin Aston proposed the term Internet of Things (IoT) [1] in the auto-id center while introducing RFID to tag different item with unique identity. Then in 2000 LG has invented first refrigerator which is connected to Internet [2]. Later on, in 2005 commercial production of smart objects was started. Also in 2008 IoT gained recognition by European union [2, 3]. At the same time Industrialization of IoT has also began. After that

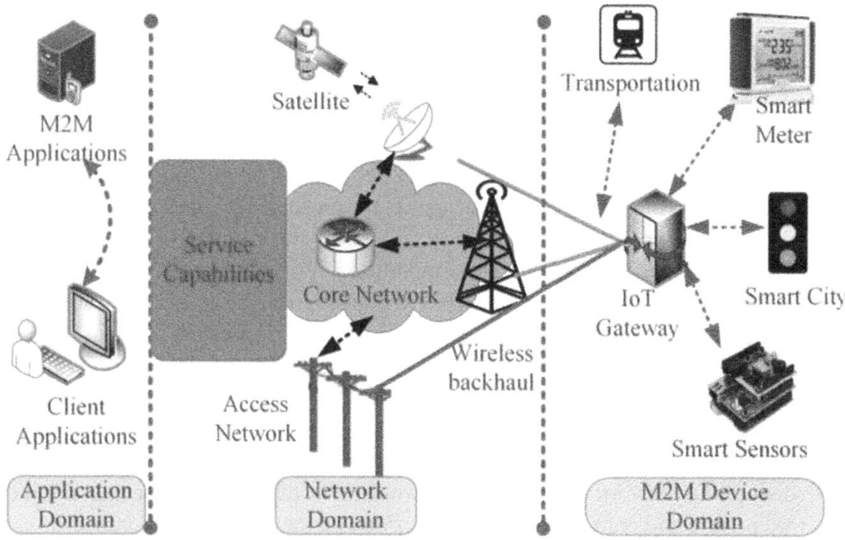

Fig. 1 High-level architecture of IoT

continuous and steady development of IoT technologies is going on and from 2016 and beyond we have numerable smart applications like connected home, connected cars, IoT enabled manufacturing plants, smart health, etc., which is using IoT as backbone. In the forthcoming generations the smart devices which are connected through IoT will become more complex. They will no more be used only for sensing and reporting tools but also for making decisions using machine learning and artificial intelligence. Irrespective of the domain starting from making smart automobile, smart homes, manufacturing robots, or a simple thermostat all the devices will be connected using IoT networks and will have greater impact on the whole society. The success of IoT has led to fusion of different communication infrastructure, which in turn has led to design smart gateways to connect sensors with the traditional Internet Devices. Like the weather sensing bio-transponder, smart automobiles, sensors for health monitoring, and so on.

B. **Advantages and disadvantages of IoT communication**: IoT has a huge impact on current technological landscape. Several advantages and disadvantages are discussed in this subsection. One of the most prominent advantages of IoT is the easy and remote access to information [1, 3, 4]. Another major advantage of IoT is automation support. Using this technology automation of day to day activities can be managed remotely and with minimum human interaction [1–6]. Among other advantages IoT gives better communication with increased efficiency in machine to machine or human to machine interaction. IoT also provides better communication since it is a network of interconnected devices. It gives quick and accurate results. Due to the advancement of technology of faster connectivity, IoT proves to be more cost effective as the one can manage the daily job from remote location [1, 3–9]. On one hand IoT systems has the ability to gather huge amount of data and on the other it can aggregate and communicate information as and when required. However, IoT systems deal with this huge amount of personal and confidential data without or minimum human intervention. This is the reason IoT comes with lot of security issues associated with it [1, 3]. The smart devices are in many cases are resource constraint in terms of computation power and energy [4]. That is why they sometimes are more vulnerable to attacks from different adversaries. The architecture of IoT is huge and very complex, different types of smart devices, networks are involved in making IoT networks. This is the reason why IoT networks are the target for malicious users and adversaries. As the devices used in the IoT often are very resource constraint and of various nature it makes easy for the adversaries to find the gap where they can attack the IoT network and disrupt the network from its normal working or misuse the information which are private and confidential making serious privacy and security concern.

C. **Applications of IoT communication**: The range of the application area of IoT is huge which is categorically depicted in Fig. 2, i.e., from health monitoring to disaster recovery it has a vast application area. In healthcare, IoT is transforming the conventional idea of health monitoring by using wearable devices, implanted devices also surveillance of patients in hospitals is being done using IoT. In armed forces or security forces make use of this technology by doing intrusion detection using various available sensors, then troop monitoring can also be done using it. For industrial purposes huge application is already in place for IoT like manufacturing control, production automation, etc. In the

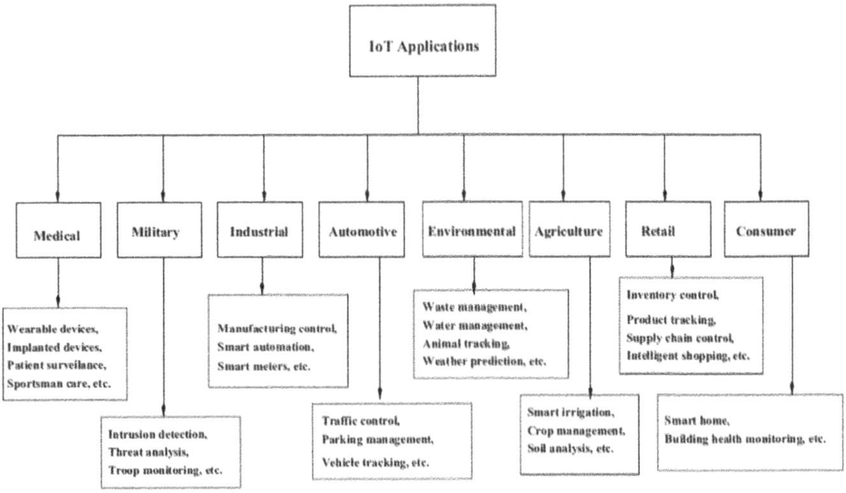

Fig. 2 IoT application

field of transportation smart cars, smart vehicles have huge application of IoT, smart traffic control, smart parking, and automobile tracking too are some major application area of IoT. In environmental front, management of waste and water control, weather prediction applications make use of various IoT sensors. One can make use of IoT even for tracking animals. Then in agriculture IoT has many applications it also makes use of different IoT sensors like humidity sensor, water sensor that tells you about the moisture condition of the soil and when to water the crop, apart from that for pest control and monitoring of crop also one can make use of this technology. In retail application IoT can be used for inventory control, tracking the products, controlling the supply chain of products, then smart shopping management, etc. Smart home and smart home appliances are another major application area of IoT where one can remotely control and monitor their own home. The concept of smart city also make use of IoT. In summary, in today's world IoT technology can be used in every aspect of our lives.

D. **Security and Privacy Issues in IoT**: IoT network always need to communicate with uncertain condition the main characteristics of IoT communication are interoperability, dynamicity and heterogeneity [1]. Thus, it is always vulnerable to various security and privacy challenges. Current IoT network is scalable and distributed in nature so, it must ensure a state-of-the-art security architecture to tackle different security and privacy challenges that could disrupt the smooth functioning of IoT smart devices and networks [3]. The traditional security goals described by the authors in [4] and they are (i) confidentiality, (ii) integrity, and (iii) availability. Here in this subsection, we present a IoT communication layer wise security and privacy challenges as proposed by authors in [5–16].

(i) ***Perception Layer***: As per the architecture of IoT network the Lower most layer is perception layer or sensing layer also it is known as device layer [16, 17]. Mainly different kinds of sensors and devices like RFID, motion sensor, acceleration sensor, etc., and actuators that are used for IoT, remain active in this layer, these sensors collect information from outside world and sends them to either gateway or edge nodes for further processing. Different types of attacks could possibly hamper the communication, thus many authors over the years have identified the probable attacks for this layer which are given in Table 1.

(ii) ***Middleware/gateway layer/Networking layer***: According to IoT communication layers the next layer after sensing layer is the gateway layer or it is also called network layer [16, 17]. The data that are sensed by the sensors eventually comes to gateway layer. The purpose of this layer is to send the information securely either to cloud server or into some machine learning applications. Adversary target this layer in different ways and try to meddle the data or try to hamper the whole communication. Authors have identified the most prominent attacks that this layer is prone to, which is given in Table 2.

Table 1 Attacks on perception/sensing layer and their description

Attacks/threats	Description
Denial of service attack	IoT sensing nodes are resource contains having very fewer computing capabilities, thus the attackers can attack these sensors and try and disrupt the service. Hence denial of service occurs [5, 6, 16]
Hardware jamming	Nodes can be damaged by the attackers by changing the parts of the node hardware [6, 7, 16]
Insertion of forged nodes	A malicious node can be placed by the attacker between the actual nodes or between server and nodes of the network to get access and control over the IoT network [8, 9, 16]
Brute force attack	As the sensing nodes are resource constraint, brute force attack can easily compromise the access control of the devices [10, 16]
Eavesdropping	İf an advarsary can get hold of the information while two nodes are communicating [11, 16]
Spoofing	If an adversary try to inject false information and try to depict that information coming from original authenticated source, then this attack occurs. Adversary can get all the permission and access by this way and can harm the system [12, 16]

Table 2 Common attacks on connection layer and their description

Attacks/threats	Description
Denial of service attack	This attck make the Servers or devices stop providing the services to the user by overloading or capturing network connectivity [5–13, 16]
Session Hijacking attacks	Through this kind of attack. Attackers can hijack the session and obtain the access to the network [5–13, 16]
Man-in-the-middle attacks	Confidential information between two sensing nodes can easily be obtained if there is no proper encryption mechanism in place. Attacker can intersect the communication channel [5, 16]
Sybil attack	More than one identity is created for a single node, the adversary targets the system by manipulating the node [5, 6, 16]
Sleep deprivation attack	This Attack keeps the sensor node awake for a portion of time which lead to batteries consumption as a result it minimizes batteries life time and causing the sensor nodes to shut down [6, 7, 16]

(iii) ***Application/cloud layer***: The topmost layer is responsible for providing the end user service. This provides an interface with which data are sent toward the cloud server or into the analytical engines for monitoring and control. The attacks that could crop up in this layer are listed in Table 3.

E. **Existing survey on security issues on IoT communication**: Several reviews and novel research works have been done for continuous improvement in respect with the relentless progress of IoT standard, the various efficiency and security

Table 3 Attacks on application layer and their description

Attacks/threats	Description
Data security in cloud computing	Cloud service provider will hold the responsibility of protecting all the Data that is collected, processed, and stored on the cloud [5–13, 16]
Malicious code injection	The adversary tries and injects a malicious code into the system and takes away user's data. This is known as Malicious code injection [5–13, 16]
DoS attack	DoS attackers terminate the availability of the service or the application [5–13, 16]
Sniffing attack	The sniffing attack takes place when the adversary tries to sniff in to the network or tries to capture data while in transit through the network [5–13, 16]

issues faced by IoT system. Here in this subsection, we will provide a picture of recent notable survey work that has been carried out in this context of security and privacy in IoT. In 2015 Granjal et al. [18] have done a survey on security protocols and the open research issues and analyzed the existing protocols, how they are solving the fundamental security problems. In 2016 Iqbal et al. [10] have done a security review of IoT where they have discussed on RFID technology and wireless sensors. This study particularly focused on suitable security architecture and risk management in IoT and also provided a guideline toward implementing smart home and smart cities. After that in 2018 Mena et al. [3] have done a thorough survey on security issues on IoT with the focus on the different technologies and architectures it also focuses on the weakness of IoT and its effect on the fundamental security and privacy requirements of IoT. A very notable work on this context is done by Ghani et al. [4] in 2019 in their survey they have given an elaborate list of security and privacy guideline in IoT communication focused on Edge nodes and reference architecture of IoT. In their work authors also identified set of implementation techniques for achieving those security requirement goals. Moreover, authors also discussed the possible challenges one has to face to implement the security guidelines and how to overcome those. In another work authors Mohanta et al. [19] in 2020 have first discussed the latest area of implementation then authors discussed the effect of latest technology like artificial intelligence, Machine learning, and blockchain on security of IoT and how those technologies could be used to achieve secure IoT communication. In the same year another survey done by authors Alferidah et al. [15] where they have done a survey of security attacks which concentrates on different layers of IoT communication and also discusses what to do, to remove these attacks and then some open research issues where work need to be done is also highlighted. Another work is done by Thilakarathne et al. [16] which summarizes the latest works done to achieve the basic security and privacy requirements. Then recently in 2021 Yue et al. [17] have done a survey on deep learning-based approaches to solve them in

IoT security and privacy concern. The comparison of existing surveys in this context, with our work is given in Table 10.4.

F. **Motivation and contribution of this work**: The attacks on IoT generally are of two types one is the attackers tries to attack in the different layers of IoT architecture, as IoT is a layered architecture and secondly the adversary's tries to cause the attack by exploiting the communication system gaps and by doing so damage the whole network. To allow greater control and autonomy it is very important to fill the security gaps in the network, so that security and privacy could be maintained in this important network scenario. This motivates us to do a survey giving layer wise solution of attacks as well as solution using state-of-the-art technology which will be helpful to fulfill the above requirement and also help in formulating optimum solution to protect the network as well as the smart devices from malicious users and adversary's and make the IoT communication more consistent. In this chapter we tried to make a draft of the major security issues in this scenario mentioned in subsection D. of Sect. 1. and also tried to do a cumulative survey on the probable solution proposed by different authors, by undertaking an extensive and latest literature survey which are available in this context. Our survey has two parts first it tries to find solution to counter the attacks on the different layers of IoT communication. And secondly it lists all possible solution to protect the security and privacy in an IoT network using state-of-the-art technologies.

G. **Organization of the chapter**: The chapter organization is as follows in Sect. 2. the state-of-the-art security solutions in IoT communication as discussed in the literature are presented. Which focuses on the traditional as well as recent solutions given by the authors to the security challenges in IoT communication. Then finally Sect. 3. Concludes the paper (Table 10.4).

2 Review of State-of-the-Art Security and Privacy Solutions

In this section, we have done a thorough review of the latest literature contributed by different authors in context of IoT security. Many surveys in context of IoT security and security implementation have been done over the years, where authors focus on the issues from end user's viewpoint with the development of IoT technology. Here in this chapter, we segregate our literature review into two parts. First, we will review the probable solution and implementation technique provided in the literature for security issues on communication layers of IoT. Then literature review will be done based on different emerging technological solution toward IoT security and privacy.

Table 4 Comparison of existing survey on security in IoT communication

Existing Survey papers & year	Discussion on Security and privacy issues	IoT communication layer wise Security issues and possible attacks	IoT communication layer wise probable solution to security attacks	Review of State-of-the-art solution to security and privacy issues	Discussion on future research issues	Key areas focused
Granjal et al. [18]	✓	×	×	✓	×	Analyzed the existing security protocols in terms of solving capabilities of the fundamental security requirements
Iqbal et al. [10]	✓	×	×	✓	✓	Analyzed security architecture and risk management approaches of different existing literature. Also focuses toward implementing smart city's and smart homes
Mena et al. [3]	✓	×	×	✓	✓	Focused on the different IoT architectures and weakness of IoT and analysis of different techniques to overcome such weakness

(continued)

Table 4 (continued)

Existing Survey papers & year	Discussion on Security and privacy issues	IoT communication layer wise Security issues and possible attacks	IoT communication layer wise probable solution to security attacks	Review of State-of-the-art solution to security and privacy issues	Discussion on future research issues	Key areas focused
Ghani et al. [4]	✓	✓	✓	×	×	Authors have given an elaborate list of security guidelines and focused on security of Edge nodes. Also identified notable implementation techniques and possible challenges
Mohanta et al. [19]	✓	×	×	✓	✓	Study is focused on different security and privacy solution using technologies like AI, Blockchain, machine learning, etc
Alferidah et al. [15]	✓	✓	×	×	✓	Survey concentrates on IoT layer wise security attacks. And discussion on techniques to make IoT communication secure from those attacks

(continued)

Table 4 (continued)

Existing Survey papers & year	Discussion on Security and privacy issues	IoT communication layer wise Security issues and possible attacks	IoT communication layer wise probable solution to security attacks	Review of State-of-the-art solution to security and privacy issues	Discussion on future research issues	Key areas focused
Thilakarathne et al. [16]	✓	✓	×	×	✓	The survey summarizes the latest security protocols to overcome basic security and privacy requirements in IoT
Yue et al. [17]	✓	×	×	✓	✓	This survey focuses on how the deep learning approaches can be used to solve security and privacy issues in IoT communication

(continued)

Table 4 (continued)

Existing Survey papers & year	Discussion on Security and privacy issues	IoT communication layer wise Security issues and possible attacks	IoT communication layer wise probable solution to security attacks	Review of State-of-the-art solution to security and privacy issues	Discussion on future research issues	Key areas focused
Proposed work 2021	✓	✓	✓	✓	✓	Have identified IoT communication layer wise possible security threats. Focused on listing layer wise existing solutions to the identified problems. Also concentrated on existing works which uses state-of-the-art technology to solve the security and privacy issues of IoT and future research direction

2.1 Security and Privacy Solution Review Based on Communication Layers

In this subsection we have analyzed the layer wise literature survey of security and privacy solutions for most of the above-mentioned layer wise problems, which is as follows:

(A) ***Sensing/perception layer Security***: According to the authors [4, 17] in the perception layer, the security requirement to tackle various attacks for this layer are Secure bootstrapping, Data authentication, access control techniques, etc. In this context a secure architecture dependent bootstrapping is discussed by the authors in [20]. Also, in the case of distributed architecture Diffie-Hellman algorithm is being used by the authors so that, the two IoT objects can make an agreement on a common secret for authentication. In this architecture two other authors in [21, 22] have shown in their work that a key exchange protocol for authentication and access control mechanism can be built without trusted party with the help of different protocols like TLS, DTLS [23], Host Identity Protocol (HIP), and IKEv2. Although the limitations of these techniques are also discussed by the authors like in case of constrained devices implementing security using the above technique is difficult. To overcome this many researchers proposed various techniques. Some notable work like Diet HIP and human notable password proposed by authors in [24, 25]. These techniques also used for the authentication purpose between IoT objects and gateway and build trust between them. To achieve authentication and access control in centralized environment key certificate from trusted agency or digital key certificates or predefined keys are used to distribute keys between communicating parties. One such technique is proposed by authors in [26] to achieve security in this layer.

(B) ***Security solutions at link layer***: In link layer or gateway layer of IoT different IoT devices communicate with each other using IEEE 802.15.4 standard which is the standard for low power and lossy network [4]. In this standard, 6loWPAN network is used for communication in IoT [27], the fundamental security and confidentiality goals are achieved by using 6lowpan [18]. Hop to hop security without any authentication, key management technique, and protection from replay attack was being used in the link layer and in IEEE 802.15.4 protocol standard. To recover from this problem a new extension of IEEE 802.15.4 was proposed by authors in [28] called IEEE 802.15.4e. Different authors have identified a very relevant problem that occurs in link layer with data packets, that data packets cannot be protected by link layer security once it exists link. Many authors have proposed different solution for the said issue. The authors [29] proposed a relevant solution in this regard by proposing key management technique for WSN and thus adding security in link layer. In [30] authors proposed a protocol where they used IPsec to secure link between a router and nodes. A new key generation technique for media access control part of link layer in IEEE 802.15.4 standard is proposed by the authors in [31].

(C) ***Security solutions at network layer***: Most of the vulnerabilities of network layers can be solved by implementing secured routing protocol, Authentication of different entities taking part in communication and access control mechanisms [17]. In this section we have discussed different techniques and protocols provided by various authors for securing 6LoWPAN and RPL routing protocol in their research papers. In this context authors in [18, 31] have developed and proposed header compression scheme and IPsec, respectively, for 6loWPAN network which helps providing security in the network layer of IoT communication. Moreover, in this context Internet key exchange protocol can be helpful for establishing key exchange between communicating parties but for the resource constraint devices this protocol cannot be used because of header size, to overcome this issue authors in [32] proposed a header compression technique for 6loWPAN environment. In this 6LoWPAN environment the routing protocol that is used is RPL routing. Many authors have worked to give security to RPL routing protocol. Some notable work has done by authors in [33, 34] where they have analyzed this routing protocol insider attacks and in [35] authors provided solutions to stop its internal attacks by using version no with message authentication code. In another work authors [36] have investigated and proposed mechanisms to stop sink-hole attacks on RPL routing.

(D) ***Security solution at transport layer***: Ensuring authentication, providing secure key management techniques, and maintaining confidentiality and integrity are the basic need in this layer. Many protocols have been devised to establish these requirements. In general TLS and SSL are used in this context for securing communication over Internet. Authors in [37] have proposed a method where SSL can be used for securing IoT communications. To establish a secure communication between remote server and IoT devices a remote lightweight protocol is proposed by the authors in [38] which uses TLS for securing the communication. But disadvantages of this method are that complete E2E solution is not achieved by this. In another work [39] authors have proposed a mutual authentication and key exchange scheme using Elliptic curve cryptography which is secure as well as lightweight. To provide security in homogeneous networks using DTLS and TLS authors in [40] have proposed two security schemes. In this context of providing security in transport layer of IoT communication, a two-way authentication scheme is proposed by the authors in [41]. The said scheme uses x.509 certificates and RSA encryption for authentication and key management.

(E) ***Security solutions at application layer***: The authors in [17] have identified the counter measures for the security attacks in application layer of IoT as key management and distribution, Intrusion detection, and access control. Many authors have given solutions in this context through their literature in some of the notable work, authors in [42, 43] have proposed a mechanism to integrate DTLS with CoAP (constraint application protocol) in the application layer, since DTLS does not support key management. Intrusion Detection system is another important part of security by which network monitoring can be done for any suspicious/malicious activity by an adversary. Authors have proposed different schemes for intrusion detection system among them in [4] authors have proposed a novel scheme for intrusion detection system which is compatible with Ipv6 environment of IoT and also gives security from black hole attack and sink-hole attack. Another notable work in this context is to track the malicious activity of the adversary authors have proposed a scheme [44] which uses information of attacks from a library and using automotive system any adverse movement can be tracked and stopped in IoT communication.

2.2 Security and Privacy Solution Review Based on State-of-the-Art Techniques

Previously many notable works related to a security issue in IoT applications infrastructure are done. In this section a thorough review of the state-of-the-art solutions provided by the authors in their literature using the emerging techniques are discussed.

(A) ***Security and privacy solution using authentication and key management***: Many authors over time have proposed many security and privacy techniques in this user-server based mutual authentication topology. Many of the previous schemes are listed in [19]. Here we have discussed only the important and latest literatures. A lightweight ECC-based two-factor authentication scheme was proposed by authors in 2016 [45]. The scheme offers a security solution for wireless sensor networks and also uses smart card for authentication procedure. While analyzing the above scheme Das et al. [46] have found that the scheme is prone to some security attacks like offline password-guessing attack and session-specific temporary information attack to counter these authors have proposed three-factor security schemes. While analyzing das et al. scheme authors [47] found that the scheme has security deficiency and is prone to various security attacks such as offline password-guessing attack and session-specific temporary information attack. Another three-factor based scheme proposed by authors in [48]. The advantages of this scheme are that this is a remote authentication scheme where the gateway is not needed between user and sensor node that means directly user can access the information from sensor node but one disadvantage of the scheme is that dynamic addition of sensor nodes cannot be done [49]. In this context another notable work using ECC and digital signature has done by authors in [50] in which authors provided a security solution by proposing secure authentication and key agreement scheme which assures user secrecy and also promise users untraced ability. In another work a scheme for hierarchical IoT networks was proposed by authors in [51]. In this scheme authors only have used one way hash function and XOR operation to make it secure and lightweight. Authors in [52–56] have proposed different schemes for securing IoT-cloud communication, for implementation of smart home they have used remote user authentication for making the communication secure against any possible attacks also they have used ECC-based schemes to make it lightweight so that it becomes useful in resource constraint environment of IoT. Very recently authors in [57] analyzed various multifactor schemes for authentication and observed the new challenges in this multi-server and multi factor authentication techniques and then proposed an all-inclusive authentication scheme which uses multifactor approach to secure the communication against probable adversaries.

(B) ***Security solution using machine learning (ML)***: With the help of latest machine learning techniques security issues in IoT can be predicted and avoided. It uses the design and test-based learning method for prediction. Authors in [58] discuss the security while applied the machine learning in smart grid and also the authors in [59, 60] used it against intrusion detection. A 3-layer Intrusion Detection System (IDS) using a supervised learning method of ML is proposed by authors in [60] to find the difference between legal and illegal network activities. Also the proposed scheme can detect network-based cyber-attacks such as DoS, Spoofing, and Replay on IoT networks. Then in 2019 using Naive Bayes classifier and Laplace estimator authors [61] presented a rule-based prediction model of mobile phone data. Then in the same year an anomaly detection system using ML is proposed by authors in [62], moreover it can also detect various cyber-attacks like command and sql injection. In [63] authors presented an initial work of neural network (NN)-based specific emitter identification on IoT devices. Then in 2020 authors [64] proposed a hybrid algorithm to detect cyber-attacks efficiently.

(C) ***Blockchain-based solution***: Due to the fact that IoT system is a heterogeneous and decentralized network of devices conventional security and privacy measures have many times failed to provide the desired level of security. Thus, researchers have concentrated toward blockchain-based security solutions that rely on cryptocurrency called Bitcoins. The main difficulty with blockchain-based techniques is it is computationally expensive with high overhead especially when low power and resource constraint devices are involved this problem is identified by authors in [65], whereas after a lot of research on this constraint in [66] authors have proposed a scheme using blockchain which is computationally lightweight and tested in home automation systems and given acceptable results. Once the lightweight blockchain comes into play researchers have concentrated to use blockchain for solving IoT security and privacy problems, since in blockchain trust model between two or more parties are not required once the transaction is done it cannot be denied. Using this afterward many security architecture like security architecture in smart city [67], Platform for IIoT [68] and, trust based transaction model is proposed by the authors. Again, in the papers [69–71] a fresh approach of how blockchain can be used in IoT security perspective is discussed and also a distributed blockchain methodology is also proposed. Different approach for solving privacy preservation and authentication issues have been proposed by authors in [72–74]. Then latest in 2020 the authors in [75] explored new benefits and design challenges of integrating blockchain with IoT Security.

(D) ***Security solution based on Artificial intelligence in IoT***: With the rise of IoT application, artificial intelligence (AI) has also got a new dimension of usage. With the combination of these two, i.e., a connected device with a sensing and acting capability makes IoT more marketable. IoT generates a huge volume of data, to process this huge data AI comes into relief. AI can help IoT to process this large volume, unstructured data to be used real time this is proposed by authors in [76]. Then to find out adversary in IoT network several methods are present but the detection rate is unsatisfactory and those are excessively dependent on the network structure. To solve this problem authors in [77] have given a detection scheme which is independent and universal. After that in [78] authors have shown a method to find out malware in IoT network. Combining Artificial intelligence with blockchain method can perform wonder in case of security that has been shown in [79] by the authors. Various applications like smart home, smart city applications using artificial intelligence combined with machine learning to process the data real time as well as discover valuable knowledge and prediction [80]. In the area of IoT data are relayed to fog and edge computing, there also several usages of AI is shown by the authors in [81]. Recently authors in [79] have proposed a IoT architecture based on blockchain and AI that provide efficient security for current applications.

(E) ***SDN (Software defined networking) based security solution in IoT***: This is a relatively new technology to be used in conjunction with IoT. The main idea of SDN was to dynamic management of the network, to give it a centralized control structure as in SDN architecture the control decision is done from the centralized SDN controller router. Gateway or switches cannot take control decision by their own. This centralized control structure of SDN can be beneficial to IoT security in case of dealing with DOS attack stated by the authors in [62]. Although SDN is useful for IoT security it has some drawback too like inability to work in a dynamic environment which is pointed out by the authors in [82]. However, in 2019 authors [83] have designed an architecture based on SDN to empower IoT system using virtualization. In the same year a novel work to protect the distributed denial of service attack has been proposed by author in [84]. Then recently in 2020 authors in [85] have proposed a methodology to protect IoT network from Man-in-the-middle attack, also authors have shown a practical approach to implement the same idea. Another chapter in 2020 authors [86] have merged the concept of blockchain with SDN to create an energy efficient SDN controller for IoT network and with a new routing protocol.

3 Conclusion

Internet of things has become a biggest phenomenon in the latest technical paradigm as it can connect billions of devices worldwide. The tremendous scalability and ability to accommodate the various heterogeneous devices make it a most sought-after technology. With this rise of IoT technology security attacks on these devices have also increased. Since IoT deals with very private and confidential data, security provisioning is extremely important in this case. In this chapter we have reviewed the important and up-to-date literature to identify and understand the latest security threats and their probable solution as given by the authors. In this survey we have first identified communication layer wise security attacks and threats that can occur in IoT networks and layer wise solution as we got from the literature. Then this chapter concentrates into the emerging state-of-the-art technology like blockchain, Machine learning, SDN, etc. which can give a new dimension to the conventional security approaches and able to address most of the existing security issue. After a thorough review of the literature we can conclude that a vast and still emerging topic like IoT communication still have some open issues like building more lightweight cryptographic protocol for embedded systems, performance and latency issues of emerging technology in IoT low power and lossy networks, standardization of security and privacy guidelines in IoT, etc. Future research should be focused in these directions.

References

1. Li L (2012) Study on security architecture in the internet of things. In: 2012 international conference on measurement, information and control (MIC), vol 1. IEEE, pp 374–377
2. Khanna A, Kaur S (2019) Evolution of internet of things (IoT) and its significant impact in the field of precision agriculture. Comput Electron Agric 157:218–231
3. Mendez D, Papapanagiotou I, Yang B (2018) Internet of things: survey on security and privacy. Inf Secur J A Glob Persp 1–16
4. Abdul-Ghani HA, Konstantas D (2019) A comprehensive study of security and privacy guidelines, threats, and countermeasures: an IoT perspective. J Sens Actuator Netw 8(2):22
5. Patel A (2017) Comprehensive survey on security problems and key technologies of the ınternet of things (IoT). Int Conf Eng Technol
6. Pasha M, Myhammad S, Pasha U (2016) Security framework for IoT systems. Int J Comput Sci Inf Secur 14(11):99–104
7. Flauzac O, Gonzalez CJ, Nolot F (2015) New security architecture for IoT network. Procedia Comput Sci 52:1028–1033
8. Efe A, Aksöz E, Hanecioğlu N, Yalman Ş (2018) Smart security of IOT against ddos attacks. Int J Innov Eng Appl 2(2):35–43
9. Zhang G, Gong W (2011) The research of access control based on UCON in the internet of things. J Soft 6(4):724–731
10. Iqbal A, Suryani MA, Saleem R, Suryani MA (2016) Internet of things (IoT): on-going security challenges and risks. Int J Comput Sci Inf Secur 14(11):671
11. Ouaddah A, Bouij-Pasquier I, Elkalam AA, Ouahman AA (2015). Security analysis and proposal of new access control model in the ınternet of thing. In: 2015 ınternational conference on electrical and ınformation technologies (ICEIT), pp. 30–35.

12. Rao TA, Ehsan-ul-Haq (2018) Security challenges facing IoT layers and its protective measures. Int J Comput Appl 179(27):31–35
13. Ali I, Sabir S, Ullah Z (2019) Internet of things security, device authentication and access control: a review
14. Borgohain T, Kumar U, Sanyal S (2015) Survey of security and privacy issues of internet of things. Int J Adv Netw Appl 6(4):2372–2378
15. Alferidah DK, Jhanjhi NZ (2020) A review on security and privacy issues and challenges in internet of things. Int J Comput Sci Netw Secur IJCSNS 20(4):263–286
16. Thilakarathne NN (2020) Security and privacy issues in IoT environment. Int J Eng Manag Res 10
17. Hosenkhan MR, Pattanayak BK (2020) Security issues in internet of things (IoT): a comprehensive review. New Paradig Dec Sci Manag 359–369
18. Granjal J, Monteiro E, Sa Silva J (2015) Security for the Internet of Things: A Survey of Existing Protocols and Open Research Issues. IEEE Commun. Surv. Tutor. 17:1294–1312
19. Mohanta BK, Jena D, Satapathy U, Patnaik S (2020). Survey on IoT security: challenges and solution using machine learning, artificial intelligence and blockchain technology. Internet Things 100227
20. Heer T, Garcia-Morchon O, Hummen R, Keoh SL, Kumar SS, Wehrle K (2011) Security challenges in the IP-based internet of things. Wirel Person Commun 61:527–542
21. Phelan T (2008) Datagram transport layer security (DTLS) over the datagram congestion control protocol (DCCP). RFC 5238, May
22. Moskowitz R, Nikander P, Jokela TH (2008) Host identity protocol; technical report for internet engineeringtask force; IETF: Fremont. CA, USA
23. Kaufman C (2005) Internet key exchange (IKEv2) protocol; technical report; internet engineering task force (IETF): Fremont. CA, USA
24. Moskowitz R (2011) HIP Diet EXchange (DEX): draft-moskowitz-hip-rg-dex-05. Internet engineering task force, status: work in progress, Technical report
25. Wook Jung S, Jung S (2015) Secure bootstrapping and reboot strapping for resource-constrained thing in internet of things. Int J Distrib Sens Netw
26. Sarikaya B, Ohba Y, Moskowitz R, Cao Z, Cragie R (2012) Security bootstrapping solution for resource- constrained devices; technical report for the internet engineering task force; IETF: Fremont, CA, USA, 22 June 2012
27. Montenegro G, Kushalnagar N, Hui J, Culler D (2007) Transmission of IPv6 packets over IEEE 802.15.4 networks; technical report for internet engineering task Force; IETF: Fremont, CA, USA
28. Watteyne T, Palattella M, Grieco L (2015) Using IEEE 802.15.4e time-slotted channel hopping (TSCH) in the internet of things (IoT): Problem Statement
29. Moskowitz R, Hummen R (2017) HIP Diet exchange (DEX); internet engineering task force (IETF): Fremont. CA, USA
30. Granjal J, Monteiro E, Silva JS (2014) Network-layer security for the internet of things using TinyOS and BLIP. Int J Commun Syst
31. Raza S, Voigt T, Jutvik V (2014) Secure communication for the internet of things—a comparison of link-layer security and IPsec for 6LoWPAN. Int J Appl Eng Res 9:5968–5974
32. Raza S, Voigt T, Jutvik V (2012) Lightweight IKEv2: a key management solution for both the compressed IPsec and the IEEE 802.15.4 security. In Proceedings of the IETF workshop on smart object security, Paris, France, 23 Mar 2012
33. Winter T, Thubert P, Brandt A, Hui J, Kelsey R, Levis P, Pister K, Struik R, Vasseur JP (2012) Alexander, R RPL: IPv6 routing protocol for low-power and lossy networks; RFC 6550; internet engineering task force (IETF) Fremont. CA, USA
34. Tsao T, Alexander R, Dohler M (2014) A security threat analysis for routing protocol for low-power and lossy networks (RPL); RFC7416; internet engineering task force (IETF): Fremont. CA, USA
35. Dvir A, Holczer T, Buttyan L (2011) VeRA—version number and rank authentication in RPL. In: Proceedings of the 8th IEEE international conference on mobile ad-hoc and sensor systems, MASS, Valencia, Spain, 17–21 Oct 2011, pp 709–714

36. Weekly K, Pister K (2012) Evaluating sinkhole defense techniques in RPL networks. In: Proceedings of the international conference on network protocols, ICNP, Austin, TX, USA, 30 Oct–2 Nov 2012, pp 1–6
37. Hong S, Kim D, Ha M, Bae S, Park S, Jung W, Kim JE (2010) SNAIL: an IP-based wireless sensor network approach to the internet of things. IEEE Wirel Commun 17:34–42
38. Fouladgar S, Mainaud B, Masmoudi K, Afifi H (2006) Tiny 3-TLS: a trust delegation protocol for wireless sensor networks. Springer, Berlin/Heidelberg, Germany
39. Granjal J, Monteiro E, Silva J (2013) End-to-end transport-layer security for Internet-integrated sensing applications with mutual and delegated ECC public-key authentication. In: Proceedings of the 2013 IFIP networking conference, Brooklyn, NY, USA, 22–24 May 2013, pp 1–9
40. Brachmann M, Keoh SL, Morchon OG, Kumar SS (2012) End-to-end transport security in the IP-based internet of things. In: Proceedings of the 2012 21st international conference on computer communications and networks (ICCCN 2012), Munich, Germany, 30 July–2 Aug 2012, pp 1–5
41. Kothmayr T, Schmitt C, Hu W, Brünig M, Carle G (2013) DTLS based security and two-way authentication for the internet of things. Ad Hoc Netw 11:2710–2723
42. Granjal J, Monteiro E, Silva JS (2013) Application-layer security for the WoT: extending CoAP to support end-to-end message security for internet-integrated sensing applications. In: Proceedings of the 11th wired/wireless internet communication, St. Petersburg, Russia, 5–7 June 2013
43. Keoh SL, Kumar SS, Garcia-Morchon O, Dijk E (2015) DTLS-Based Mul-Ticast security for low-power and lossy; technical report for the internet engineering task force. IETF, Fremont, CA, USA, pp 1–22
44. Hartke K (2014) Practical issues with datagram transport layer security in constrained environments; DICE working group. Fremont, CA, USA
45. Chang C-C, Le H-D (2016) A provably secure, efficient, and flexible authentication scheme for ad hoc wireless sensor networks. IEEE Trans Wirel Commun 15(1):357–366
46. Das AK, Goswami A (2015) A robust anonymous biometric-based remote user authentication scheme using smart cards. J King Saud Univ Comput Inf Sci 27(2):193–210
47. Kumari S, Li X, Wu F, Das AK, Arshad H, Khan MK (2016) A user friendly mutual authentication and key agreement scheme for wireless sensor networks using chaotic maps. Futur Gener Comput Syst 63:56–75
48. Dhillon PK, Kalra S (2017) A lightweight biometrics based remote user authentication scheme for IoT services. J Inf Secur Appl 34:255–270
49. Souri A, Norouzi M (2019) A state-of-the-art survey on formal verification of the internet of things applications. J Serv Sci Res 11(1):47–67
50. Challa S, Wazid M, Das AK, Kumar N, Reddy AG, Yoon EJ, Yoo KY (2017) Secure signature-based authenticated key establishment scheme for future IoT applications. IEEE Access 5:3028–3043
51. Fakroon M, Alshahrani M, Gebali F, Traore I (2020) Secure remote anonymous user authentication scheme for smart home environment. Internet Things, 100158
52. Wazid M, Das AK, Odelu V, Kumar N, Conti M, Jo M (2018) Design of secure user authenticated key management protocol for generic iot networks. IEEE Internet Things J 5(1):269–282
53. Sharma G, Kalra S (2018) A lightweight multi-factor secure smart card based remote user authentication scheme for cloud-IoT applications. J Inf Secur Appl 42:95–106
54. Shuai M, Yu N, Wang H, Xiong L (2019) Anonymous authentication scheme for smart home environment with provable security. Comput Secur 86:132–146
55. Sowjanya K, Dasgupta M, Ray S, Obaidat MS (2019) An efficient elliptic curve cryptography-based without pairing KPABE for internet of things. IEEE Syst J
56. Sadhukhan D, Ray S, Biswas GP, Khan MK, Dasgupta M (2020) A lightweight remote user authentication scheme for IoT communication using elliptic curve cryptography. J Supercomput
57. Wang D, Zhang X, Zhang Z, Wang P (2020) Understanding security failures of multi-factor authentication schemes for multi-server environments. Comput Secur 88:101619

58. Hossain E, Khan I, Un-Noor F, Sikander SS, Sunny MSH (2019) Application of big data and machine learning in smart grid, and associated security concerns: a review. IEEE Access 7:13960–13988
59. Chaabouni N, Mosbah M, Zemmari A, Sauvignac C, Faruki P (2019) Network intrusion detection for IoT security based on learning techniques. IEEE Commun Surv Tutor 21(3):2671–2701
60. Anthi E, Williams L, Słowińska M, Theodorakopoulos G, Burnap P (2019) A supervised intrusion detection system for smart home IoT devices. IEEE Internet Things J 6(5):9042–9053
61. Sarker IH (2019) A machine learning based robust prediction model for real-life mobile phone data. Internet Things 5:180–193
62. Gonzalez C, Charfadine SM, Flauzac O, Nolot F (2016) SDN-based security framework for the IoT in distributed grid. Proc Int Multidiscip Conf Comput Energy Sci SpliTech Split Croatia 13–15:1–5
63. McGinthy JM, Wong LJ, Michaels AJ (2019) Groundwork for neural network-based specific emitter identification authentication for IoT. IEEE Internet Things J 6(4):6429–6440
64. Shafiq M, Tian Z, Sun Y, Du X, Guizani M (2020) Selection of effective machine learning algorithm and Bot-IoT attacks traffic identification for internet of things in smart city. Futur Gener Comput Syst 107:433–442
65. Dorri A, Kanhere SS, Jurdak R (2016) Blockchain in internet of things: challenges and solutions. arXiv:1608.05187.
66. Dorri A, Kanhere SS, Jurdak R, Gauravaram P (2017) LSB: a lightweight scalable blockchain for IoT security and privacy, pp 2–17
67. Biswas K, Muthukkumarasamy V (2016) Securing smart cities using blockchain technology. In: Proceedings of the 18th IEEE international conference on high performance computing and communications, 14th IEEE international conference on smart city and 2nd IEEE international conference on data science and systems, HPCC/SmartCity/DSS, Sydney, Australia, 12–14 Dec 2016, pp 1392–1393
68. Bahga A, Madisetti VK (2016) Blockchain platform for industrial internet of things. J Softw Eng Appl 9:533–546
69. Banerjee M, Lee J, Choo KKR (2018) A blockchain future for internet of things security: a position paper. Digit Commun Netw 4(3):149–160
70. Minoli D, Occhiogrosso B (2018) Blockchain mechanisms for IoT security. Internet Things 1:1–13
71. Satapathy U, Mohanta BK, Panda SS, Sobhanayak S, Jena D (2019) A secure framework for communication in internet of things application using hyperledger based blockchain. In: 2019 10th international conference on computing, communication and networking technologies (ICCCNT). IEEE, pp 1–7
72. Xu J, Xue K, Li S, Tian H, Hong J, Hong P, Yu N (2019) Healthchain: a blockchain-based privacy preserving scheme for large-scale health data. IEEE Internet Things J 6(5):8770–8781
73. Hammi MT, Hammi B, Bellot P, Serhrouchni A (2018) Bubbles of trust: a decentralized blockchain-based authentication system for IoT. Comput Secur 78:126–142
74. Lin C, He D, Huang X, Choo KKR, Vasilakos AV (2018) BSeIn: a blockchain-based secure mutual authentication with fine-grained access control system for industry 4.0. J Netw Comput Appl 116:42–52
75. Dedeoglu V, Jurdak R, Dorri A, Lunardi RC, Michelin RA, Zorzo AF, Kanhere SS (2020) Blockchain technologies for iot. In: Advanced applications of blockchain technology. Springer, Singapore, pp 55–89
76. Ghosh A, Chakraborty D, Law A (2018) Artificial intelligence in Internet of things. CAAI Trans Intell Technol 3(4):208–218
77. Wang S, Qiao Z (2019) Robust pervasive detection for adversarial samples of artificial intelligence in IoT environments. IEEE Access 7:88693–88704
78. Zolotukhin M, Hämäläinen T (2018) On artificial intelligent malware tolerant networking for IoT. In 2018 IEEE conference on network function virtualization and software defined networks (NFV-SDN). IEEE, pp 1–6

79. Singh SK, Rathore S, Park JH (2020) Blockiotintelligence: a blockchain-enabled intelligent IoT architecture with artificial intelligence. Futur Gener Comput Syst 110:721–743
80. Falco G, Viswanathan A, Caldera C, Shrobe H (2018) A master attack methodology for an AI-based automated attack planner for smart cities. IEEE Access 6:48360–48373
81. Zou Z, Jin Y, Nevalainen P, Huan Y, Heikkonen J, Westerlund T (2019) Edge and fog computing enabled AI for IoT-an overview. In: 2019 IEEE international conference on artificial intelligence circuits and systems (AICAS). IEEE, pp 51–56
82. Kouicem DE, Bouabdallah A, Lakhlef H, Kouicem DE, Bouabdallah A, Lakhlef H (2018) Internet of things security: a top-down survey. Comput Netw 141:199–221
83. Zarca AM, Bernabe JB, Trapero R, Rivera D, Villalobos J, Skarmeta A, Gouvas P (2019) Security management architecture for NFV/SDN-aware IoT systems. IEEE Internet Things J 6(5):8005–8020
84. Abou El Houda Z, Hafid A, Khoukhi L (2019) Co-IoT: a collaborative DDoS mitigation scheme in IoT environment based on blockchain using SDN. In: 2019 IEEE Global Communications Conference (GLOBECOM). IEEE, pp 1–6
85. Al-Hayajneh A, Bhuiyan ZA, McAndrew I (2020) Improving internet of things (IoT) security with software-defined networking (SDN). Computers 9(1):8
86. Yazdinejad A, Parizi RM, Dehghantanha A, Zhang Q, Choo KKR (2020) An energy-efficient SDN controller architecture for IoT networks with blockchain-based security. IEEE Trans Serv Comput

Causality and Its Applications

Pramod Kumar Parida

Abstract The diagnosis of cause–effect relations boosts the both predictive and prescriptive models and their analysis. The recent trend and growth in machine learning, leading the field of predictive analytics, base the certain case for causal analysis. Causal analysis supports the study of causes and effect as they are observed and their underneath relationship directing such trends, are analysed for predictive modelling. The relation in cause–effect provides much needed information on interdependency in the various features. The interdependency helps to identify the valuable or influential features which supports for any certain trend, which is of interest. In this chapter, machine learning and necessity for causal analysis is discussed in depth. The main focus point is to provide different scenario where causal integration into machine learning is useful but lacks the process of inference. A depth view is shown for different situations where causal analysis can produce and largely effect the standard industrial process of predictive modelling. Next to this trend, a more advanced step is prescriptive analysis. Going one step further of predicting what is expected, is prescriptive analytics focus on what to avoid to achieve the required target. In this case, causal analysis can be used to find feature dependency model, leading to the prescriptive future model. As seen in machine learning modelling, much effort is put upon setting up the predictive modelling. But most industry require a better understanding of feature dependency and a model leading to business success while avoiding losses, a prescriptive modelling. In deep learning, much effort is lost on building the most accurate predictive modelling. But neural networks never seen in the lights of models which are explainable. The weights produced in most accurate models of deep learning, does not provide interpretability or their effectiveness in producing final results. With the rise of explainable AI, there is now efforts put onto the explain ability of the deep networks. This new trend/flow now opened the windows for causal explanations in deep neural network models. The task to explain deep learning models, require understanding on hidden weights and their influence on final neural outputs. Causal influence of input features through

P. K. Parida (✉)
Department of Systemics, School of Computer Science, University of Petroleum and Energy Studies, Dehradun, UK, India
e-mail: Pramodkumar.parida@ddn.upes.ac.in

S. Chakraverty (ed.), *Soft Computing in Interdisciplinary Sciences*,
Studies in Computational Intelligence 988,
https://doi.org/10.1007/978-981-16-4713-0_11

the weights inform their contribution towards neural outputs. Through explainable neural network weights, the better models can be built to overpower or underpower the effects of input features. In this note, the chapter provide how causal influence is useful and their significance for causal influence analysis. The criteria for causal influence analysis in deep neural models are discussed with examples.

1 Introduction

1.1 An Introduction to Causality

Any observation in natural phenomena, experiments or the effects in reality is due to a cause. Most of the time though the components or factors are responsible for such causes, the exact relations in them are unknown. There is the possibility that the factors just may not affect each other in a linear process, such as 1 affecting 2, then 2 affecting 3 and then so on. The process may be a more complicated one, where 1 may affect 2, then 3 and 4 in combination with 2 create a new process 5. It should not be forgotten that these processes may extend to a nonlinear form, making it even more complex to trace their relationships. For such processes and the curiosity to understand the cause–effect relations, the term Causality has been coined. In causality, the main purpose relies on finding each relation in factors to explain the process in the best possible way. Not involving in the complexity of representation of multiple features in multiple dimensions, the easiest way of representation is a two-dimensional graphical model with connected and directed edges to provide visualization to such processes. This will be easy to understand that connected edges show whether causality exists in the factors and directions to signify what affects what.

The meaning of causal is very clear in physics by only considering a force and its effect. In other fields, the derivation of variable relations is strictly dependent on characteristics of data and its domains. A more special case to causality, the Granger causality has been widely researched and the time factor relating to it helps to analyse it in simple steps, while the general view of causality is far more complicated and underdeveloped.

Defining a causal model for an irreversible system, it is... "*a directed acyclic graph with nodes representing the variables, connected by weighted directed edges to show the relation from one to another and weights for the connection strength or the flow of information*" [6]. It is not necessary for a causal model to be acyclic, though it can be cyclic for a reversible system. The depth of information is provided in the following sections.

1.2 A Representation of Causal Model

In multiple factors, if one affects the other, then there must be something such as energy, property or some influence which is passed from the cause to the effect. A unified name would be the 'information', which results in effect when transferred from the cause. Obviously, this information carries some weight, as there are not always the same changes or the same amount of effect created by the same cause–effect relation. So, information varies each time in case of even the same factor relations. This information can be represented in a model by providing the weights for the directed edges, showing the amount of information passed from the source to the receiver. All of these define basic properties of causality and it can be represented in a model as shown in Fig. 1.

Figure 1 shows a causal model, where factors are represented by nodes $\{N_1, N_2, N_3,.., N_n\}$, an existing causal relation in the nodes are shown by the directed edges and the weights $\{W_{12}, W_{13}, W_{14}, ..., W_{(n-1)n}\}$ on the edges shows the information that is passed on. To make it easier to understand, the source/cause is called the Parent and the effect/receiver as the Child. So W_{12} shows the amount of information passed from parent N_1 to the child node N_2 and in a similar way for all the other nodes as shown in the Fig. 1. It can be noticed that the causal structure is arranged from the primary cause to the extended effects as shown in Fig. 1. This arrangement is termed a causal ordering and represented as $O(.)$. In this case for Fig. 1, the orders are $O(N_1)$

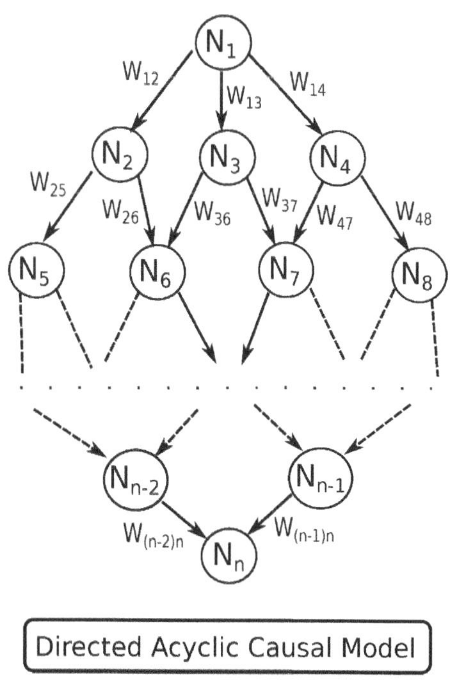

Fig. 1 Multivariate directed acyclic causal model with n number of nodes

> $O(N_2)$,$O(N_3)$ > $O(N_6)$ and similarly for all other nodes as they are represented. In particular, an order is derived from how the factors are related and how they are observed. This will be discussed later in the sections with examples.

Consider a more reasonable question that should arise by now: are the causal models cyclic or acyclic? Most natural processes are cyclic because of the interdependency in all the natural systems. Most of the artificial processes as caused by humans are acyclic. As in case of a causal model which depends on factors, it can be both cyclic and acyclic. Until now all have been busy considering the causal models as acyclic, or more precisely, Directed Acyclic Graphs (DAGs) as represented and solved in most of the papers. Why is this so? Because it is easier to analyse and represent, considering the huge amount of complexity one has to solve in the cyclic processes. In an acyclic model, the feature set complexity is less as information only flows in one direction, whereas in the case of cyclic models, the information flow becomes more complicated as the paths cross each other.

This can be explained in this way: in a single-lane road, if all cars are going in the same direction, then there is less chance of an accident. But if in a single lane cars are going in both directions, then there is a high chance of an accident. The accident can be understood as an effect identification problem in the case of causality. Whenever the relations in factors become cyclic, detecting the direction of information flow and effect identification becomes more complicated.

The above concepts provide the hints of problems which are going to be dealt with while solving causal models. Some major issues that have never been addressed in this regard are discussed in the following section.

2 Causal Identification

The following best explains the problems to be faced in contributing towards the development of causal analysis.

Causality is the general case of the study of causal analysis and inference without the consideration of time. Although the general theory of causality does not explain it explicitly, there are very few methods available which explicitly emphasizes causal inference in observation sets and construction of causal models.

2.1 The Quantitative Analysis of Causality

For a long time, what all the researchers have provided is the quantitative analysis of the causality (without any concern for it), although the only purpose of causal inference is about qualitative analysis. The first step in any analysis is to find a suitable model for the system. The next section presents models which are available or used for causal inference.

2.1.1 Structural Equation Model

The first choice for graphical analysis is Structural Equation Models (SEMs) proposed by Bollen [1]. The linearity of SEM makes it easier to implement for path estimations. But path finding is not what is considered for causality: if this were true then all the other models such as linear or nonlinear ones would do the same. A more common and simple functional relation that can explain both linear and nonlinear cases can be given as

$$n_j = f(n_i, e), \tag{2.1}$$

where the direction/path $\{n_i \rightarrow n_j\}$ is estimated, with the system containing a noise e_j. While the particular representation for linear and nonlinear cases will be significantly different for the above bivariate model, the purpose remains the same: to estimate the parameters in the model. This type of model is better suited for nonlinear causal models and more efficient than the linear models.

In the case of linear system, a bivariate model for estimation of causal direction in between (n_i, n_j) can be given as follows:

$$\mathrm{n_j = c_{ij} n_i + e_j (i \neq j)}, \tag{2.2}$$

where c_{ij}, $(i \neq j)$ is the connection strength for the directed edge $\{n_i \rightarrow n_j\}$ and e_j is a noise in the system. For any graph with m number of nodes, there will be $({}^m c_2)$ number of path coefficients and m number of noises for estimation. In every model, the nodes are arranged in a definite order where none of the later nodes have directed edges towards the earlier observed ones. This implies every prior node has a directed path to reach any of the later child nodes. This ordering from the parent node to successive descendant nodes can follow an ascending or descending order depending on the causal influence factors.

Example Consider a case where a particularly prominent factor is significantly improved through the information passed from parent to child nodes. This can be understood in the case of medical observations where the primary host of a particular disease may not be severely affected by it. But as it spreads from primary host to secondary and then to others, the case may become more infectious with worsening behaviours. In this case, the influence becomes more effective from the parent stage to the descendant stages, which will favour an ascending ordering for the node arrangements.

The problem is reduced where the only requirement is to find the significant paths responsible for causal evaluation.

The above bivariate model was used for causal analysis by [3, 7, 8, 10, 9, 11, 13] with different parameter estimation methods. But the problem arises whether the bivariate model is sufficient to analyse causal inference. This question is explained in the following section. Some of the propositions and definitions useful for causal inference are discussed below.

Graph d-Separable: The condition of d-separable makes it easier to analyse any graph by observing or blocking the flow of information. For the structure $\{x \leftrightharpoons y \leftrightharpoons z\}$ the following d-separable conditions are equivalent:

i. The structures $\{x \rightarrow y \rightarrow z\}$ and $\{x \leftarrow y \leftarrow z\}$ are d-separable when y is not observed.
ii. In $\{x \leftarrow y \rightarrow z\}$, the graph is d-separated for unobserved y.
iii. The graph $\{x \rightarrow y \leftarrow z\}$ is d-separated whenever y or the descendant of y is observed.

Both (i) and (ii) have the same conditional independence where $(x \perp z|y)$. But in (iii), the observation of y makes the nodes (x, z) become mutually dependent while before both are mutually independent of each other. The graph (iii) is of a common effect type referred to as V-structure and it is the only directed graph which can be used to solve the causal inference in sub-structural levels. The causal analysis from the point of probabilistic inference requires conditional probabilities of child nodes with respect to their parent nodes.

The Markov Condition: Consider a graph G with vertex set $\{V_1, V_2, \ldots, V_n\}$ with the probability distribution P. Applying the Markov Condition on the graph G, the joint distribution factorization of conditional probabilities can be represented in an order from prior to later ones. A mathematical representation can be given as

$$\mathrm{P}(\mathrm{V}_1, \mathrm{V}_2, \ldots, \mathrm{V}_n) = \prod_{\mathrm{i}} \mathrm{P}(\mathrm{V}_\mathrm{i}\mathrm{Pa}(\mathrm{V}_\mathrm{i})), \tag{2.3}$$

where $Pa(V_i)$ represent the parent of V_i.

Causal Markov Condition: Consider the above graph G and the subset of the vertex set.

$\{v \in V\ |v \subseteq V\}$ and let $v = \{x \rightarrow y \leftarrow z\}$. Then the Causal Markov Condition says that the subset v is independent of all the other variables in the set V which are not the direct causes. Then for the subset v, for any set $\{S \subseteq V\ |S \nrightarrow v\}$ the Causal Markov Condition can be written as

$$P(y|S, Pa(y)) = P(y|Pa(y)) = P(y|x, z). \tag{2.4}$$

These definitions are the foundations for later causal development and analysis. How the bivariate models fail to hold on to these basic requirements is discussed in following sections.

Markov Equivalent Classes: The sub-structural analysis through d-separable V-structures requires a minimal arrangement to learn the feature relation. In this regard, the Markov equivalent classes are very helpful which offer the criteria for the minimal arrangement of V-structures. It says the connected causal models with the same number of node sets are Markov equivalent in their respective classes if they have the same connected edge sets irrespective of the edge directions in them.

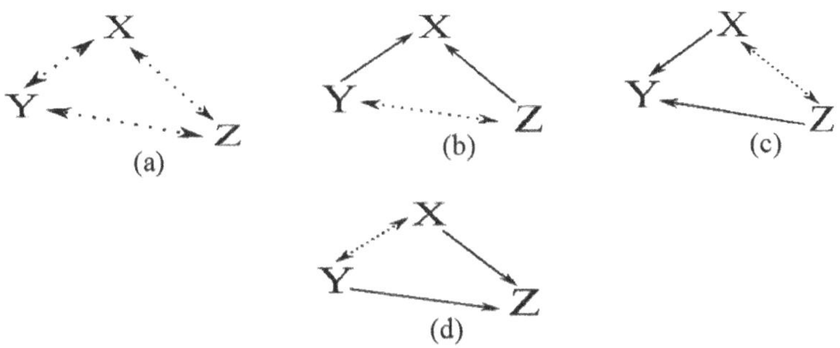

Fig. 2 All possible causal relations in {X,Y, Z} as shown in (**a**) and three different Markov equivalent models shown by (**b**), (**c**) and (**d**)

This is explained in Fig. 2 where Fig. 2a represent a completely connected causal model with all possible causal directions and Fig. 2b–d show the possible Markov equivalent classes for the Fig. 2a.

The d-separable V-structures which are Markov equivalent are shown in Fig. 2b–d for the causal model in Fig. 2a with node set $\{X, Y, Z\}$. The chapter by He and Geng [2] provides a better understanding for the use of Markov equivalent classes on causal inference using graphical sub-structural analysis.

Notice that by comparing the causal relations in Markov equivalent classes the correct directions can be found, however, for this comparison, the causal influence values are to be estimated. Then by selecting the highest causal influence, the causal directions can be confirmed. This process is explained in the qualitative analysis section of the causal inference.

2.1.2 The Insufficient Bivariate Models

The primary bivariate model is analysed using probability and conditional probability for causal relations. The solution of conditional independence given using probability measures is definitely a quantitative case. Consider the below bivariate model:

$$x_j = x_i + e_j. \tag{2.5}$$

Here, the causal direction $\{x_i \rightarrow x_j\}$ is estimated, where the order of node is given as

$O(i) > O(j)$ and e_j is the noise/additive noise in the system for x_j.

The probability representation of the exact case $\{x_i \rightarrow x_j\}$ using the causal Markov condition can be given as

$$P(x_j|Pa(x_j)) = P(x_j|x_i). \tag{2.6}$$

This turns out to be the conditional probabilities of child node for observed parent sets. The question arises whether the conditional probability helps for causal identification. The representation of conditional probability is discussed in the below.

Observation from Conditional Probability: The main criterion for checking conditional probability is to find out whether two nodes which are observed on condition are dependent or independent. The probability value inform whether these two observed nodes are conditionally dependent or independent. The conditional dependence case in between the node.

$\{x, y\}$ can be represented as

$$(x/\perp\perp y) = (y/\perp\perp x),$$
$$P(x|y) = P(y|x). \tag{2.7}$$

Any conditional independence case of a node $\{w \in V\, |w \nrightarrow y\}$, has a representation of

$$(w\perp y) = (y\perp w),$$
$$P(w|x) = P(w), P(x|w) = P(x). \tag{2.8}$$

It is evident that the effect of conditional analysis for causal calculation does not make any sense. Both conditional dependence and independence in the bivariate case make both the nodes dependent or independent at the same time. The example below explains this in a simple case study.

Example The conditional dependence or independence problem in the bivariate case can be better understood from this example. Consider the case of a DNA test for mother, father and child. The DNA samples A, B and C are unlabelled, but in real case A is the father, B is mother and C is the child. Without knowing which is a mother and a child's DNA, let us assume that there is 50% similarity in B and C's DNAs test results. This also means there is a 50% difference/dissimilarity in both DNAs. Can it be confirmed which one is a parent and which one is a child from this result? The answer is 'No'. To resolve this problem, the father's DNA sample is required. Comparing the DNA samples in A with C, may result in a 45% similarity. Then without any doubt, one can conclude that A and B are parents and C is the child, where one obvious case is that DNAs of A and B are completely dissimilar. But even then from similarity test, it is hard to confirm which one is the father and which one is the mother for the multivariate case. And for bivariate, it is even more complicated to distinguish between mother, father and child from similarity tests. These case scenarios for the test can be changed and the results will remain the same. In no case for two DNA comparisons, the parent can be identified.

It is clear that for the bivariate case, the parent and child features cannot be concluded from conditional dependence or independence test results. These help to clarify the claim that bivariate models are insufficient for causal analysis. The previous statement need to be corrected, because until now, what the conditional independence is trying to do is to find out the dependence or independence in two features for the bivariate case. So, it is nowhere nearly discussing the causal inference in all the above cases, till the conditional probability is used.

Unused D-separation Criteria: Let us add one more point to this claim. If it is noticed in the discussion that, the definition of graph d-separation is not applied anywhere in the case of the bivariate model. To use d-separation condition at least three nodes are needed, but this is impossible in case of bivariate models. This is the point where one primary proposition for causal analysis has just been violated. How is that true? Arguing that after analysing the structure for the bivariate model and after getting all the possible causal directions, the graph d-separation still can be used for causal construction. That is not possible, because d-separation is a condition which enables us to identify the parent and child nodes in the process of separation. What it means is that, if in the primary stage while causal directions are solved, if it did not include the d-separation (or not analysing three variables at a time for directions) then, the directions estimated are not the causal directions (as in the case of bivariate models). So, our method of estimation and model should use the d-separation criteria whenever the directions are estimated. This is the turning point where bivariate models seem to be using the fundamental conditions wrongly and are found to be insufficient. A complete review of most noticeable works on causal analysis is given in Parida et al. [4], which discuss the advantages and disadvantages of proposed works in different method categories.

For this case, Fig. 2 can be followed which provides the d-separable V-structures necessary for sub-structural learning using Markov equivalent classes. But while using a bivariate model, the d-separable V-structures cannot be found for sub-structural analysis.

2.2 *Causal Inference: A Qualitative Analysis*

Much have been discussed about the problem of bivariate models for causal analysis, ex- plaining why they are insufficient and those leading to the quantitative inference in the above sections. So, what exactly does qualitative analysis means, how is it different from the conditional independence or dependence cases and how to identify it?

The multivariate additive noise proposed by Parida et al. [5] provide a solution to the insufficient bivariate model for causal analysis.

2.2.1 A Multivariate Additive Noise Model

It is clear that insufficient bivariate models lead to the development of multivariate models. But how many dependent nodes are needed to be considered or how many parents should be taken into consideration for the estimation of causal directions in parent–child relation? Taking a closer look at the d-separation condition which is not used in the bivariate case, it takes at least three nodes to find causal directions. That means exactly two parents and one child node are required. In terms of mathematics, two nodes are needed to represent the third node in a linear or nonlinear equation format with an added noise (as additive noise model suggests). Why three, why not four or five or more? The easiest one to analyse for d-separation are three node sets and adding more nodes to it needs more skills and restrictions to find the directions in them. So, it is always easy to go with three nodes which preserve the primary assumptions. This can be represented by the following equation for the case of triplet $\{x \rightarrow y \leftarrow z\}$ as

$$y = f(x) + f(y) + e_y. \tag{2.9}$$

Equation 5.9 can be a linear or nonlinear one, where e_y represents the additive noise in the system. A linear form of the Eq. 2.9 with connection strength and additive noise can be given as

$$y = c_{xy}.x + c_{zy}.z + e_y. \tag{2.10}$$

where c_{xy}, c_{zy} are the connection strength values for directions $\{x \rightarrow y, z \rightarrow y\}$ and e_y is the additive noise in the system. Analysing the importance of this equation reveals that the d-separation condition is fully implementable in this case. Also, while estimating for causal directions, the equation can be used to impose the d-separation criteria.

The model properties used in case of bivariate models also hold here. So for a node set $V = \{v_1, v_2, \ldots, v_n\}$, the order of the nodes can be given as $\{O(v_1) > O(v_2) > \cdots > O(v_n)\}$ following a parent to child structural arrangement. But in previous cases, the order of the nodes depends on how they are connected in the graph or more specifically, using the conditional independent order over the conditional dependent ones. As argued before, the conditional probability could not justify the ordering of the nodes in the causal model, so a new type of independence is introduced for causal analysis, called causal independence.

2.2.2 Causal Independence

Causal independence is a qualitative signifier of the causal direction. While the basics of conditional independence provide the quantitative support, these do not specify the directions in feature sets. The causal independence/dependence are very specific and exclusive to identify the directions in the observed feature sets. Causal independence also signifies the order of a node in the causal structure. While most causal independent nodes are arranged at the top of the structure, the least independent ones are arranged in descending order in the causal model. Let us start from the conditional dependence case as given for the observed feature set $\{x \rightarrow y \leftarrow z\}$ as

$$P(y|x) = P(x|y) \Rightarrow (y/\perp\perp x) = (x/\perp\perp y) \tag{2.11}$$

But in the case of causal dependence, the same conditional dependence has a different meaning and it can be seen in the following:

$$\begin{aligned} P(y|x) \Rightarrow (x \rightarrow y) \quad &= x \text{ is causal independence of } y \\ &= y \text{ is causal dependence on } x, \\ P(x|y) \Rightarrow (y \rightarrow x) \quad &= y \text{ is causal independence of } x \\ &= x \text{ is causal dependence on } y. \end{aligned} \tag{2.12}$$

It is evident why the conditional independence/dependence is not very useful for direction detection, whereas the causal independence/dependence exclusively informs about the direction in the feature set. So, all the criteria and primary assumptions defined for causal inference can be used in Eq. 2.9 with causal independence criteria to find the causal directions.

Different methods can be used to analyse the Eq. 2.9 to find the directions in the triplets. The question is, what parameter values are needed to be estimated for a successful causal inference? Following the structure of causal model, the connection strength values are required to detect the causal directions and the error values added into the model. As it has been discussed, the qualitative analysis of causal model requires causal influence values, let us discuss the role of causal influence.

2.2.3 Causal Influence

The role of causal influence is to verify the goodness of information passed on from parent node to the child node. The connection strength values show the weight/amount of information transferred, but the potential of that information only can be verified using the causal influence condition. The goodness of information depends on the quality of information passed and the level of noise added into the system. The meaning of this is that, if there is more noise/error added into the system then the influence of transmitted information becomes less effective. Therefore, the more the noise the system contains, the less influential the connection strength becomes. But causal influence does not only depend on the information but also depends on the feature value. So, the parent and its connection strength combined provide the causal influence on the child node.

In Fig. 2, 3 minimal d-separable Markov equivalent V-structures were provided, which are useful for the detection of causal directions from the comparison of their causal influence values. By solving Eq. 2.10, the values of the unknowns can be estimated easily, i.e.

$\{c_{xy}, c_{zy}, e_y\}$ for the case $\{x \rightarrow y \leftarrow z\}$ as shown in Fig. 2c. Using Eq. 2.10, the causal influence in the features can be defined by

$$CI_{xy} = \frac{c_{xy}E(x)}{E(y)} < \frac{E(e_y)}{E(y)}, CI_{zy} = \frac{c_{zy}E(z)}{E(y)} < \frac{E(e_z)}{E(y)} \tag{2.13}$$

The values $\{CI_{xy}, CI_{zy}\}$ represent the causal influence/causal influence factors for direction sets $\{x \rightarrow y, z \rightarrow y\}$ and value $E(.)$ represents the expectations of the indicated variables. The values of CI indicate the goodness of causal inference for the direction set $\{x \rightarrow y \leftarrow z\}$ and the same process is applied for the other Markov equivalent classes as shown in Figs. 2b and d. The values of CI can be represented using a percentage range from [0,100%] or the probability range of [0, 1].

The causal influence values indicate the existence of direction in the observed features and how influential/effective the information is for constructing the child node. After examining all possible direction sets for a dataset, these can be arranged using the causal independence condition. The most causally independent node will be the top of the causal structure, where nodes are arranged in descending order, depending on their causal independence.

2.2.4 Causal Level

The causal structures are arranged using the causal influence values which represent the causal independence of an observed feature. The effective arrangement of nodes can be done in a way such that the graphical arrangement follow a tree like but connected and cyclic/acyclic arrangement in an ascending or a descending order using the causal independence of the node set. This kind of ordering produces the levels where nodes are grouped using their causal independence. Figure 3 provides an example of causal levels in the shown causal structure.

Figure 3 represents a 3-causal levelled model with seven nodes. In Fig. 3, the nodes $\{V_1, V_3\}$ are the two parent sets and represent the parent causal level. The

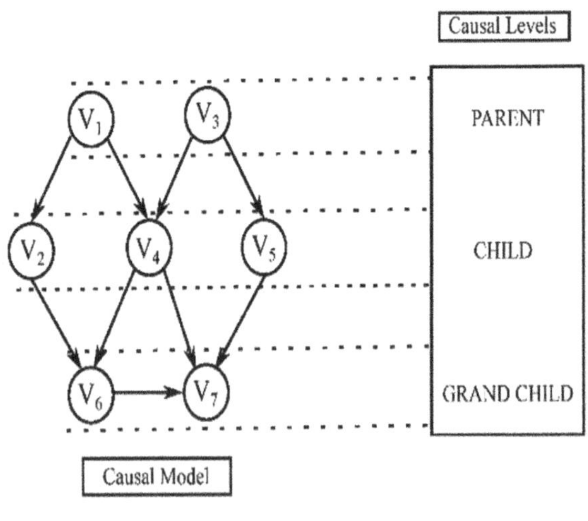

Fig. 3 A 7-node directed acyclic causal model with three causal levels

node set $\{V_2, V_4, V_5\}$ are the children of parent set and are shown in the child causal level. In the next level, the two nodes.

$\{V_6, V_7\}$ which are the grandchildren of parents, are represented in grandchild causal level. This kind of causal level represents the kinship relation in the causal structure. Furthermore, the causal levels can be shown using different relation types as is present in the feature set.

Here, a genealogy of the feature set is used to level the structure where nodes with the same relations are staged together.

Causal levelling can be seen as clustering of nodes having the same shared relation for the same types of ancestral and descendant structures. Not only does it help to improve the structural arrangement of the nodes in the causal model but also helps to extract the necessary information of causal influence and causal independence.

Causal Model Construction: A complete causal structure can be constructed after getting the hierarchy of causal independent and dependent sets represented in causal levels. A complete and orderly arranged causal level represents a complete causal model. Also, the causal models can be assembled using causal sub-structures found from causal inference, although this process is somewhat complicated and time-consuming. It requires more construction skills to arrange sub-structures into a perfectly arranged complete causal model. So, one of the easiest methods is to find the causal levels and arrange them in top-down or bottom-up graphical order to find the complete causal model:Causality and Machine Learning, Deep Learning.

3 Machine Learning, Deep Learning and Causal Reasoning

The recent developments in machine learning in both supervised and unsupervised cases, mostly concern on analysing, making predictions and fine-tuning the models. Leaving the case of regression the most machine learning methods can't be interpretable when it comes to find the cause and effect inference in the system. Most of the time, accuracy is highly a concern when it comes to general predictability of the trained model. But again the parameters and selected features for trained accurate model can not explain why it is responsible. Which features are responsible for what kinds of changes in the prediction? If we can infuse causality with machine learning models to bring out the causal influence factors or the causal levels, it will be more than sufficient to provide every possible unanswered questions in the prediction model.

3.1 Deep Learning and the Black Box

Modern computations makes it possible for complex algorithms and designs to run comfortably. The deep learning models are good examples of complex models which runs on big data to achieve specific goals. From large image classification, tagging, to new image generations, Language translations and interpretations to speech recognition are become possible using deep learning architect models. As models and algorithms become complex and deep the parameters, number of features and the models layers become larger than before. Which brings to the question of what exactly these parameters mean and what are all those features that are extracted inside the complex deep models and how they are effecting through the deep layers to achieve the accuracy we seek for. And the term for the process of input data into the deep models then magically generating and learning through the networks/layers to output the results become a black box. There is no interpretability as such in deep learning. The designs are changed to serve the purpose but not so simple explanations to support the result. Even the learned and extracted features does not make full justification of the process and time consumed to train them.

3.2 Predictive Versus Prescriptive Analysis

Predictive analysis, most certainly the most commonly used model. In this, we are trying to predict the outcome in the observed features which are incorporated into the trained model from our understanding of the environment or the process itself. But are we certain of our selection of right features which going to predict the model, although we have carefully selected most of the features based on their predictability towards the target. But it might also possible most influential feature from causal point of view may have very less influence on the target which diminishes through the causal levels. So causal influence is not quite detectable when we are trying to accumulate features to provide a better prediction.

Now if exactly we know the causal effects then, should we do a prediction? Let's understand this in details. Causal analysis provides us with information like causal influence and arranging them we can find causal levels. Now we can definitely arrange the causal levels such that our target is at the top or bottom, remember that we are removing subsequent childs after the target to retain a clear picture of causal levels and parental structure only. From causal influence, we can find the contribution of any feature towards the target through the causal levels. With all this information in hand, we can now perform a prescriptive analysis. The prescriptive analysis will be one step further of predictive analysis. In prescriptive analysis, we know how the features are responsible towards what is happening with the target. So no need to fine tune, train or make models more accurate to achieve the better target. Now we can prescribe what is necessary and even sufficient to do to get the target outcome.

3.3 *Integrating Causality into Machine Learning and Deep Learning*

Machine Learning fails to generalize the model with addition of new data and needs further fine tuning and optimizations. Most of all the addition of new data or new developments in the environments where machine learning is deployed, the models unable to cover all the data distributions. This is the reason why different machine learning methods are used to tackle this problem but none the less fails. Integrating causality in to feature dependency identification can substantially change how machine learning models are tuned. Causal learning is independent of data additions and does not require large training process. So the change in environment can be explained in priory observed statistical dependencies.

In case of deep learning the features selection process is specific to the requirements goal or target. So multiples of different representations of the same object is required to capture the underlying feature structures and to get the desired accuracy. With causal introductions into deep learning, the models won't require large data to learn different representations of the same object, but it tries to learn the feature dependency towards the desired goal state. Causal analysis can capture different feature relations and generalize them to find their contributions towards the prediction.

3.4 *Causal Applications in Machine Learning, Deep Learning*

Causal inference can be much of help in machine learning tasks without giving to much concerns on model training, testing, parameter tuning and optimizations for accurate predictions.

3.4.1 Causal Inference in Classification Task

Machine learning models for classifications singularly depend on training and testing of the models for specific data sets to achieve high accuracy. And in the end the models with selected feature sets and acquired parameters does not produce explanations for their specific contributions towards the target class. It is often not interpretable to say which features are significant for the change in target. Also adding new data changes the models ability to generalize for the new data and significantly drops the performance of the old model.

Inducting causal inference into classification task, we can explain the feature dependency and causal influence can provide us with the much required information of how they are related. We can find which features are responding highly towards the target class. So the exhaustive process of model tuning are not required while using causal inference.

3.4.2 Deep Learning Network Analysis Using Causal Inference

Consider a specific case of convolution network where we are extracting feature by using the convolution functions. Then large no of parameters are generated and increases with each layer and taking a stack of all convolutions we can find the output. Now we definitely know what convolutions operators to use to extract specific features from images, but the features relations are lost in the layer. At the end we don't know how the deep models are learning and what is learned in the process.

Here, the causal analysis can help to track the dependency through the layers which transfers into different features and the target itself. The causal influence values with the in the same path of the network connections but providing much required links to feature convolution and transferring causal influence in each layer.

Other case of LTM and RNN, where the deep connections and dependency of features are stored in memory to learn the positional values. But required huge computational power and time and large volume of data with different representations to achieve admissible accuracy. In such cases, causal influence can be used to learn all dependency structures and then it can be used in different data samples to study the generalization of the observed causal influence structure.

References

1. Bollen KA (1989) Structural equations with latent variables. Wiley
2. He YB, Geng Z (2008) Active learning of causal networks with intervention experiments and optimal designs. J Mach Learn Res 9:2523–2547
3. Hyvärinen A, Smith SM (2013) Pairwise likelihood ratios for estimation of non- gaussian structural equation models. J Mach Learn Res 14:111–152
4. Parida PK, Marwala T, Chakraverty S (2016) An overview of recent advancements in causal studies. Arch Comput Methods Eng 24(2):319–335
5. Parida PK, Marwala T, Chakraverty S (2018) A multivariate additive noise model complete causal discovery. Neural Netw 103:44–54
6. Pearl J (2009) Causal inference in statistics: an overview. Statistical Surveys, 3:96–146. Pellet JP and Elisseeff A (2008). Using markov blankets for causal structure learning. J Mach Learn Res 9:1295–1342
7. Peters J, Mooij MJ, Janzing D, Schölkopf B (2014) Causal discovery with continuous additive noise models. J Mach Learn Res 15(1):2009–2053
8. Petrovi\'{c}, L. and Dimitrijevi\'{c}, S. (2011) Invariance of statistical causality under convergence. Statist Probab Lett 81(9):1445–1448
9. Shimizu S, Inazumi T, Sogawa Y, Hyvärinen A, Kawahara Y, Washio T, Hoyer PO, Bollen K (2011) Directlingam: a direct method for learning a linear non-gaussian structural equation model. J Mach Learn Res 12:1225–1248
10. Shimizu S, Hyvärinen A, Kano Y, Hoyer PO (2005) Discovery of non-gaussian linear causal models using ICA. In: Proceedings of the 21st conference on uncertainty in artificial intelligence, pp 526–533
11. Shpitser I, Pearl J (2008) Complete identification methods for the causal hierarchy. J Mach Learn Res 9:1941–1979
12. Spirtes P, Glymour C, Scheines R (1993) Causation, Prediction, and Search. Springer Verlag

13. Sun X, Janzing D, Schölkopf B (2006) Causal inference by choosing graphs with most plausible Markov Kernels. In: Proceeding of the 9th international symposium art international and mathematics, Fort Lauderdale, FL

Hybrid Evolutionary Computing-based Association Rule Mining

Ganghishetti Pradeep, Vadlamani Ravi, and Gutha Jaya Krishna

Abstract This chapter proposes three novel association rule mining algorithms based on hybrid evolutionary techniques, which obviate the necessity of pre-specifying the minimum support and minimum confidence, unlike Apriori and FP-Growth. Further, these rules are obtained in a single run of these algorithms. These features signal a clear departure from the traditional algorithms without sacrificing the power of the rules. Their effectiveness is tested on a real-life commercial bank dataset from India and five other standard datasets. This chapter opens up a new line of research, wherein hybrid evolutionary techniques can be exploited to their fullest potential in mining association rules from databases.

Keywords Firefly optimization algorithm · Threshold accepting algorithm · Particle swarm optimization · Association rule mining · Combinatorial global optimization

1 Introduction

Data mining uncovers hidden patterns from the huge mass of data. The tasks of data mining include association rule mining, classification, clustering, regression, and outlier analysis. It found diverse applications in varied domains. Association rule mining extracts useful correlations among the items from databases. It is paramount in decision-making in a wide variety of areas ranging from market basket analysis, medical diagnosis, fraud detection in web, click stream mining among others.

Agrawal et al. [1] and Agarwal and Srikant [2] were the first to propose an algorithm for mining association rules from large customer transactional datasets. An association rule is of the form A → B, where A and B represent item sets (I) or products, and an item set includes all possible items $\{i_1, i_2, \ldots, im\}$ in a transactional database. The algorithm works in two stages. The first phase generates frequent itemsets using a user-defined parameter called support count (SUP), while the second

---

G. Pradeep · V. Ravi (✉) · G. J. Krishna
Center of Excellence in Analytics, Institute for Development and Research in Banking Technology (IDRBT), Masab Tank, Castle Hills Road #1, Hyderabad 500057, India

S. Chakraverty (ed.), *Soft Computing in Interdisciplinary Sciences*,
Studies in Computational Intelligence 988,
https://doi.org/10.1007/978-981-16-4713-0_12

generates the rules using another user-defined parameter namely minimum confidence. Another algorithm FP-Growth, proposed by Han et al. [3] requires two scans on the database. Then FP-Growth starts to mine the FP-tree for each item whose support is larger than the minimum support by recursively building its conditional FP-tree.

In this chapter, binary firefly optimization (BFFO), binary firefly optimization—threshold accepting (BFFO-TA) and binary particle swarm optimization-threshold accepting (BPSO-TA) association rule miners are proposed to be executed in a single run in order to obtain the rules. These algorithms are tested for their effectiveness on 6 datasets and compared with traditional approaches like Apriori and FP-Growth.

The remainder of the chapter is structured as follows. Section 2 presents related work in the field of evolutionary algorithm-based association rule mining. Section 3 introduces various algorithms used in our proposed approaches. Then, Sect. 4 presents various steps in the devised strategies. The results obtained on various datasets are discussed in Sect. 5 followed by conclusions.

2 Literature Review

The application of evolutionary algorithms to mine association rules has been a relatively new area. Saggar et al. [4] proposed Genetic Algorithm (GA) to optimize the rules yielded by Apriori. Waiswa et al. [5] proposed Pareto-based GA-based multi-objective evolutionary algorithm to extract the association rules. Ghosh and Nath [6] also did the same by considering three measures—comprehensibility, interestingness, and predictive accuracy. Gupta [7] used weighted particle swarm optimization (WPSO), while Asadi et ai. [8] used PSO for finding the minimum support and minimum confidence values for the Apriori. Minaei-Bidgoli et al. [9] proposed multi-objective GA for mining numerical association rules considering confidence, interestingness, and comprehensibility. Nandhini et al. [10] devised PSO and domain ontology-based association rule mining algorithm. Alatas et al. [11], used pareto-based multi-objective differential evolution (DE) for mining accurate and comprehensible numeric association rules. Hadian et al. [12] proposed a Cluster-Based Multi-Objective Genetic Algorithm (CBMOGA). Kuo et al. [13] devised a particle swarm optimization algorithm for the extraction of rules for large databases. Further, Maheshkumar et al. [14] proposed PSO-TA hybridized algorithm for unconstrained continuous optimization. Most recently, Sarath and Ravi [15] proposed binary particle swarm optimization (BPSO) to extract association rules without specifying minimum support and confidence. The motivation for current work is that BPSO must be run M times to get top M rules, which is not an attractive proposition in practical applications. In order to overcome this drawback, hybrid evolutionary algorithms for association rule mining which generate the top M in just a single run of the algorithms are proposed. This feature makes these algorithms attractive in practical applications. Cheng [16] applied GA for selecting items from a database of a certain chain convenience store in Shanghai.

Shenoy et al. [17] analyzed a dynamic transaction database using GA. Kuo and Shih [18] employed ACO to mine the association rules efficiently. Chien and Chen [19] built an associative classifier to discover trading rules from a GA-based algorithm on numerical data. Christian and Martin [20] obtained association rules with Apriori, on the search space reduced by GA. Hansen et al. [21] considered the retail shelf allocation problem with non-linear profit functions, vertical and horizontal location effects, and product cross elasticity. Khademolghorani [22] proposed ICA for mining interesting and comprehensible association rules. Yang et al. [23] suggested the evolutionary associative classification method for both adjustments of the order of rules as well as the refinement of each single rule. Bhugra et al. [24] generated and optimized association rules. Birtolo et al. [25] searched for product bundles that are optimal using GA. Cunha and Castro [26] built association rules by evolutionary and AIS algorithms. Luna et al. [27] extracted both numerical and nominal association rules in a single step. Ganghishetti and Vadlamani [28] developed three multi-objective evolutionary association rule miners, namely, MO-BPSO, MO-BPSO-TA, and MO-BFFO-TA on XYZ Bank datasets. However, the algorithms presented here need to be run only once to get all the rules.

3 Association Rule Mining Using Firefly Optimization, Particle Swarm Optimization, Threshold Accepting-based Techniques

3.1 Firefly Optimization Algorithm (FFO)

The Firefly algorithm proposed by Yang [29], is based on the natural behavior of fireflies. It is a population-based technique to obtain the global optimal solution. For further details, please refer to Yang [29], as it has become too popular to describe in detail here.

3.2 Threshold Acceptance (TA)

The threshold Accepting algorithm [30] is a faster variant of simulated annealing. For more details, the reader is referred to Ganghishetti and Vadlamani [28] (Tables 1

Table 1 Parameters chosen for BPSO Rule Miner on Bakery and Clickstream Dataset

Dataset	Inertia	C1	C2	N	Max. Iterations
Bakery	100.8	2	2	50	100
Clickstream	0.8	2	2	50	50

Table 2 Rule representation

I_1		I_2		I_3		…	I_N	
V_{11}	V_{12}	V_{21}	V_{22}	V_{31}	V_{32}	…	V_{N1}	V_{N2}

Table 3 Parameters chosen for BFFO Rule Miner

Dataset	Beta_0	Gamma	Alpha	n	Delta	Bias	Max Iterations
Books	2	2	0.4	30	0.97	−0.2	50
Foods	2	2	0.4	30	0.97	−0.2	50
Grocery	1	0.01	1	10	0.97	−0.25	1000
Bank	0.1	1	0.6	10	0.97	−0.4	50
Bakery	0.1	2.5	0.5	10	1	−0.4	400
Clickstream	0.1	2.5	0.5	10	1	−0.4	100

and 2).

3.3 Binary Firefly Optimization (BFFO)

The firefly optimization algorithm described in Sect. 3.1 is well suited for continuous problems alone. For the details of BFFO, the reader is referred to Ganghishetti and Vadlamani [28]. Binarization of FFO is accomplished by the use of $S(x) = \frac{1}{1+e^{-x}} + bias$. The parameter bias is a user-defined small value lying between 0 and 1. It assists the BFFO algorithm to converge faster. The bias component is not needed for BFFO-TA algorithm. For BFFO and BFFO-TA, the parameters tuned for the datasets considered in this chapter are presented in Tables 3 and 4.

3.4 Particle Swarm Optimization (PSO)

Introduced by Kennedy and Eberhart [31], PSO belongs to the class of swarm intelligence and is inspired by behavioral models of bird flocking. PSO is a population-based stochastic approach for solving continuous and discrete optimization problems. It is too well-known to be described here in detail.

3.5 Binary PSO

Sarath and Ravi [15] developed binary particle swarm optimization-based association rule miner. The BPSO is employed to extract M rules in exactly M runs. For more

Table 4 Parameters chosen for BFFO-TA rule miner

	β_0	Γ	α	N	Δ	Prob_TA (%)	Max Iterations	GI	II	Acc	Thresh	Eps	Thr. Tol
Books	2	2.5	0.5	10	1	20	50	5	100	1.2E-06	2	0.06	1.8
Foods	2	2.5	0.5	10	1	30	50	5	100	1.2E-06	2	0.06	1.8
Grocery	2	2.5	0.5	10	1	30	50	5	200	1.2E-06	2	0.06	1.8
Bank	2	2.5	0.5	10	1	4	50	5	100	1.2E-06	2	0.06	1.8
Bakery	2	2.5	0.5	10	1	90	50	25	100	1.2E-06	2	0.06	1.8
Clickstream	0.1	2.5	0.5	10	1	5	100	25	100	1.2E-06	2	0.06	1.8

details of BPSO, the reader is referred to Sarath and Ravi [15]. The various parameters used in BPSO for all datasets are presented in Sarath and Ravi [15]. The parameters tuned for the two additional datasets considered in this chapter are presented in Table 1. For the BPSO-TA, the parameters tuned for the datasets considered in this chapter are presented in Table 5.

3.6 Feature Selection

This is step is to be followed as preprocessing step when the number of features exceeds 50. This scheme is devised to reduce the number of features to be considered for rule generation. Since our objective function is to maximize both support and confidence, if the individual item support values are extremely low, they are unlikely to be included in the final set of rules generated. Here, all the features whose item support is less than α are discarded, where α is very low. This helps in obviating the unnecessary computations for the algorithm. The steps involved are depicted in Figs. 1, 2, 3 and 4. The rule representation is presented in Table 2. This issue is further discussed in Sect. 4.7.

4 BFFO/BFFO-TA/BPSO-TA for Association Rule Mining

The proposed hybrid algorithms are divided into two phases. In data preprocessing phase, the data is transformed into a set of binary records. In preprocessing phase, in the case of datasets with sparse transactions, feature selection is performed based on item support values. In the second phase, BFFO/BFFO-TA rule miner is applied to extract association rules. First, binary encoding of firefly positions is performed. Next, a population of fireflies whose fitness values are greater than 0.001 is generated. The mining of rules continues until the stopping criterion of a maximum number of iterations is reached.

4.1 Binary Transformation

In order to extract top M rules from the transactional database, the data is transformed into a binary-encoded format where each record is stored in terms of 0s and 1s as in Wur and Leu [32]. For more details, the reader is referred to Ganghishetti and Vadlamani [28].

Table 5 Parameters chosen for BPSO-TA rule miner

Dataset	Inertia	C1	C2	N	Prob_of invoking TA (%)	Max Iterations	GI	II	Acc	Thresh	Eps	Thr Tol
Books	0.8	2	2	30	5	50	5	50	1.2E−06	2	0.01	1.8
Food	0.8	2	2	30	5	50	5	50	1.2E−06	2	0.01	1.8
Grocery	0.8	2	2	30	10	100	5	50	1.2E−06	2	0.01	1.8
Bank	0.8	2	2	50	10	50	5	50	1.2E−06	2	0.01	1.8
Bakery	100.8	2	2	70	10	100	25	300	1.2E−06	0.5	0.0001	0.3
Click stream	0.8	2	2	50	5	100	25	50	1.2E−06	0.02	0.0001	0.001

GI-global iterations; II-inner iterations

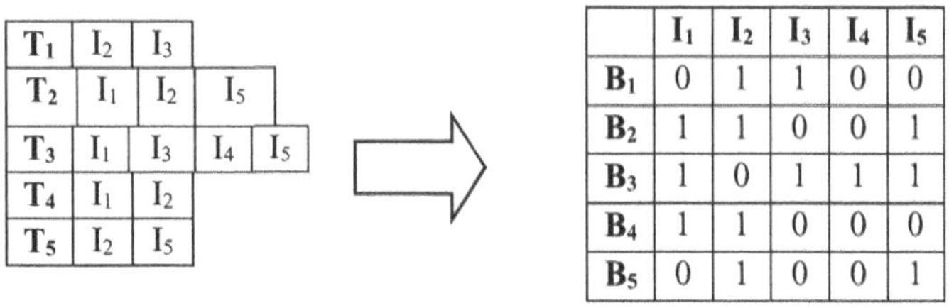

Fig. 1 Binary transformation of the original dataset

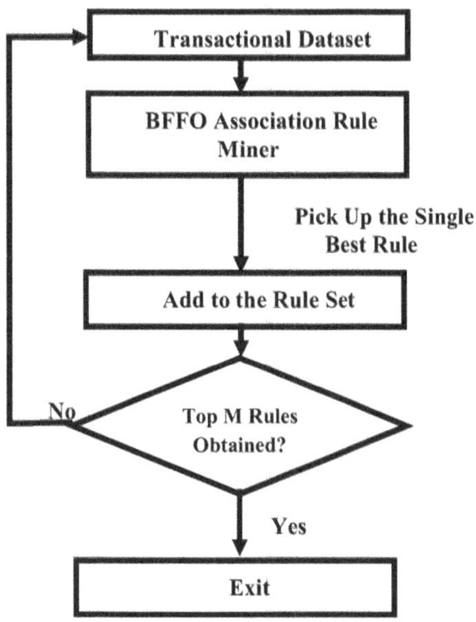

Fig. 2 Block diagram of our proposed BFFO method

4.2 Rule Representation

While applying optimization techniques to association rules, the first step involved is to represent an association rule as a firefly position. In this chapter, the Michigan approach is followed. Let there be N number of items in the dataset. Each item is represented by two bits and each bit can take the values either 1 or 0.

For further details, please refer Ganghishetti and Vadlamani [28]. This same procedure is followed in Kennedy and Eberhart [31, 33]. The rule encoding is needed for selecting items in the antecedent and consequent part of the association rule.

4.3 Objective Function

The product of support and confidence is chosen as the objective function as follows:

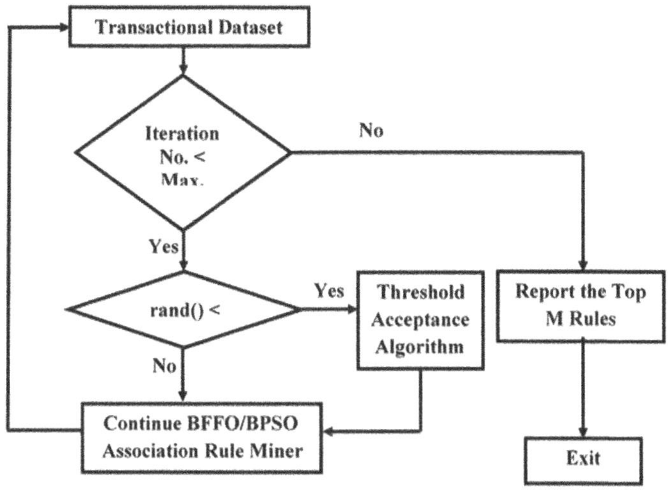

Fig. 3 Block diagram of the BFFO-TA/BPSO-TA method

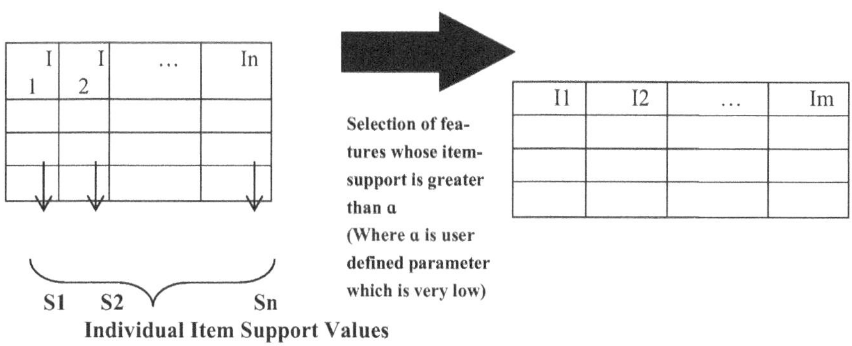

Fig. 4 Reduction of feature for sparse dataset

$$\text{Objective Function} = \text{maximize}\{\text{Support }(A \rightarrow C) \times \text{ Confidence }(A \rightarrow C)\} \quad (1)$$

Support is an indicator of how frequently the item appears in the database. Confidence indicates the number of times the if/then statements have been found to be true. Taking support alone as the objective function doesn't serve our purpose. This is due to the rare item problem as the items that occur sparsely/infrequently in the dataset are pruned, although they would be extracting potentially interesting and valuable rules. The rare item problem discussed in Li et al. [34] is important for transaction data due to the fact that they usually have an uneven distribution of supports for individual items.

Similarly, confidence alone cannot be considered as an objective function as it is sensitive to the frequency of the consequent part in the database. From the definition of confidence, consequents with higher support will automatically produce higher confidence values although there may be no associations among items.

Therefore, the product of support and confidence is considered as our objective function. The rationale is that, because both support and confidence lie between 0 and 1, if their product is maximized, they are individually maximized.

For the steps involved in BFFO, BFFO-TA, and BPSO-TA, the reader is referred to Ganghishetti and Vadlamani [28].

4.4 Special Cases

The following special mechanisms are devised for both BFFO-TA and BPSO-TA:

Case 1: When a dataset has more than 50 features (as in clickstream dataset), then perform feature selection to reduce the number of features.

Case 2: When a dataset is extremely sparse having features with less item support count and has more than 50 features (as in Bank dataset), perform feature selection as well as perform re-initializations. Re-initialization step is described as follows: In the process of movement of particles/fireflies from one position to another, whenever encountered with situations such as the rule containing all antecedents or all consequents or all zeros indicating that the rule is out of desired range, the position of the fireflies/particles is reinitialized. This re-initialization step adds diversity to the population.

4.5 Advantages of the Proposed Approaches

Rules are generated without having to specify the minimum support and minimum confidence. Redundant rules are not produced, unlike Apriori and FP-Growth. The rules extracted are of varied length and are independent of frequent Item sets as against fixed rule lengths obtained in Apriori and FP-Growth. Overall, the BFFO-TA takes less computational time to mine association rules compared to these traditional strategies. In case of BFFO-TA, it is found to produce top M rules in just a single run. However, in our approaches, the user needs to specify the number of best rules he/she wants to generate from a given dataset.

5 Results and Discussion

In this chapter, six datasets are analyzed for demonstrating the effectiveness and to perform comparative analysis for various algorithms. The datasets vary in terms of the number of features, transaction count, and sparsity in the data. The first one is Books dataset taken from http://www.solver.com/xlminer-data-mining. The second one is the Food dataset taken from http://www.ibm.com/software/analytics/spss. The third one is the grocery dataset taken from http://www.sas.com/technologies/analytics/datamining/miner. The study analyzed a real-world dataset from XYZ commercial bank. It is a sparse dataset with 12,191 customers' transactional records and 134 different product and service offerings. The fifth one is the Bakery dataset taken from https://wiki.csc.calpoly.edu/datasets/wiki/ExtendedBakery. The last one is the Anonymous Web dataset taken from http://archive.ics.uci.edu/ml/datasets/Anonymous+Microsoft+Web+Data. The results obtained on various datasets are discussed in the following sections.

5.1 Books Dataset

For books dataset, kept the population size (p) as 30, absorption coefficient (γ) as 2, attractiveness ($\beta 0$) as 2, and the iterations count (I) as 50 for the BFFO algorithm. The top 10 association rules extracted using BFFO after executing it 10 times are presented in Table 6. It can be observed that BFFO could produce rules with more support and confidence values. To increase the performance and efficiency of the BFFO association rule miner, it is hybridized with TA Algorithm. The effectiveness of this hybrid is discussed below

On books dataset, the chosen attractiveness is 2 and absorption coefficient is 2.5 for the BFFO-TA. The population size is kept at 10 and the number of iterations is fixed at 50. In every iteration of BFFO, the TA algorithm is called with a probability of 20%. By hybridizing TA with BFFO, were able to extract all the Top 10 rules just in a single run. It is a noteworthy achievement of our proposed approach. Also, BFFO-TA Rule Miner takes computationally less time. This increase in the robustness of our

Table 6 Results of books datasets using BFFO, BFFO-TA, BPSO-TA, and BPSO

Rule number	Antecedent	Consequent	Support	Confidence
1	Child books	Cook Books	25.6	60.52
2	Geog books	Child Books	19.5	70.65
3	Geog books	Cook Books	19.25	69.74
4	Doity books	Cook Books	18.75	66.48
5	Doity books	Child Books	18.4	65.24
6	Cook books, Geog Books	Child Books	14.95	77.66
7	Child books, Doity Books	Cook Books	14.6	79.34
8	Art books	Cook Books	16.7	69.29
9	Italcooks	Cook Books	11.35	100
10	Youth books	Child Books	16.5	66.67

algorithm is because of the invocation of TA. The rules produced by this approach are the same top 10 rules obtained from BFFO and BPSO presented in Table 6. On books dataset, for the BPSO-TA algorithm, the chosen inertia is 0.8, the population size is kept at 30, and the number of iterations is fixed at 50. In every iteration, the TA algorithm is called with a probability of 5%. Also, BPSO-TA Rule Miner takes computationally less time and produces the same top 10 rules obtained from BFFO, BFFO-TA, and BPSO as presented in Table 6.

5.2 *Food Dataset*

The population size of BFFO algorithm for Food item dataset is fixed at 30, the absorption coefficient at 2, attractiveness at 2, and iterations count at 50. The results are presented in Table 7.

While applying BFFO-TA on Food items dataset, the number of fireflies in the population is chosen as 10 and the number of iterations is fixed at 50. In every iteration, the TA algorithm is called with a probability of 30%. The parameters $\beta 0$ and γ are chosen as 2 and 2.5. To extract the top 10 rules, BFFO-TA is run just once. BFFO-TA is found to take computationally less time compared to BFFO as it needs less population size and one run. The top ten rules produced by BFFO-TA are the same rules produced by BFFO, BPSO and are presented in Table 7.

While applying BPSO-TA on food dataset, the chosen inertia is 0.8, the number of particles is kept at 30, and the number of iterations is fixed at 50. The TA algorithm is called with a probability of 5% in every iteration. Also, BPSO-TA Rule Miner takes computationally less time. This increase in the robustness of our algorithm is because of the inclusion of the invocation of the TA algorithm. The rules produced by this approach are the same top 10 rules obtained from BFFO, BFFO-TA, and BPSO as presented in Table 7.

Table 7 Results of food dataset using BFFO, BFFO-TA, BPSO-TA, and BPSO

Rule Number	Antecedent	Consequent	Support	Confidence
1	Canned Veg, Beer	Frozen Meal	14.6	87.42
2	Frozen meal	Canned Veg	17.3	57.28
3	Beer	Frozen Meal	17	58.02
4	Beer	Canned Veg	16.7	56.99
5	Confectionery	Wine	14.4	52.17
6	Fish	Fruit Veg	14.5	49.65
7	Canned Veg, Beer, Fish	Frozen Meal	4.4	91.66
8	Fruit Veg, Canned Veg, Beer	Frozen Meal	4	86.95
9	Wine	Canned Veg	9.7	33.79
10	Canned Meat, Frozen Meal, Beer	Canned Veg	3.6	90

Table 8 Results of grocery dataset using BFFO, BFFO-TA, BPSO-TA, and BPSO

Rule number	Antecedent	Consequent	Support	Confidence
1	Cracker	Heineken	36.56	75
2	Cracker, Soda	Heineken	23.37	93.22
3	Artichoke	Heineken	25.17	82.62
4	Soda	Heineken	25.67	80.81
5	Soda	Cracker	25.07	78.93
6	Artichoke, Avocado	Heineken	19.88	94.31
7	Baguette, Herring	Heineken	21.37	85.94
8	Baguette	Heineken	26.07	66.58
9	Turkey	Olives	22.07	78.09
10	Corned_B, Olives	Herring	20.17	85.23

5.3 *Grocery Dataset*

In case of Grocery dataset, fix p as 10, γ as 0.01, β_0 as 1, and I as 1000 for BFFO algorithm. The top rules by this approach are presented in Table 8.

The BFFO-TA algorithm is applied on this dataset with p as 10 and I as 50. Due to the nature of the dataset, the probability of calling the TA algorithm in each iteration is increased to 30%. The other parameters β0 and γ are taken as 2 and 2.5, respectively. The BFFO-TA produced the top 10 rules in a single run itself. These are the same rules produced by BFFO and BPSO as presented in Table 8. The current BFFO-TA based approach is found to take less computationally time compared to BPSO.

For the BPSO-TA algorithm, the parameter inertia, population size, maximum iterations are fixed as 0.8, 30, and 100, respectively. In every iteration, the TA is called with a probability of 10%. By hybridizing TA with BPSO Rule Miner, were able to extract all the Top 10 rules only in a single run. Also, BFFO-TA Rule Miner takes computationally less time compared to every other strategy and this can be attributed to the hybridization of TA with BPSO. The rules produced by this approach are the same top 10 rules obtained from BFFO, BFFO-TA, and BPSO as presented in Table 8.

5.4 *XYZ Bank Dataset*

When applied the BFFO Rule Miner on the real-life bank data set taken from India named as XYZ bank dataset, it was found to be extremely sparse in nature with respect to support and confidence values. In this dataset, specifically, the objective function values of all the rules whose rule length is 3 or more is found to be almost equal to zero. The objective function values of most of the rules whose rule length is

2 is also close to zero. Due to this nature of the dataset, the proposed BFFO approach was not found to be effective in extracting the best rules even after running 10,000 iterations and took huge computational time. The dataset is unusually deterring the algorithm in reaching the global optimum as almost all the fireflies in the population are reaching a fitness function value of zero. To speed up the process and to improve the effectiveness of our proposed strategy, a new strategy is devised as follows:

The objective function considered in our algorithm is taken as the product of support and confidence as in Eq. 1. Our algorithm's primary goal is to maximize support and confidence, and therefore, the rules produced will not contain items which have too small support values. In order to overcome the situation created by sparse datasets of this kind, and to remove the unnecessary computations, performed feature/product selection i.e., identified and dropped irrelevant features/products from the bank dataset. In this procedure, we considered only those features whose item support values are greater than 0.99, a user-defined parameter. By this process, the original set of 134 features got reduced to 25. The feature selection process is depicted in Fig. 4.

For banks dataset, the population size is kept at 50 and number of iterations as 2000 during the application of BFFO algorithm. The top rules produced are presented in Table 9. However, due to the sparsity of the dataset, the support and confidence values are low.

Prior to application BFFO-TA to Banks dataset, reduce the number of features to 25 as in BFFO to overcome sparsity in the data. The p is kept at 10 and the I is fixed at 50. Taken $\beta 0$ as 2 and γ as 2.5. In each of the iterations, the TA algorithm is called with a probability of 30%. Here, the Top 10 rules are extracted in just a single run and are presented in Table 10. This also takes computationally very less time compared to BFFO. Furthermore, the support and confidence values produced are comparatively higher than BFFO.

As regards the BPSO-TA algorithm, the chosen inertia is 0.8, the population size is kept at 50, and the number of iterations is fixed at 50. In every iteration, the TA

Table 9 Results of XYZ bank dataset using BFFO

Rule number	Antecedent	Consequent	Support (%)	Confidence (%)
1	AGL1	SB2	1.87	72.15
2	LAD1	FD1	1.61	74.52
3	LAD2	FD1	1.53	77.82
4	AG1, AGL2	CC1	1.07	86.75
5	SB3, GL3	GL1	1.27	54.58
6	FDOGP	FD1	1.29	52.15
7	GL1	SB3	2.21	28.86
8	CC1	FD1	1.39	39.91
9	CC1	AG1	1.26	36.15
10	CC1, AGL2	FD1	1.13	39.54

Table 10 Results of XYZ Bank using BFFO-TA, BPSO-TA, and BPSO

Rule number	Antecedent	Consequent	Support (%)	Confidence (%)
1	FD1	SB1	6.6	46.72
2	CC1	AGL2	2.86	81.92
3	GL1	GL3	3.97	51.93
4	GL2	GL3	3.18	57.74
5	FD2	FD1	2.26	68.24
6	AGL1	SB2	1.87	72.15
7	GL1, GL2	GL3	1.67	76.12
8	LAD1	FD1	1.61	74.52
9	LAD2	FD1	1.53	77.82
10	GL3	SB1	3.27	35.62

algorithm is called with a probability of 10%. Also, BPSO-TA Rule Miner takes computationally less time compared to BPSO, BFFO, Apriori, and FP-Growth. The rules produced by this approach are the same top 10 rules obtained from BFFO-TA and BPSO as presented in Table 10.

5.5 *Bakery Dataset*

It is found that it has very good item support count values and did not require any feature selection. However, the dataset is found to have very low support values when different combinations of items are considered. This forced the need to increase the number of global iterations in TA to get good particles/fireflies with high fitness values. This increased the execution time for this particular dataset.

Association rules via BFFO for Bakery dataset were obtained by fixing population size (p) as 10, absorption coefficient (γ) as 2.5, attractiveness ($\beta 0$) as 0.1, and the number of iterations (I) as 400, respectively. The top 10 rules, thus extracted using BFFO after executing 10 times are presented in Table 11.

While applying BFFO-TA on Bakery dataset, the population size is kept at 10 and the number of iterations is fixed at 50. In every iteration, the TA algorithm is called with a probability of 90%. The parameters $\beta 0$ and γ are kept as 2 and 2.5 and the top 10 rules are extracted only in a single run. BFFO-TA is found to take computationally less time compared to BFFO as it needs a fewer iterations and a reduced number of runs. The top ten rules produced by BFFO-TA are the same rules produced by BPSO, BPSO-TA and are presented in Table 12. It is clear that BFFO-TA produced superior results compared to BFFO.

In case of BPSO algorithm on Bakery dataset, the population size and maximum iterations are both fixed at 50. To extract the top 10 rules, BPSO is run 10 times. BPSO is found to take computationally less time compared to BFFO, BFFO-TA.

Table 11 Results of bakery using BFFO

Rule	Antecedent	Consequent	Support	Confidence
1	Truffle	Gongolais	5.8	56.32
2	Napoleon	Strawberry	4.9	54.44
3	Coffee, Almond_T3	Apple_T1	2.7	90.00
4	Casino	Chocolate_T1	4	55.56
5	Coffee, Hot	Apple_T1	2.4	68.24
6	Blackberry, Single,	Coffee,	2.3	95.83
7	Opera	Cherry_T1	4.1	52.56
8	Lemon_T2	Lemon_T1	4	52.63
9	Casino	Chocolate_T1, Chocolate_T6	3.8	52.78
10	Cherry_T1	Opera	4.1	48.81

Table 12 Results of bakery using BFFO-TA, BPSO, and BPSO-TA

Rule	Antecedent	Consequent	Support	Confidence
1	Apple_T2, Apple_T4	Apple_T3	4	97.56
2	Casino, Chocolate_T6,	Chocolate_T1	3.8	97.44
3	Opera, Cherry_T1	Apricot_T3	3.8	92.68
4	Truffle	Gongolais	5.8	56.31
5	Apricot_T2, Hot	Blueberry_T1	3.2	100.00
6	Marzipan	Tuile	5.3	58.89
7	Apple_T3, Cherry_T2	Apple_T2	3.1	93.94
8	Raspberry_T1, Raspberry_T2	Lemon_T3	2.9	100.00
9	Apricot_T3	Cherry_T1	4.6	61.33
10	Lemon_T4, Raspberry_T2	Raspberry_T1	2.8	96.55

The top 10 rules produced by BPSO are presented in Table 12 and are also found to be superior compared to that of BFFO.

The BPSO-TA algorithm is applied on Bakery dataset with particles count kept at 70 and the number of iterations is fixed at 50. The TA algorithm is called with a probability of 10% in every iteration. The parameters inertia, C1, and C2 are kept as 100.8, 2, and 2. The top ten rules are extracted by running the BPSO-TA algorithm only once. BPSO-TA is found to take computationally less time compared to BFFO, BFFO-TA and the rules produced by them are presented in Table 12.

5.6 Clickstream Dataset

Similar to Bank dataset, clickstream is found to be extremely sparse and needed feature selection. Here, the number of features is reduced by considering only those features whose support count is greater than 0.3. The original dataset which contained 294 features was finally reduced to 18 features.

The top 10 association rules for the Clickstream dataset (http://archive.ics.uci.edu/ml/datasets/Anonymous+Microsoft+Web+Data) were obtained using BFFO by choosing the population size (p) as 10, absorption coefficient (γ) as 2.5, attractiveness ($\beta 0$) as 0.1, and the iterations count (I) as 100. The rules thus obtained are presented in Table 13.

The population size is kept ate 10 and iterations count as 100 for BFFO-TA algorithm on the Clickstream dataset, wherein TA is invoked with a probability of 5% in every iteration. The top 10 rules are extracted only in a single run by keeping $\beta 0$ and γ as 0.1 and 2.5. The top ten rules produced by BFFO TA are the same rules produced by BPSO, BPSO-TA and are presented in Table 14. It is evident that BFFO-TA produced superior results and takes less computational time compared to BFFO.

While applying BPSO on the Clickstream dataset, the particles count is kept as 50 and the number of iterations as 50. The top 10 rules are extracted in 10 runs. BPSO is found to take computationally less time compared to BFFO, BFFO-TA. The top ten rules produced by BPSO are the same ones produced by BPSO-TA, BFFO-TA and are presented in Table 14. Further, BPSO takes computationally less time compared to BFFO, BFFO-TA.

For BPSO-TA on the Clickstream dataset, the number of particles in the population is kept at 50 and the number of iterations is fixed at 100. In every iteration, the TA algorithm is called with a probability of 5%. The parameters inertia, C1, and C2 are chosen as 0.8, 2, and 2 respectively. To extract the top 10 rules, BPSO-TA is run

Table 13 Results of clickstream using BFFO

Rule	Antecedent	Consequent	Support	Confidence
1	Windows Family of Oss Link	Free Downloads Link	7.79	55.08
2	Knowledge Base Link	Support Desktop Link	5.52	60.85
3	Isapi Link	Free Downloads Link	7.31	44.84
4	Support Desktop Link	isapi Link	5.94	43.68
5	Knowledge Base Link	isapi Link	4.69	51.69
6	Products Link	Free Downloads Link	6.12	39.21
7	Free Downloads, L1035 Link	isapi Link	2.46	90.56
8	Support Desktop, Knowledge Base Link	isapi Link	3.32	60.19
9	Knowledge Base, L1035 Link	isapi Link	2.12	87.60
10	Support Desktop Link	Microsoft.com Search Link	4.86	35.70

Table 14 Results of clickstream using BFFO-TA, BPSO, and BPSO-TA

Rule	Antecedent	Consequent	Support	Confidence
1	Internet Explorer Link	Free Downloads Link	16.08	56.06
2	Windows Family of Oss Link	Free Downloads Link	7.79	55.08
3	Windows95 Support Link	isapi Link	4.61	84.14
4	Knowledge Base Link	Support Desktop Link	5.52	60.85
5	Isapi Link	Free Downloads Link	7.31	44.84
6	Windows 95 Link	Windows Family of OSs Link	3.24	91.47
7	Support Desktop Link	isapi Link	5.94	43.68
8	Windows Family of OSs, L1035 Link	isapi Link	2.87	87.52
9	Knowledge Base Link	isapi Link	4.69	51.69
10	Products Link	Free Downloads Link	6.12	39.21

only once and is found to take computationally less time compared to BPSO, BFFO, BFFO-TA. The top ten rules produced by BFFO-TA are presented in Table 14. It can be observed that BPSO-TA also produced superior results compared to BFFO.

The computational time obtained by various algorithms on all the seven datasets is presented in Table 15. The computational time of BFFO on Books dataset is around 2 min and it is around 1 min for Food dataset. The reduction in time for Food dataset is mainly due to less number of transactional records involved in calculations of support and confidence values during the computation of objective function. The time taken for grocery dataset is around 6 min as it requires 1000 iterations due to the nature of the dataset. For all the datasets, though BFFO was found to mine effectively the important rules, computationally it took more time as it needs to be run 10 times. In case of BFFO-TA and BPSO-TA, the running times for smaller datasets are comparable with the traditional approaches. But as regards the XYZ bank dataset, where the number of transactions is relatively high, the BFFO-TA algorithm yielded the least computational time across algorithms including BPSO. Further, it could yield all top 10 rules in just a single run, which is a remarkable outcome of the present study. This superior performance is primarily due to the integration BFFO

Table 15 Computational times for various approaches

Data set	BFFO-TA	BFFO	BPSO-TA	BPSO
BOOKS	4 s	2 m13 s	3 s	3 s
FOOD	4 s	1 m7 s	2 s	3.5 s
GROCERY	13 s	5 m56 s	7 s	27 s
XYZ Bank	52 s	2 m45 s	1 min 3 s	120 s
Bakery	2 m 53 s	3 m 29 s	53 s	82 s
Clickstream	4 m 47 s	8 m 7 s	2 m 48 s	6 m 9 s

with TA. For the sake of brevity, the comparison of the results of the proposed hybrid algorithms BFFO-TA and BPSO-TA with that of the Apriori and the FP-Growth algorithms is not presented.

6 Conclusions

In this chapter, we proposed three new hybrid evolutionary techniques to extract association rules that need to be run just once to obtain all the association rules. The striking advantage of these algorithms is that minimum support and minimum confidence need not be specified upfront, unlike popular algorithms. In case of BFFO rule miner, it extracts all the top 10 rules by ensembling the best rules produced in 10 runs of the algorithm. However, the BFFO-TA and BPSO-TA Rule Miners are able to generate all the top 10 rules in just a single run. It is a spectacular benefit achieved over BFFO and a previously published BPSO. This is a dramatic achievement of the present study. These hybrid methods also outscore the traditional ones in not generating redundant rules. It can be concluded that the proposed algorithms show superior performance compared to other traditional ones. In future, we would like to extend the current work to a multi-objective optimization framework by bringing in other measures of strength. In future, optimization algorithms with new operators and objective functions can be devised for association rule mining. Further, better rule encoding than the present one can be employed. This work presented here can be further applied to high utility itemset mining, sequential rule mining, periodic pattern mining, episode mining, etc. Further, class association rule mining can also be performed on various biomedical classification datasets available in [35–37].

References

1. Agrawal R, Imieliński T, Swami A (1993) Mining association rules between sets of items in large databases. In: Buneman P, Jajodia S (eds) ACM SIGMOD international conference on management of data. ACM, Washington D.C. USA, pp 207–216
2. Agrawal R, Srikant R (1994) Fast algorithms for mining association rules. In: Proceedings 20th international conference VLDB, pp 487–499
3. Han J, Pei J, Yin Y (2000) Mining frequent patterns without candidate generation. SIGMOD record. ACM, Dallas, Texas USA, pp 1–12
4. Saggar M, Agrawal AK, Lad A (2004) Optimization of association rule mining using improved genetic algorithms. In: IEEE international conference on systems, man and cybernetics. IEEE, The Hague, Netherlands, pp 3725–3729
5. Wakabi-Waiswa PP, Baryamureeba V, Sarukesi K (2011) Optimized association rule mining with genetic algorithms. In 7th international conference on natural computation, ICNC 2011. IEEE, Shanghai, China, pp 1116–1120
6. Ghosh A, Nath B (2004) Multi-objective rule mining using genetic algorithms. Inf Sci (Ny) 163:123–133. https://doi.org/10.1016/j.ins.2003.03.021
7. Gupta M (2012) Application of weighted particle swarm optimization in association rule mining. Int J Comput Sci Inform ISSN (PRINT 1:2231–5292)

8. Asadi A, Afzali M, Shojaei A, Sulaimani S (2012) New Binary PSO based method for finding best thresholds in association rule mining. Life Sci J 9:1097–8135
9. Minaei-Bidgoli B, Barmaki R, Nasiri M (2013) Mining numerical association rules via multi-objective genetic algorithms. Inf Sci (Ny) 233:15–24. https://doi.org/10.1016/j.ins.2013.01.028
10. Nandhini M, Janani M, Sivanandham SN (2012) Association rule mining using swarm intelligence and domain ontology. In: International conference on recent trends in information technology (ICRTIT). IEEE, Chennai, Tamil Nadu, India, pp 537–541
11. Alatas B, Akin E, Karci A (2008) MODENAR: multi-objective differential evolution algorithm for mining numeric association rules. Appl Soft Comput J 8:646–656. https://doi.org/10.1016/j.asoc.2007.05.003
12. Hadian A, Nasiri M, Minaei-Bidgoli B (2010) Clustering based multi-objective rule mining using genetic algorithm. Int J Digit Content Technol Its Appl 4:37–42. https://doi.org/10.4156/jdcta.vol4.issue1.5
13. Kuo RJ, Chao CM, Chiu YT (2011) Application of particle swarm optimization to association rule mining. Appl Soft Comput 11:326–336. https://doi.org/10.1016/j.asoc.2009.11.023
14. Maheshkumar Y, Ravi V, Abraham A (2013) A particle swarm optimization-threshold accepting hybrid algorithm for unconstrained optimization. Neural Netw World 23:191–221
15. Sarath KNVD, Ravi V (2013) Association rule mining using binary particle swarm optimization. Eng Appl Artif Intell 26:1832–1840. https://doi.org/10.1016/j.engappai.2013.06.003
16. Cheng Y (2005) Genetic algorithm for item selection with cross-selling. In: Proceedings of 2005 international conference on machine learning and cybernetics, pp 18–21
17. Shenoy PD, Srinivasa KG, Venugopal KR, Patnaik LM (2005) Dynamic association rule mining using genetic algorithms. Intell Data Anal 9:439–453
18. Kuo RJ, Shih CW (2007) Association rule mining through the ant colony system for national health insurance research database in Taiwan. Comput Math Appl 54:1303–1318. https://doi.org/10.1016/j.camwa.2006.03.043
19. Chang Chien YW, Chen YL (2010) Mining associative classification rules with stock trading data-A GA-based method. Knowledge-Based Syst 23:605–614. https://doi.org/10.1016/j.knosys.2010.04.007
20. Christian AJ, Martin GP (2010) Optimization of association rules with genetic algorithms. In: 2010 XXIX international conference of the Chilean computer science society, pp 193–197
21. Hansen JM, Raut S, Swami S (2010) Retail shelf allocation: a comparative analysis of heuristic and meta-heuristic approaches. J Retail 86:94–105. https://doi.org/10.1016/j.jretai.2010.01.004
22. Khademolghorani F (2011) An effective algorithm for mining association rules based on imperialist competitive algorithm. In: 2011 sixth international conference on digital information management, pp 6–11
23. Yang GF, Mabu S, Shimada K, Hirasawa K (2011) An evolutionary approach to rank class association rules with feedback mechanism. Expert Syst Appl 38:15040–15048. https://doi.org/10.1016/j.eswa.2011.05.042
24. Bhugra D, Goel S, Singhania V (2013) Association rule analysis using biogeography based optimization. In: 2013 international conference on computer communication and informatics, pp 1–5
25. Birtolo C, De Chiara D, Losito S, et al (2013) Searching optimal product bundles by means of GA-based engine and market basket analysis. In: IFSA world congress and NAFIPS annual meeting (IFSA/NAFIPS), 2013 Joint, pp 448–453
26. da Cunha DS, de Castro LN (2013) Bioinspired algorithms applied to association rule mining in electronic commerce databases. In: 2013 BRICS congress on computational intelligence and 11th Brazilian congress on computational intelligence, pp 189–194
27. Luna JM, Romero JR, Ventura S (2013) Grammar-based multi-objective algorithms for mining association rules. Data Knowl Eng 86:19–37. https://doi.org/10.1016/j.datak.2013.01.002
28. Ganghishetti P, Ravi V (2014) Association rule mining via evolutionary multi-objective optimization. In: Murty MN, He X, Chillarige RR, Weng P (eds) Multi-disciplinary trends in artificial intelligence (MIWAI). Springer, Cham, Bangalore, India, pp 35–46
29. Yang XS (2010) Nature-inspired metaheuristic algorithms. Luniver press

30. Dueck G, Scheurer T (1990) Threshold accepting: a general purpose optimization algorithm. J Comput Phys 90:161–175. https://doi.org/10.1016/0021-9991(90)90201-B
31. Kennedy J, Eberhart R (1995) Particle swarm optimization. In: International conference on neural networks (ICNN'95). IEEE, Piscataway, NJ, pp 1942–1948
32. Wur SH, Leu Y (1999) An effective Boolean algorithm for mining association rules in large databases. 6th international conference on database systems for advanced applications (DASFAA). Institute of electrical and electronics engineers Inc., Hsinchu, Taiwan, Taiwan, pp 179–186
33. Kennedy J, Eberhart RC (1997) A discrete binary version of the particle swarm algorithm. In: IEEE international conference on systems, man, and cybernetics. Comput Cybern Simul. IEEE, pp 4104–4108
34. Li Y, Ning P, Wang XS, Jajodia S (2003) Discovering calendar-based temporal association rules. Data Knowl Eng 44:193–218. https://doi.org/10.1016/S0169-023X(02)00135-0
35. Das Himansuand Naik B, BHS (2018) Classification of diabetes mellitus disease (DMD): A data mining (DM) approach. In: Pattnaik PK, Rautaray SS, DH and NJ (ed) Progress in computing, analytics and networking. Springer Singapore, Singapore, pp 539–549
36. Das H, Naik B, Behera HS (2020) An experimental analysis of machine learning classification algorithms on biomedical data. In: Kundu S, Acharya U, De C, Mukherjee S (eds) 2nd international conference on communication, devices and computing. Springer, Haldia, India, pp 525–539
37. Das H, Naik B, Behera HS (2020) Medical disease analysis using neuro-fuzzy with feature extraction model for classification. Inform Med Unlocked 18. https://doi.org/10.1016/j.imu.2019.100288

Toward Sarcasm Detection in Reviews—A Dual Parametric Approach with Emojis and Ratings

Aanshi Rustagi, Annapurna Jonnalagadda, Aswani Kumar Cherukuri, and Amir Ahmad

Abstract Detection of sarcasm is very crucial in today's world where social media become a major platform of expressing emotions. Sarcastic statements are the statements where the sentiment polarity and contextual meaning are completely contrary. It affects the efficiency and accuracy of present sentiment analysis systems (SAS). Most of the currently available sarcasm detection models such as vector space models, CNN, RNN, etc. consider the raw review text in order to determine the sentiment which ignores the presence of negation, lexical ambiguity, and irony created by general facts. Due to this, detecting sarcasm may not be very accurate. To improve the accuracy of sarcasm detection and to better understand the context, the model proposed integrates the ratings, reviews, and emojis. The performance of the proposed system is evaluated using annotator-agreement methods with the metrics such as F1 score, Precision, Recall, and Accuracy. The performance shows that integrating more features enhances the accuracy by a considerable margin as compared to previously defined methodologies.

Keywords Emojis · Ratings · Sarcasm detection · Sentiment analysis · Tokenization · Web scraping

1 Introduction

Sentiment analysis plays a very important role in today's world in determining the sentiment of the people over a period of time or on a particular topic. For example,

---

A. Rustagi · A. Jonnalagadda (✉)
School of Computer Science and Engineering, VIT University, Vellore 632 014, India

A. K. Cherukuri
School of Information Technology and Engineering, VIT University, Vellore 632 014, India
e-mail: cherukuri@acm.org

A. Ahmad
Department of Information Technology, College of Information Technology, UAE University, Al Ain, UAE

S. Chakraverty (ed.), *Soft Computing in Interdisciplinary Sciences*,
Studies in Computational Intelligence 988,
https://doi.org/10.1007/978-981-16-4713-0_13

collecting the data by companies about their products from various review sites and predicting the sentiment of the people toward their product from the users may help the companies to provide better recommendations to their customers or to improve their service or to improve the quality of the product. Hence, it is necessary to know the exact emotions and reactions given by the users in order to predict the sales or improvise the product, etc. Due to the extensive use of sarcasm and irony in reviews these days it has become extremely difficult to compute the exact sentiment using those conventional methodologies that are based on sentiment polarities and analysis. The model incorporates the contextual references as well to decipher the exact emotion and meaning of the review given by the user. Word ambiguities also pose a major challenge to the accuracy of conventional sentiment analysis models because the referred meaning of the word can be understood by analyzing the context and background of the review. Same word can be used as various figures of speech as well. Consider the review, "Product is very cheap." This review ambiguously uses the word "cheap," it can be considered as the measure of quality or cost but the fact is that if used for quality then the emotion is negative else if it is used for cost, then it is positive comment. This can be understood by taking into consideration other factors and context of the comment given. Upcoming is numerical sarcasm, that is, sarcasm is reflected in the numerical values provided in the statement like, "It's really very hot—at –5 °C.", this statement is clearly sarcastic and can be detected by using the value of the temperature given by the comment. So, in social media world where the corpus is more dynamic, the current sentiment analysis models have to be upgraded and improved. The abovementioned challenges motivate to change the conventional methodologies and move to more accurate and sophisticated approaches which capture the contextual knowledge of the comment as well. Such one challenge is sarcastic comments which give a positive sentiment polarity but are actually negative in the given context. These issues are leading to the failure of one of the core reasons of industrial sentiment analysis which is helping the industry understand the response of their customers about the products they launch. Such faulty analysis of sarcastic, ironical, and slang language leads to high misunderstanding of the sentiment behind the statement. Hence, it has become very important to address this issue and challenging to solve this problem.

Progress has been made for this issue in the past few years using various technologies and approaches. One of the main drawbacks for these methods is the use of only the review text for sarcasm detection which directly affects the accuracy and efficiency of the model. While extraction of opinion words, generally the repeated words are removed and then converted to vectors but the removal of repeating words leads to ignoring the challenge of word ambiguity and hence cannot understand the meaning of the sentence, like in the sentence, "She bought a silver ring in a silver car," in this sentence if only unique words are considered, it changes the context of the sentence. Moreover, all the describing figures of speech should be used while opinion word extraction [7]. After experimentation with machine learning techniques, pattern-based approaches, and opinion word analysis, gradually shift was made toward neural networks, which did improve the accuracy of the existing models drastically but lacked the integration of any other feature with the review text.

In today's slang-driven world, knowing and feeding all the words in our lexicon dictionary is nearly impossible. So, relying on one feature may not yield satisfactory results in sarcasm detection [18]. We intend to propose a model integrating other features as well like ratings and emoji classification in order to improve performance of the system. This factor ensures the performance stability of the proposed model in a real-world scenario. So, in the proposed model, ratings, emojis, and review text are integrated for sarcasm detection.

The proposed model extracts reviews from a movie review-aggregating website and detects sarcasm in real time without any pre-defined dataset which is used for training purposes. Extraction of ratings and review text occurs separately and then emojis are separated from reviews and processed separately. From the review text, all the opinion words are extracted and sentiment polarity for each word is computed. This helps to calculate the sentiment vector of the review based on the occurrence and polarity of each token of the extracted words from the review. These processed features are integrated to detect sarcasm in real-time streaming data.

The remaining of the report is organized as follows: Sect. 2 explains the related work, Sect. 3 elaborates on dataset preparation, the methodology is explained in Sect. 4 followed by Results and Discussions in Sect. 5. We then conclude the article providing some future directions.

2 Related Works

Sentiment analysis touches every aspect of the natural language processing which can be beneficial for generalizing customer feedbacks, [5] by annotating reviews with respect to aspect terms and their polarity for a given domain. Reviewing process can also be considered similar to that of student survey form analysis [3] that was done by hood, Karoline et al. [3], so it can be concluded that affective computing and sentiment analysis is very important for the customer-relationship management [4] as mentioned by Cambria [4]. Data is collected from a review aggregation website, as the method suggested by goel, bansal et al., to use beautiful soup to extract data and urllib to open the link [1]. In [6], authors have extracted adjectives for them being able to describe the emotion in a given review. The methodology of Baid, Palak et al. [7], divides the algorithm in phases like cleaning data, feature selection, and classification [7] using various other tools. To determine the polarity, the study of Kim, Hyopil et al. [9], shows the use of vector space models [9] which compute the point polarity of the word with respect to its usage in the given text. Semiautomatic polarity expansion algorithm was also used by Fernandez, Milagros for the classification of sentiments, it includes a unsupervised dependency parsing-based text classification method involving many natural language toolkit methodologies. The paper of Redhu and Swati reflects on all the steps of the general algorithm [10] adopted in any natural language processing project, and they also discuss about various methods used in opinion mining.

It is commonly observed that a lot of sarcasm is used to describe the movies in disappointment. Online critic websites such as Rotten Tomatoes, IMDB, etc. have become a common platform for people to express their emotions, as people watch movies out of passion for actors. It also reveals a lot of patterns [2] in likes and dislikes of the audience who is watching the movie as suggested by Daniel and Walid. Reviewing process can also be considered similar to that of student survey form analysis [3] that was done by hood, Karoline et al. [3], so can be concluded that affective computing and sentiment analysis is very important for the customer-relationship management [4] as mentioned by Cambria [4].

In the work of Bouazizi, Mondher et al. [11], they discuss about various forms of sarcasm present in social media or opinion aggregation websites, and they also speak of various patterns like punctuations [11], emoticons, use of literature jargons, etc. to express sarcasm. The work of Kochinski and Y. Alex shows the importance of determining the human relationship with the text [12] given to determine the context of the statement given and detect sarcasm. Conneau, Alexis et al. also highlighted the use of recurrent neural networks and LSTMs for sentiment analysis. The deployment 29 convolutional layers [13] in the analysis improves performance drastically. Sivashankari and Valarmathi introduce a novel-approach NLP-modified token-based frequencies of left right [14] which is very effective to determine the multi-word product names and dominant multi-word product name from the customer review corpus. Facebook sarcasm detection [17] was attempted by Kuo, Po chen et al. through emoticons, slangs to reveal a pattern in those and detect. They also use annotator agreement to validate their experiment results. Justo, R., Corcoran, T. et al. provide alternative sets of features obtained according to different criteria and test range of different feature sets using two different classifiers, which shows that sarcasm detection [19] is improved by considering linguistic and semantic factors. Deliens, Antoniou, et al. work investigates perspective-taking process [25] in sarcasm detection task and confirms that perspective-shifting is egocentrically grounded. Ahuja et al. [25] have used 12 classification algorithms on four datasets whose split ratio was varied to check the accuracy of models. Also, gradient was used to boost the performance. Work of Ren et al. [26] suggested that context-augmented neural models can effectively decode sarcastic clues from contextual information, and give a relative improvement in the detection performance. Filik et al. [28] used functional magnetic resonance imaging (fMRI) with the aim of mapping the neural networks involved in the processing of sarcastic and non-sarcastic irony. Proposed work by Sandhu et al. [28] showed using sentimental analysis that sarcasm led to over-categorization of positive tweets, which altered the results by suggesting that the public viewed partisanship on the Supreme Court favorably. Sulis et al. [29] suggest novel data-driven methods for irony and sarcasm differentiation. Finally, result is a novel set of sentiment, structural, and psycholinguistic features evaluated in binary classification experiments. Matsui et al. [30] provide evidence that the left inferior frontal gyrus, particularly BA 47, is involved in integration of discourse context and utterance with affective prosody in the comprehension of sarcasm. Kumar and Harish [31] use two-stage feature selec-

S.no.	Approach	Reference	Methodology
1	Word embedding	Joshi, Aditya, et al. "Are word embedding-based features useful for sarcasm detection?." arXiv preprint arXiv:1610.00883 (2016).	enhancement using semantic similarity/discordance between word embeddings, experiments with four types of word embeddings.}&{Word2Vec and dependency weight-based features outperform LSA and GloVe, in terms of their benefit to sarcasm detection.
2	Hyperbolic Sarcasm Detection	Bharti, S.K., Vachha, B., Pradhan, R.K., Babu, K.S. and Jena, S.K., 2016. Sarcastic sentiment detection in tweets streamed in real time: a big data approach. Digital Communications and Networks, 2(3), pp.108-121	Used parsing based lexicon generation algorithm, interjection start of a sentence, positive sentiment with an antonym pair to detect saracsm in real time streaming tweets.
3	Hashtags analysis	Maynard, D.G. and Greenwood, M.A., 2014, March. Who cares about sarcastic tweets? investigating the impact of sarcasm on sentiment analysis. In LREC 2014 Proceedings. ELRA.	Sentiment analysis of all the hashtags present in the tweet. Except hashtags no other feature used.
4	Machine Learning Algorithms	Prasad, A.G., Sanjana, S., Bhat, S.M. and Harish, B.S., 2017, October. Sentiment analysis for sarcasm detection on streaming short text data. In 2017 2nd International Conference on Knowledge Engineering and Applications (ICKEA) (pp. 1-5). IEEE.	Detection performed with various algorithms. Best algorithm selected with novel introduction of slang and emoji dictionary
5	Naïve Bayes and fuzzy clustering algorithms	Mukherjee, S. and Bala, P.K., 2017. Sarcasm detection in microblogs using Naïve Bayes and fuzzy clustering. *Technology in Society*, *48*, pp.19-27.	Provides knowledge to the system to detect sarcasm which is relatively independent of text, like n-grams and function words. Acuuracy of 65 perc

Fig. 1 Various approaches for sarcasm detection

tion process with second stage using k-means clustering. Classifiers used were SVM and random forest and outperform the previous works on Amazon review dataset (Fig. 1).

As for the previous works, one common gap was identified in all the works. Only one feature was used for sarcasm detection whether it was in real-time streaming tweets, consumer reviews, or literature works. Moreover, in the work proposed by

Maynard and Greenwood [23], only hashtags were used for sarcasm detection and resulted in 90% accuracy but with a big drawback of detecting only in tweets containing a hashtag. Thus, it cannot be generalized for all the tweets but only a certain category of them. Approach of Prasad et al. [24] requires a huge amount of effort for creating dictionaries so that previous knowledge can be provided for detection but this is impractical in a generation where hundreds of slangs are used and created on a daily basis. Methodology proposed by Mukherjee and Bala [20] used a different category of features but yielded a low accuracy of 65%.

So, in the current proposed approach, review text with emojis and ratings is used to improve its efficiency to detect sarcasm in real time with no previous knowledge required. Our methodology is evaluated against corpus which is annotated by five annotators. The experiment dataset is also obtained in real time by scraping the review website. Metrics used are Precision, Recall, F1 score, and Accuracy for four different experimental datasets.

3 Dataset Preparation

In this sub-module, the process involved in dataset extraction and pre-processing is explained.

3.1 Dataset Extraction

The experimental datasets used are the set of movie reviews from Rotten Tomatoes (https://www.rottentomatoes.com/). Review dataset is obtained using web scraping. For loading the URL, request module is used and then HTML is parsed using beautiful soup. Due to the fact the reviews for a movie are distributed over various pages because of the paging property in the website, beautiful soup is used to scrape data out of a single page and Python Selenium WebDriver to scrape data from different pages. Like this all the review data for a movie are scraped. The data collected from the web scraping is raw review text and star ratings (0–5) in HTML format. After collecting the data, it is stored in two different lists for further processing.

3.2 Data Pre-processing

After the data is collected successfully, the data is pre-processed for further procedure. Firstly, emojis are separated from the review text to process them. Then, review text is tokenized using regex tokenizer, in which all words less than three in length are removed so that the tagging has shorter list of useless words to tag and process becomes comparatively faster. So, for n reviews (r1 to rn) for each review ri, all the stopwords are removed which removes all the unnecessary words not required for sentiment detection like conjunctions, preposition, and the non-English words. Further, lemmatize the words which converts the words to linguistically correct lemmas, also perform stemming which converts all the words to their root forms like running to run, etc. After this, tag the parts of speech and also all the emojis are converted if present in the review to its substitute expression by using emoji package in Python. After the processing of review texts, all the ratings are converted from HTML tag formats to numeric format so that the polarities and ratings can be combined for the detection.

4 Methodology

4.1 Overview

The dataset used for detection is retrieved from Rotten Tomatoes website through web scraping as explained in the above section. It returns features like review text, emoji, and rating which are then analyzed further using NLTK sentiment analysis. The data after pre-processing is used as an input. All the important features required in the sentiment detection of the reviews are extracted and collected. The process is shown in the figure below.

4.2 Rating Extraction

Ratings are an important factor in sarcasm detection as even if the sentiment comes out to be positive, through ratings it's always known whether the user is being sarcastic or happy. After the pre-processing of the HTML ratings is done, a list of ratings in numbers is obtained. The list thus obtained is zipped with that of the reviews and emoji lists to ease the further processing of the features.

4.3 Emoji Extraction

Emojis help us furthermore to clarify and confirm the sentiment of the review given to us for sentiment processing. In data pre-processing, emojis are converted to their substitute text for the reviews which had them while for others the list was initialized with the value zero.

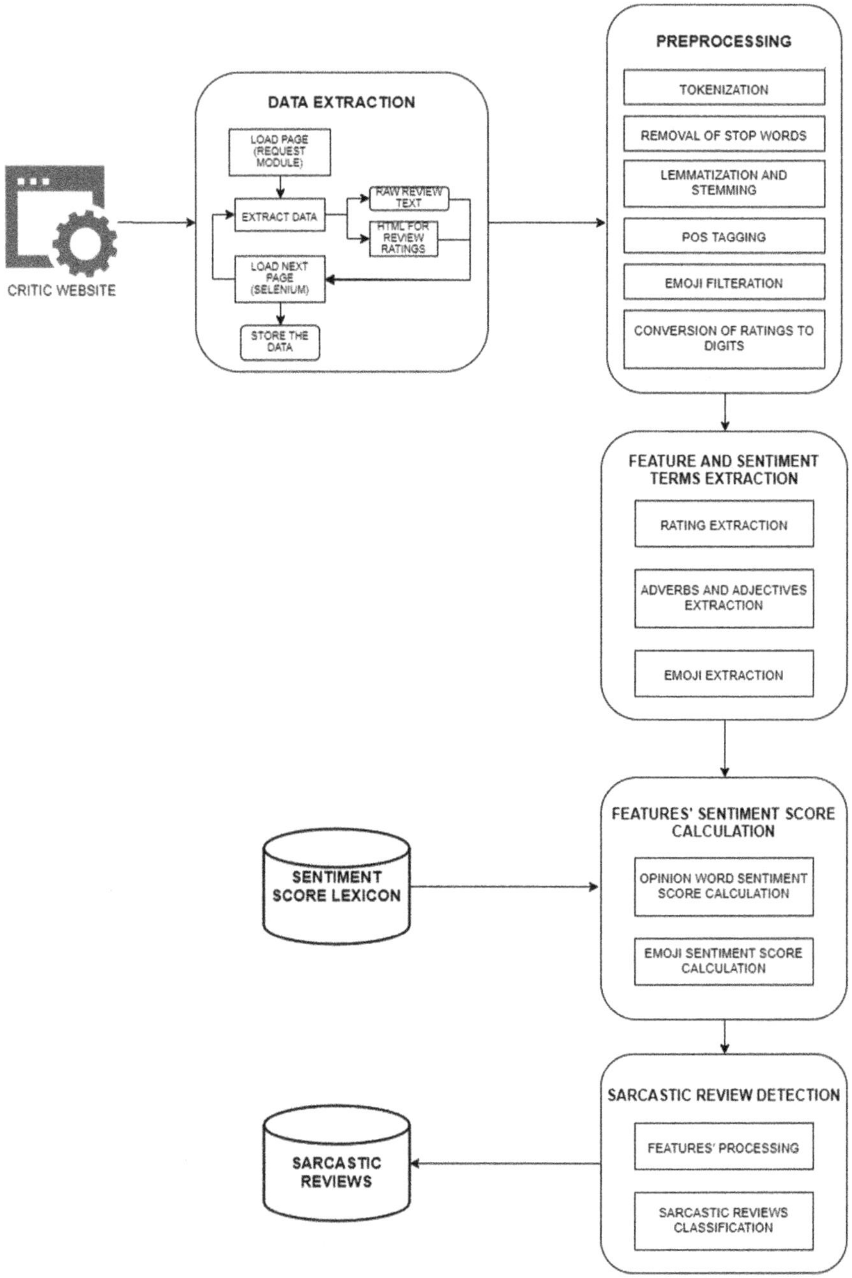

Fig. 2 Methodology for the proposed approach

4.4 Adverbs and Adjectives Extraction

In the pre-processing step, the reviews are POS tagged, so for feature extraction the required parts of speech were adverbs and adjectives. All these words are considered because these are used to describe the properties of verbs and nouns and the emoji expressions, and hence can be used to analyze the user feelings regarding the movie through their written review. All the required features are extracted and stored in list for further processing.

4.5 Feature Sentiment Score Calculation

In this sub-module, the features extracted are processed and their sentiment score is calculated for final sarcasm detection. This is done using VADER library in NLTK which calculates the valence score for a sentence by summing up valence scores of each sentiment bearing token as listed in the VADER source code dictionary. These scores are then normalized by using the following formula to map the value between –1 and 1:

$$valencescore = \frac{x}{\sqrt{x^2 + \alpha}},$$

where x is the sum of valence scores and α is the normalization parameter. Its scores are calculated considering five other heuristics which are punctuation marks, capitalization, degree modifiers, occurrence of "but," and capturing negation.

4.6 Opinion Word Sentiment Score Calculation

This step involves the sentiment analysis of the extracted adverbs and adjectives. For that to happen, the sentiment lexicon is taken as input which helps in determining the sentiment percentage score according to the occurrence of words in that particular review. It compares the opinion word list with sentiment lexicon and gives the percentage of positive, negative, and neutral sentiments as a list. These values will further be combined with other features to detect sarcasm in reviews.

4.7 Emoji Sentiment Score Calculation

Similar to the opinion word sentiment score calculation, the substitute text of the emojis is used and compared with the sentiment lexicon given as input. This gives the percentage score calculation of the various emojis.

Table 1 Statistics for experimental datasets

Sets	No. of reviews	No. of annotators
1	23	5
2	39	5
3	27	5
4	33	5

4.8 Sarcastic Review Detection

After the complete processing of review text, emojis, and ratings, they are combined using logical operators. The logic behind the use of operators is the basic definition of sarcasm, i.e., a sentence which uses positive words (implying good meaning) to give a negative comment about the concerned movie. So, when logical operators are used, reviews are checked for containing maximum positive sentiment but rating less than half or vice versa, from such a situation it is inferred that the user review is making positive comment but the users are not happy as seen through the rating given. Hence, it can be concluded that the review is giving a sarcastic sense.

5 Results and Discussion

5.1 Data

Sarcasm is a form of expression which is often used but not so easily recognized even by the human brain. So, for the validation and verification of the results obtained by the algorithm shown in figure, any existing labeled dataset is not used but instead its completely based on the human mind for the validation. For the above mentioned, four sets of data containing varying number of reviews (maximum being 39 in total) are used. To label the data obtained, each set is assigned to various annotators for the labeling. Table below shows number of reviews in all the sets and the number of annotators assigned (Table 1).

5.2 Ground Truth

In this section, the methodology for marking the data is done using the human psychology. So, for true and efficient validation of the algorithm mentioned in Fig. 2, inter-annotator agreement is used in three levels in which single annotator (agree 1), two annotators (agree 2), and three annotators (agree 3) are taken into account. For

SET	AGREE 1				AGREE 2				AGREE 3			
METRICS ->	ACCURACY	F1 SCORE	PRECISION	RECALL	ACCURACY	F1 SCORE	PRECISION	RECALL	ACCURACY	F1 SCORE	PRECISION	RECALL
SET1	73.9	69.9	77.7	63.6	65.2	59.9	66.6	54.5	56.5	50	55.5	45.5
SET2	82.1	75.8	78.5	73.3	74.6	64.2	69.2	60	66.7	51.8	58.3	46.6
SET3	77.7	70	70	70	62.9	54.5	60	50	55.5	45.4	50	41.6
SET4	78.7	71.9	75	69.2	66.7	55.9	58.3	53.8	63.6	53.8	58.3	50

Fig. 3 Experimental results

each agreement level, the four metrics—Precision, Recall, F1 score, and Accuracy—are calculated. To compute all the metrics, the confusion matrix for multi-feature classification is constructed, features being Positive, Negative, and Sarcastic which were previously labeled by the annotators in all the datasets. Precision gives the percentage of all true sarcastic reviews; recall gives the percentage of reviews labeled as sarcastic by our model actually being sarcastic; F1 score is the metric that defines the balance between recall and precision; and accuracy is the percentage of correctly detected sentiments in reviews irrespective of it being positive, negative, or sarcastic.

The next discussed about the corner and edge test cases which were used to check the limitations and advantages of the model and to which extent t can be suitable for the real-life scenarios.

5.3 *Results and Discussion*

In this section, we take into consideration the corner test cases that were mentioned as the limitation of other models but can be predicted correctly by the proposed model. The model proposed addresses the issue discussed in Sect. 1 with the example, "Product is very cheap" by taking into consideration the rating aspect of the review. The model analyzes the context in an effective manner because the emotion of user will be reflected in the rating given. Similarly, other test cases were used to verify the metrics shown in the below table and to test the proposed approach thoroughly so that its limitations and advantages are clearly identified (Fig. 3).

As the results are shown above in table, most accurate results were found using set 2 data, accuracy being 82.1 and precision being 78.5; it was due to the fact that it had the highest number of data instances present. As all the metrics involve proportion calculation, even one wrong detection leads to drastic fall in the results which is cause of low percentages under the agree three category. This fact is also reflected in the comparatively high accuracy of set 4, accuracy being 78.7. The table highlights the three highest metrics achieved with the use of four different sets of data scraped and marked for testing using inter-annotator agreement.

6 Conclusion and Future Work

This work tries to highlight the importance of human interaction on opinion-aggregating platforms, and how the rating patterns and review readings played a major part in developing this method for sarcasm detection. This method is majorly based on how human psychology works and the common sense factor of humans. There is still a lot of work to be done in this field but to best of our knowledge this is the first work using ratings of reviews and emojis for detecting sarcasm. Extension of this work can be the development of region-based customized sentiment analyzers which can be language and culture specific. Moreover, study can be done to predict the movie which is more likely to get sarcastic comments from users, which can be used for business benefits.

References

1. GOEL S et al (2019) Web Crawling-based Search Engine using Python. In: 3rd international conference on electronics. Communication and Aerospace Technology (ICECA), IEEE, p 2019
2. Martens D, Maalej W (2019) Release early, release often, and watch your users' emotions: lessons from emotional patterns. IEEE Softw 36(5):32–37
3. Hood K, Kuiper PK (2018) Improving student surveys with natural language processing. In: 2018 second IEEE international conference on robotic computing (IRC). IEEE
4. Cambria E (2016) Affective computing and sentiment analysis. IEEE Intell Syst 31(2):102–107
5. Pontiki M et al (2016) Semeval-2016 task 5: aspect based sentiment analysis. In: Proceedings of the 10th international workshop on semantic evaluation (SemEval-2016)
6. Manek AS et al (2017) Aspect term extraction for sentiment analysis in large movie reviews using Gini Index feature selection method and SVM classifier. World Wide Web 20(2):135–154
7. Baid P, Gupta A, Chaplot N (2017) Sentiment analysis of movie reviews using machine learning techniques. Int J Comput Appl 975:8887
8. Fernández-Gavilanes M et al (2016) Unsupervised method for sentiment analysis in online texts. Expert Syst Appl 58:57–75
9. Kim Y, Shin H (2017) Finding sentiment dimension in vector space of movie reviews: an unsupervised approach. J Cogn Sci 18(1):85–101
10. Redhu S et al (2018) Sentiment analysis using text mining: a review. Int J Data Sci Technol 4(2):49
11. Bouazizi M, Ohtsuki TO (2016) A pattern-based approach for sarcasm detection on twitter. IEEE Access 4:5477–5488
12. Kolchinski YA, Christopher P (2018) Representing social media users for sarcasm detection. arXiv:1808.08470
13. Conneau A et al (2016) Very deep convolutional networks for natural language processing 2. arXiv:1606.01781
14. Sivashankari R, Valarmathi B (2018) NLP-MTFLR: document-level prioritization and identification of dominant multi-word named products in customer reviews. Arab J Sci Eng 43(2):843–855
15. Mozetič I, Grčar M, Smailović J (2016) Multilingual Twitter sentiment classification: the role of human annotators. PloS One 11(5)
16. Nisioi S et al (2017) Exploring neural text simplification models. In: Proceedings of the 55th annual meeting of the association for computational linguistics (Vol 2: Short Papers)
17. Kuo PC, Alvarado FH, Chen YS (2018) Facebook reaction-based emotion classifier as cue for sarcasm detection. arXiv:1805.06510

18. Justo R, Corcoran T, Lukin SM, Walker M, Torres MI (2014) Extracting relevant knowledge for the detection of sarcasm and nastiness in the social web. Knowl-Based Syst 69:124–133
19. Mukherjee S, Bala PK (2017) Sarcasm detection in microblogs using Naïve Bayes and fuzzy clustering. Technol Soc 48:19–27
20. Joshi A et al (2016) Are word embedding-based features useful for sarcasm detection?. arXiv:1610.00883
21. Bharti SK, Vachha B, Pradhan RK, Babu KS, Jena SK (2016) Sarcastic sentiment detection in tweets streamed in real time: a big data approach. Digit Commun Netw 2(3):108–121
22. Maynard DG, Greenwood MA (2014) Who cares about sarcastic tweets? investigating the impact of sarcasm on sentiment analysis. In: LREC 2014 Proceedings. ELRA
23. Prasad AG, Sanjana S, Bhat SM, Harish BS (2017) Sentiment analysis for sarcasm detection on streaming short text data. In: 2017 2nd international conference on knowledge engineering and applications (ICKEA). IEEE, pp 1–5
24. Deliens G, Antoniou K, Clin E, Kissine M (2017) Perspective-taking and frugal strategies: evidence from sarcasm detection. J Pragmat 119:33–45
25. Ahuja R, Bansal S, Prakash S, Venkataraman K, Banga A (2018) Comparative study of different sarcasm detection algorithms based on behavioral approach. Procedia Comput Sci 143:411–418
26. Ren Y, Ji D, Ren H (2018) Context-augmented convolutional neural networks for twitter sarcasm detection. Neurocomputing 308:1–7
27. Filik R, Ţurcan A, Ralph-Nearman C, Pitiot A (2019) What is the difference between irony and sarcasm? an fMRI study. Cortex 115:112–122
28. Sandhu M, Vinson CD, Mago VK, Giabbanelli PJ (2019) From associations to sarcasm: mining the shift of opinions regarding the supreme court on twitter. Online Soc Netw Media 14
29. Sulis E, Farías DIH, Rosso P, Patti V, Ruffo G (2016) Figurative messages and affect in twitter: differences between # irony, # sarcasm and # not. Knowl-Based Syst 108:132–143
30. Matsui T, Nakamura T, Utsumi A, Sasaki AT, Koike T, Yoshida Y, Harada T, Tanabe HC, Sadato N (2016) The role of prosody and context in sarcasm comprehension: behavioral and fMRI evidence. Neuropsychologia 87:74–84
31. Kumar HK, Harish BS (2018) Sarcasm classification: a novel approach by using content based feature selection method. Procedia Comput Sci 143:378–386

The manufacturer's authorised representative in the EU is Springer Nature Customer Service Centre GmbH, Europaplatz 3, 69115 Heidelberg, Germany. If you have any concerns regarding our products, please contact ProductSafety@springernature.com

Printed and bound by CPI Group (UK) Ltd, Croydon, CR0 4YY
30/09/2026
02225246-0009